Transition to Agricultural Market Economies

The Future of Kazakhstan, Russia and Ukraine

Transition to Agricultural Market Economies

The Future of Kazakhstan, Russia and Ukraine

Edited by

Andrew Schmitz

University of Florida, Gainesville, Florida, USA

and

William H. Meyers

University of Missouri, Columbia, Missouri, USA

www.cabi.org

CABI is a trading name of CAB International

CABI
Nosworthy Way
Wallingford
Oxfordshire OX10 8DE
UK

Tel: +44 (0)1491 832111
Fax: +44 (0)1491 833508
E-mail: info@cabi.org
Website: www.cabi.org

CABI
38 Chauncy Street
Suite 1002
Boston, MA 02111
USA

Tel: +1 800 552 3083 (toll free)
E-mail: cabi-nao@cabi.org

A catalogue record for this book is available from the British Library, London, UK.

Library of Congress Cataloging-in-Publication Data

Transition to agricultural market economies : the future of Kazakhstan, Russia, and Ukraine / Andrew Schmitz and William H. Meyers, editors.
pages cm
Includes bibliographical references and index.
ISBN 978-1-78064-535-3 (hbk : alk. paper) 1. Grain trade--Former Soviet republics. 2. Grain trade--Russia (Federation) 3. Grain trade--Ukraine. 4. Grain trade--Kazakhstan. 5. Agriculture--Economic aspects--Former Soviet republics. 6. Agriculture--Economic aspects--Russia (Federation) 7. Agriculture--Economic aspects--Ukraine. 8. Agriculture--Economic aspects--Kazakhstan. I. Schmitz, Andrew, editor. II. Meyers, William H. (William Henry), 1941- editor.

HD9045.F62T73 2015
338.10947--dc23

2015011145

ISBN-13: 9781780645353

Commissioning editor: Claire Parfitt
Editorial assistant: Emma McCann
Production editor: James Bishop

Typeset by SPi, Pondicherry, India.
Printed and bound in the UK by CPI Group (UK) Ltd, Croydon, CR0 4YY.

Table of Contents

Introduction

The genesis of this book originated at an S-1043 regional research meeting in Washington, DC (USA), in 2010, when the editors discovered they had a common interest in the grain economies of Kazakhstan, Russia and Ukraine (KRU) and a shared view that the KRU region would be important to the global grain economy of the future. At that time, at least from an academic perspective, little was known about the agricultural economies of the KRU region. The first outcome of this conversation was that a panel of experts met to discuss the topic 'Competition from Ukraine and Russia: Making Agriculture More Productive and Profitable' at the 18th Annual Moose Jaw Conference on 'Farming for Profit?' in June of 2011. Later, as both authors travelled to Russia and Ukraine, and other symposia were organized by them at an Agricultural and Applied Economics Association (AAEA) annual meeting, a Southern Agricultural Economics Association (SAEA) annual meeting and an International Agricultural Trade Research Consortium (IATRC) annual meeting, a growing body of research was generated and networks developed that made it possible to attract a very comprehensive and diverse set of high-quality papers.

These chapters include a wide range of topics, such as agricultural policies that affect production and trade in the KRU region, the privatization of land markets, the optimal use of fertilizer and other inputs, export grain embargoes and the transition from livestock to grain production. The authors of these chapters include scholars from Europe, the USA and Canada who are working with collaborators in those countries. In addition, many of these scholars have conducted extensive graduate research in the USA. This variety of talent and experience and the wide scope of topics make this book a unique collection of information and analysis, and a valuable resource for anyone working on agricultural issues in Kazakhstan, Russia or Ukraine, including the global grain economy.

As this book goes to press, troubling conflicts are unfolding in this region that increase the political and economic volatility of all three countries, especially of Ukraine. The chapters in this volume were completed before the political conflicts emerged, so they did not anticipate the possible impacts of these developments. The longer-term implications of these events for the Black Sea and for global agricultural markets cannot be assessed at this time. Our feeling is that the current political and economic conflict increases the need for a deeper understanding of this region. Our book provides this understanding and the basis for a later comparison of the impact of the crisis on agriculture in the KRU region. It may be a long time before the impact can be assessed, and it would be premature to attempt any assessment, given that so many uncertainties exist. However, even though statements are highly speculative, one can be reasonably certain of the likely impact of the crisis in this region. For example, in the short and medium term, livestock and dairy prices are likely to decline in Ukraine and rise in Russia and Kazakhstan after the ban by Russia on imports of these products from Ukraine. This will lead to increased livestock and dairy production at least in Russia if the trade embargo continues. This will be enhanced further if the Russian embargo on the importation of US livestock products continues.

Professor Andrew Schmitz, University of Florida
Professor William H. Meyers, University of Missouri-Columbia

Acknowledgements

The publication of this book was supported with contributions from the following:

Institute of Food and Agricultural Sciences, University of Florida, Gainesville, Florida
CAFNR International Programs, University of Missouri, Columbia, Missouri
Leibniz Institute of Agricultural Development in Transition Economies (IAMO), Halle, Germany
Economic Research Service, United States Department of Agriculture, Washington, DC

Foreword

A long, secular decline in world food prices, starting in the 1974–2007 period, with relative stability around the downward trend, engendered some complacency among policy makers in many countries. Investments in agriculture stagnated, and lower priority was accorded to efficiency-enhancing agricultural policy reforms. The food crisis of 2007/08, with large price spikes in key global food prices, changed the perceptions of all stakeholders. The World Bank's global food price index increased by 60% in just a few months in 2008, with the prices of maize, rice and wheat increasing by 70%, 180% and 120%, respectively, compared to mid-2007. These changes caused major hardships to vulnerable groups globally, particularly in poorer, food-importing developing countries. It was estimated that tens of millions of people were pushed into poverty, and many more that were already poor suffered worsening malnutrition, hunger and related harms to health.

Domestic political turmoil ensued in dozens of countries around the globe, and attention focused on agricultural development and policy issues. After 2008, world food prices spiked again in early 2011, with prices reaching 2008 levels. In mid-2012, maize prices rose to an all-time high, and wheat prices jumped by over 50% within 2 months. In both post-2008 spikes, adverse weather in major grain-exporting countries played a significant role in generating a contraction of supply that pushed up prices. However, export restrictions in these countries and related short-term domestic policies played a role too, albeit not as significant as in 2008, when export bans and pro-cyclical, short-term trade policies were responsible for 30% and 45% of the increases in rice and wheat prices, respectively.

The parallel increases in food prices and broader agricultural prices are perceived by many scholars to be indicative of longer-term structural, rather than temporary, weaknesses in the global food supply system. Furthermore, higher volatility of food prices is viewed as a long-term phenomenon. Consequently, it is expected that price spikes and volatility will continue in the foreseeable future.

The evolution of supply capacities in major food exporting countries and the short- and long-term policies and investments that affect these countries' supply are very relevant when assessing global food market prospects in the coming years and decades. Among the major grain-exporting countries, the Kazakhstan, Russia and Ukraine (KRU) region has become a major player in the global food market: between 2006 and 2012, the KRU region contributed 14% of total world grain exports and 21% of world exports of wheat. Furthermore, the US Department of Agriculture (USDA) projects that by 2021, the KRU region's share in total world grain and wheat exports will rise to 22% and 29%, respectively. The KRU region has the potential to expand grain production by increasing both grain area and yields.

Given the prominence of the KRU region within the global grain trade, a thorough understanding of their agricultural production capacities, structural constraints and policy environment are essential to ascertaining the future course of global food markets. This book makes a major and timely contribution to our knowledge about the agricultural economy of these key countries. The analysis cannot be confined to the grain sector alone, because the natural and fiscal resources available to agriculture are ultimately limited, and developments and policies in the non-grain sector have implications for the grain sector and for its export potential. In this respect, the book covers a wide range of relevant issues, dwelling on the constraints to productivity growth in the grain sector, the impacts of trade and price policies and the interconnections of the grain sectors with the meat, dairy, fruit and vegetable sectors. In particular, this book provides insights as to why the KRU region began emerging in 2000 as a major grain exporter and assesses its potential to expand grain production and exports further. By broadening our knowledge and understanding of the constraints, potentials and policy trade-offs that will determine the agricultural development and trade performance of the KRU countries, this book will serve as an important resource for global stakeholders focused on food security and food markets and for the scholarly community that analyses the workings and evolution of these markets.

Gershon Feder
International Food Policy Research Institute
Washington, DC

Contributors

Asci, Serhat, University of Florida, PO Box 110240, Gainesville, FL 32611, USA. E-mail: sasci@ufl.edu

Boersch, Marlene, Mercantile Consulting Venture, Inc, 605 Des Meurons, Winnipeg, Manitoba R2H 2R1, Canada. E-mail: mboersch@mercantileventure.com

Brink, Lars, 4 Brink Road, Chelsea, Quebec J9B 2C6, Canada. E-mail: lars.brink@hotmail.com

Burkitbayeva, Saule, LICOS KU Leuven, LICOS Centre for Institutions and Economic Performance, Waaistraat 6 – Bus 3511, B-3000 Leuven, Belgium. E-mail: Sab868@mail.usask.ca

Chastikova, Elena, North-West Institute of Economics and Organization of Agriculture, Pushkin, Shosse Podbelskogo 7, St Petersburg, Russia 196608. E-mail: Chasticova_lena@mail.ru

Djuric, Ivan, Leibniz Institute of Agricultural Development in Transition Economies, Theodor-Lieser-Strasse 2, D-06120 Halle (Saale), Germany. E-mail: Djuric@iamo.de

Epstein, David, North-West Research Institute of Agricultural Economics, Podbelskogo 7, St Petersburg, Russia 196608. E-mail: epsteindb@gmail.com

Erokhin, Vasily, Moscow University of Finance and Law, 8, Building 1, Presnenskaya Embankment, Moscow 115114, Russia. E-mail: basilic@list.ru

Feder, Gershon, IFPRI, 2033 K Street, Room 4015, Washington, DC 20006, USA. E-mail: gfeder@hotmail.com

Gagalyuk, Taras, Ukrainian Agribusiness Club, 3 Floor, 146 Zhylianska Str, Kyiv, Ukraine. 01032. (Current affiliation: World Bank Group, 2121 Pennsylvania Ave., NW, Washington, DC 20037, USA. E-mail: tgagalyuk@worldbank.org)

Glauben, Thomas, Leibniz Institute of Agricultural Development in Transition Economies, Theodor-Lieser-Strasse 2, D-06120 Halle (Saale), Germany. E-mail: glauben@iamo.de

Golovina, Svetlana, Kurgan State Agricultural Academy, Ketovo District, Kurgan Region, Russia. E-mail: s_golovina@yahoo.com

Götz, Linde, Leibniz Institute of Agricultural Development in Transition Economies, Theodor-Lieser-Strasse 2, D-06120 Halle (Saale), Germany. E-mail: goetz@iamo.de

Hahlbrock, Konstantin, VitalFields.com (Head of Product), Am Kaiserkai 47, 20457 Hamburg, Germany. E-mail: konstantin@vitalfields.com

Hockmann, Henrich, Leibniz Institute of Agricultural Development in Transition Economies, Theodor- Lieser-Strasse 2, D-06120 Halle (Saale), Germany. E-mail: hockmann@iamo.de

Kerr, William A., University of Saskatchewan, 3D20 – 51 Campus Drive, Saskatoon, SK Canada S7N 5A2. E-mail: william.kerr@usask.ca

Kirby, Sandra, University of Winnipeg, 515 Portage Avenue, Room 3D10, Winnipeg, MB R3B 2E9, Canada. E-mail: s.kirby@uwinnipeg.ca

Kobuta, Iryna, FAO Regional Office for Europe and Central Asia, 34, Benczur utca, Budapest, Hungary, 1068. E-mail: iryna.kobuta@fao.org

Lapa, Volodymyr, Ukrainian Agribusiness Club, 3 Floor, 146 Zhylianska Str, Kyiv, Ukraine, 01032. E-mail: lapa@ucab.ua

Lerman, Zvi, Department of Agricultural Economics and Management, Hebrew University of Jerusalem, Rehovot, Israel. E-mail: Zvi.lerman@mail.huji.ac.il

Liefert, Olga, USDA/ERS, 355 E Street SW, Washington, DC 20024, USA. E-mail: oliefert@ers.usda.gov

Liefert, William, USDA/ERS, 355 E Street SW, Washington, DC 20024, USA. E-mail: wliefert@ers.usda.gov

Loveridge, Scott, Michigan State University, 446 West Circle Drive, Suite 66 (Morrill), East Lansing, MI 48824-1039, USA. E-mail: Loverid2@msu.edu

Meyers, William H., FAPRI, University of Missouri, 101 Park DeVille Drive, Suite E, Columbia, MO 65211, USA. E-mail: meyersw@missouri.edu

Moss, Charles B., University of Florida, PO Box 110240, Gainesville, FL 32611, USA. E-mail: cbmoss@ufl.edu

Murova, Olga, Department of Agricultural and Applied Economics, Ag Sciences Bldg, Room 317, Texas Tech University, 15th and Boston, Lubbock, TX 79409, USA. E-mail: olga.murova@ttu.edu

Nikulina, Yulia, North-West Institute of Economics and Organization of Agriculture, Pushkin, Shosse Podbelskogo 7, St Petersburg, Russia 196608. E-mail: julia.nikylina@mail.ru

Nilsson, Jerker, Swedish University of Agricultural Sciences, PO Box 7013, SE-750 07 Uppsala, Sweden. E-mail: jerker.nilsson@slu.se

Nizalov, Denys, Kiev School of Economics, 1 Ivana Mazepy Str, 01010 Kiev, Ukraine. E-mail: nizalov@kse.org.ua

Oshakbaev, Dauren, Bogenbay Str 30, Apt 81, Astana, 010000, Kazakhstan. E-mail: lunoskok@gmail.com

Ostapchuk, Igor, Ukrainian Agribusiness Club, 3 Floor, 146 Zhylianska Str, Kyir, Ukraine. E-mail: ostapchuk@ucab.ua

Petrick, Martin, Leibniz Institute of Agricultural Development in Transition Economies, Theodor-Lieser-Strasse 2, D-06120 Halle (Saale), Germany. E-mail: petrick@iamo.de

Ponomarev, Mikhail, North-West Institute of Economics and Organization of Agriculture, Pushkin, Shosse Podbelskogo 7, St Petersburg, Russia 196608. E-mail: m.a.ponomarev@gmail.com

Saigusheva, Uliana, North-West Institute of Economics and Organization of Agriculture, Pushkin, Shosse Podbelskogo 7, St Petersburg, Russia 196608. E-mail: skarapejas@yandex.ru

Schedrin, Evgeny, North-West Institute of Economics and Organization of Agriculture, Pushkin, Shosse Podbelskogo 7, St Petersburg, Russia 196608. E-mail: eshedrin@gmail.com

Schmitz, Andrew, University of Florida, PO Box 110240, Gainesville, FL 32611, USA. E-mail: aschmitz@ufl.edu

Schroeder, Kateryna G., The World Bank Group, 1818 H Street Northwest, Washington, DC 20433, USA. E-mail: Kateryna.schroeder@gmail.com

Seale, James L. Jr, University of Florida, PO Box 110240, Gainesville, FL 32611, USA. E-mail: jseale@ufl.edu

Sedik, David, Food and Agricultural Organization of the United Nations, 34 Benczar ut, Budapest, Hungary 1068. E-mail: david.sedik@fao.org

Surovtsev, Vladimir N., North-West Institute of Economics and Organization of Agriculture, Pushkin, Shosse Podbelskogo 7, St Petersburg, Russia 196608. E-mail: vnsurovtsev@gmail.com

Thornsbury, Suzanne, USDA/ERS, 355 E Street SW, Washington, DC 20024, USA. E-mail: sthornsbury@ers.usda.gov

Vorotnikova, Ekaterina, University of Idaho, Agricultural Economics and Rural Sociology, 875 Perimeter Drive, MS 2334, Moscow, ID 83844, USA. E-mail: evorotnikova@uidaho.edu

Wolz, Axel, Leibniz Institute of Agricultural Development in Transition Economies, Theodor-Lieser-Strasse 2, D-06120 Halle (Saale), Germany. E-mail: wolz@iamo.de

Woods, Mollie, Michigan State University, Room 85, Morrill Hall of Agriculture, East Lansing, MI 48824-1039, USA. E-mail: Willi751@msu.edu

Zadorozhna, Olha, Lazarski University, Świeradowska 43, 02-662, Warsaw, Poland. E-mail: ozadorozhna@kse.org.ua

1 Overview of Agriculture in Kazakhstan, Russia and Ukraine

William M. Liefert* and Olga Liefert
Economic Research Service, US Department of Agriculture, Washington, DC, USA

Abstract

This overview chapter examines the main systemic and policy changes for agriculture in Kazakhstan, Russia and Ukraine (KRU) that were part of these countries' transition from a planned economy to a market economy, and how these changes altered the commodity mix and volumes of KRU agricultural production, consumption and trade. Key systemic and policy changes involved budget subsidies, price formation, trade and the allocation of industrial and agricultural inputs and outputs. These changes drove the main commodity developments, which included the major contraction of the KRU livestock sector; the elimination of the large Soviet era imports of grain, soybean and soybean meal; the substantial growth (especially by Russia) in imports of meat and other livestock products; and the emergence of the KRU region as a major grain exporter. These outcomes appear to be a result of the economically rational restructuring of KRU agricultural production and of trade that is consistent with these countries' cost/price competitiveness (comparative advantage) in world markets. This chapter also examines the major changes in KRU farm-level organization, structure and management after the dissolution of the Soviet Union in 1991, which also affected production.

Acknowledgements

Disclaimer: The views expressed in this chapter are those of the authors and may not be attributed to the Economic Research Service or the US Department of Agriculture.

Introduction

Kazakhstan, Russia and Ukraine (KRU), the main agricultural countries of the former Soviet Union, are becoming increasingly important in world agricultural markets. The two main developments are that this region has become a large grain exporter, especially of wheat, and Russia, specifically, has become a large importer of meat.

The emergence of the KRU region as a major grain exporter is important for both world commercial agriculture and world food security. The surge in world agricultural and food commodity prices in 2006–2008, and again in 2011/12, raised concerns about the world's ability to feed an ever-growing population (Godfray *et al.*, 2010). Increasing the production of grain is central to meeting this challenge, both to provide sufficient food grain and to satisfy the demand for animal feed, especially as income growth in emerging market economies raises demand for meat and

*Corresponding author; e-mail: wliefert@ers.usda.gov

other livestock products. Many observers see the KRU as a region with strong (and perhaps the most) potential to strengthen world food security by expanding grain production and exports (EBRD/FAO, 2008).

This overview chapter provides:

1. Background data on the KRU countries and agriculture's importance within their economies.
2. A review of KRU agricultural trade.
3. A discussion of how the systemic and policy changes that were part of the KRU countries' transition from centrally planned economies to market economies restructured their commodity agriculture by altering the mix and volume of agricultural production, consumption and trade.
4. An examination of how the transition and economic reform has affected various aspects of the KRU agrofood systems: farm-level operations (the nature and size of farms and their operations), upstream and downstream operations, institutional and physical infrastructure and the introduction of large agroholdings.
5. The outlook for KRU agricultural policy.
6. Conclusions.

The chapter draws heavily from Liefert *et al.* (2010) and Liefert and Liefert (2012).

Background Information on the KRU Countries and Agriculture

Table 1.1 presents background information on the KRU countries and the importance of agriculture to their economies. One of the reasons why the per capita gross domestic product (GDP) for Russia and Kazakhstan is higher than that for Ukraine is because Russia and Kazakhstan are large producers and exporters of oil, natural gas and metals, which allows them to run large trade surpluses. Table 1.1 also shows that of the three KRU countries, Ukraine has the greatest agricultural share in GDP and total land area, while Kazakhstan has the largest share in total labour force.

KRU Agricultural Trade

The KRU region is a larger agricultural importer than it is an exporter, with the value of its imports in 2010 about double that of exports (Table 1.2). This is primarily because of the large agrofood trade deficit of Russia, whose 2010 imports of US$33 billion were about four times its exports. Ukraine is a net agricultural exporter, while in any year Kazakhstan can be a small net importer or exporter.

The main KRU agricultural imports are meat (beef, pork and poultry), fruit, vegetables and processed foods. The European Union (EU) is the KRU region's dominant foreign agrofood supplier, followed by Brazil and the USA (Global Trade Information Services, 2011).

The KRU region's main agricultural exports are grain and sunflower seeds, with grain and associated products accounting for 45% of total KRU agricultural exports in 2010. Wheat is the dominant KRU grain export, comprising more than 70% of total grain exports, followed by maize and barley (USDA/PSD, 2013). During

Table 1.1. KRU region and agriculture indicators, by country, 2008–2012. (From Kazakhstan Republic Statistical Agency, 1987–2012; Russian Federal Service of State Statistics, 1987–2012; Ukrainian State Statistics Service, 1987–2012; CIA, 2008–2012; World Bank, 2008–2012.)

Item	Russia	Ukraine	Kazakhstan
(1) Total country area (thousand km^2) in 2012	17,080	604	2,725
(2) Population (millions) in 2012[a]	143	45	17
(3) GDP (US$ billion) in 2012[a]	2,504	335	231
(4) Per capita GDP (US$) in 2012[a]	17,700	7,600	13,900
(5) Agriculture's share in GDP in 2008–2012[b]	4%	10%	6%
(6) Agricultural area (thousand km^2) in 2010[c]	2,210	416	937
(7) Agricultural area's share in total country area in 2010[c]	13%	69%	34%
(8) Agriculture's share in total labour force in 2010[c]	10%	15%	28%

Notes: [a](2)–(4) are in 2012; [b](5) covers 2008–2012; [c](6)–(8) are in 2010.

Table 1.2. KRU agriculture trade, 1997–2010 (in US$ billion), and item's share (%) in 2008. (From Global Trade Information Services, 2011; FAOSTAT, 2013.)

	Exports (US$ billion)				Imports (US$ billion)			
Year	Russia	Ukraine	Kazakhstan	KRU	Russia	Ukraine	Kazakhstan	KRU
1997	1.4	1.8	0.9	4.1	12.8	0.9	0.6	14.3
2000	1.3	1.7	0.7	3.7	7.0	1.0	0.4	8.4
2002	2.2	2.4	0.6	5.1	9.7	1.1	0.5	11.4
2004	2.6	3.5	0.8	6.9	12.8	1.9	0.9	15.6
2006	4.8	4.7	1.0	10.6	20.4	3.2	1.7	25.2
2008	8.4	10.8	3.0	22.2	33.3	6.5	3.0	42.8
2010	8.0	10.0	2.0	20.0	33.0	5.8	2.3	41.1
Item	Exports (Share (%))				Imports (Share (%))			
2008	Russia	Ukraine	Kazakhstan	KRU	Russia	Ukraine	Kazakhstan	KRU
Meat	0.2	0.7	0.1	0.4	21.6	13.1	4.2	19.1
Vegetables, fruit and nuts	1.7	2.3	4.6	2.4	18.7	10.4	4.4	16.5
Grain and products	42.2	35.9	84.2	44.7	1.9	3.0	2.5	2.1
Fats and oils	11.0	18.0	0.6	13.0	4.8	9.5	9.2	5.8

2006–2012, the KRU region contributed 14% of total world grain export volume and 21% of world wheat export volume. The main foreign markets for KRU grain are the EU, North Africa, the Middle East, certain Asian countries and other countries of the former Soviet Union.

The negative KRU agrofood balance is due mainly to the KRU region exporting bulk crops and importing high-value products. For example, in 2010, KRU agrofood imports totalled US$41 billion (with the bulk being imported by Russia). These imports are about two-thirds those of China, despite the fact that the KRU population is about one-sixth that of China (Global Trade Information Services, 2011). Even though both KRU agricultural exports and imports have grown substantially since 2000 (Table 1.2), exports have increased at the greater rate. Thus, the major focus is on the KRU agricultural export trade, especially of grain, rather than on the import trade.

How the Transition Restructured KRU Commodity Agriculture and Trade

The main objective of Soviet agricultural policy in the 1970s and 1980s was to expand the livestock sector, mainly to improve the standard of living by increasing meat and dairy consumption. Using large budget subsidies to both producers and consumers, along with controlled prices and trade, the regime succeeded in increasing meat production between 1970 and 1990 by over 60% (Liefert and Swinnen, 2002). By 1990, Soviet per capita consumption of meat and dairy products was close in volume to that in wealthy developed countries, while their per capita GDP was less than one-half in comparison (Sedik, 1993). Because the Soviet Union could not produce enough animal feed to support its growing livestock herds, it became a large importer of feed grain, soybean and soybean meal, to the benefit of bulk crop producers such as the USA, Canada and Australia.

Immediately following the dissolution of the Soviet Union in 1991, the KRU countries began their transition from planned economies to market economies. After just a few years, the KRU region had largely ended central planning, state allocation of resources and state monopoly over foreign trade. Reform and market liberalization substantially reduced the large subsidies enjoyed by KRU agriculture. During the planned period, Soviet agriculture had been subsidized heavily three ways. The first was through state budget subsidies to the sector, which in 1990 equalled about 10% of Soviet GDP. The second was through price policy, whereby prices for agricultural inputs were set low relative to their production cost and to agricultural output prices. The third was also through price policy, in that output prices for many agricultural products, in particular livestock goods, were set high relative to world prices. For example, in 1986, Soviet producer prices for beef and poultry were about one-quarter and two-thirds above world prices, respectively (Liefert *et al.*, 1993).

Due to the collapse in state revenue after the 1991 Soviet dissolution, the large Soviet era budget subsidies to KRU agricultural producers and consumers were largely eliminated. Price liberalization also terminated the indirect subsidies that producers had received from the Soviet price system. When domestic prices jumped to reflect the real cost of producing goods, agricultural input prices rose relative to output prices, such that agricultural producers' terms of trade (output versus input prices) worsened dramatically. Trade liberalization then resulted in domestic prices for many agricultural goods moving downward toward world prices, further exacerbating producers' terms of trade vis-à-vis their domestic input suppliers. For example, from 1991 to 1997, the domestic terms of trade of Russian agricultural producers fell by about 75% (OECD, 1999). In 1992, Russian wheat producers had to sell on average 0.3 tonnes (t) of output to purchase 1.0 t of nitrogen fertilizer, whereas by 1997, they had to sell 1.4 t of wheat to purchase the same 1.0 t of nitrogen fertilizer (Russian Federal Service of State Statistics, 1987–2012).

Transition to a market economy reduced KRU agricultural output severely, especially in the livestock sector (Table 1.3). The main reason for the production collapse was that market liberalization decreased substantially or eliminated

Table 1.3. Volume indices of agricultural output, KRU countries, selected years, 1990–2012. (From Kazakhstan Republic Statistical Agency, 1987–2012; Russian Federal Service of State Statistics, 1987–2012; Ukrainian State Statistics Service, 1987–2012.)

	1990	1991–1995	1996–2000	2001–2005	2006–2012
Russia					
Total agricultural output	100	81	60	66	75
Crops	100	90	75	86	97
Livestock products	100	76	51	53	60
Ukraine					
Total agricultural output	100	76	54	60	67
Crops	100	81	63	74	87
Livestock products	100	71	46	47	51
Kazakhstan					
Total agricultural output	100	77	46	58	70
Crops	100	74	48	69	81
Livestock products	100	80	43	48	59

Note: The table gives indices with 1990 = 100.

the huge direct and indirect subsidies that Soviet agriculture, and particularly the livestock sector, received under the planned economy system. Higher relative prices for inputs generated a severe drop in their purchase and use, which in turn reduced production.

Table 1.3 shows that during 1996–2000, average annual agricultural output in Russia, Ukraine and Kazakhstan equalled 60%, 54% and 46% of the level in 1990, respectively, while output of livestock products was only 51%, 46% and 43% of the 1990 volume, respectively. Average annual meat production (beef, pork and poultry) fell in all three KRU countries by more than one-half, with total KRU meat output over 1996–2000 (average annual) down by 52% compared to 1987–1991 (Table 1.4). KRU livestock herds also contracted commensurately.

The huge downsizing of one-half or more of the KRU livestock sectors substantially reduced domestic demand for animal feed, such that the large Soviet era imports of grain virtually disappeared (Table 1.4), along with those of soybean and soybean meal. Domestic feed grain requirements declined so much that KRU grain production also fell. Average annual KRU grain output dropped from 160 million tonnes (Mt) during 1987–1991 to 100 Mt during 1996–2000. (Note: in this chapter, KRU grain production and trade exclude rice, buckwheat, sorghum and pulses.)

Rather than importing animal feed to maintain a large and costly livestock sector, during the 1990s Russia increased its imports of meat (Table 1.4). Net average annual meat imports by Russia grew from 1.9 Mt in 1989–1991 to 2.5 Mt in 1996–2000.

The KRU agricultural establishments regarded the severe contraction of the livestock sector as a disaster, to be reversed when the state finally had the resources to do so. Yet, the downsizing can be viewed largely as a necessary part of the market-driven reallocation of resources away from an uncompetitive high-cost sector that had expanded during the planned period to levels that could not be maintained under market conditions. Put another way, given the high real cost of KRU livestock production and consumers' real incomes in the late Soviet period, the country was producing and consuming much more meat and other livestock products than it would with a market economy.

By 2000, KRU agricultural production had begun to rebound. From 1996–2000 to 2006–2012, average annual total agricultural output increased in Russia (25%), Ukraine (24%) and Kazakhstan (52%), with both the crop and livestock sectors expanding (Table 1.3). Over this same period, average annual KRU meat production increased by about one-third, from 5.7 Mt to 7.7 Mt (Table 1.4).

KRU grain production also rose substantially after 2000. From 1996–2000 to 2006–2012,

Table 1.4. KRU grain and meat production and trade (Mt), selected years, 1987–2012.[a] (From USDA/PSD, 2013.)

Years	Grain (Million tonnes)		Meat (Million tonnes)	
	Production	Net trade	Production	Net trade
Total KRU				
1987–1991	160	(16.2)	11.9	(1.2)
1992–1995	138	(2.1)	8.7	(1.1)
1996–2000	100	4.0	5.7	(2.5)
2001–2005	126	20.9	5.6	(2.9)
2006–2012	142	40.5	7.7	(3.4)
Russia				
1987–1991	95	(20.9)	7.2	(1.9)
1992–1995	84	(7.9)	5.4	(1.4)
1996–2000	63	(3.0)	3.6	(2.5)
2001–2005	76	8.1	3.7	(2.8)
2006–2012	81	15.9	5.3	(2.9)
Ukraine				
1987–1991	43	0.1	3.7	0.4
1992–1995	35	0.0	2.5	0.1
1996–2000	26	2.7	1.6	0.1
2001–2005	35	8.2	1.4	(0.0)
2006–2012	40	16.5	1.7	(0.3)
Kazakhstan				
1987–1991	21	4.6	1.1	0.2
1992–1995	19	5.7	0.9	0.1
1996–2000	11	4.3	0.5	(0.0)
2001–2005	14	4.6	0.5	(0.1)
2006–2012	18	8.1	0.7	(0.2)

Note: [a]Figures are average annual values during the period identified in the first column. Figures for grain follow the marketing year (July–June), while figures for meat follow the calendar year (January–December). For meat production and trade, the first row for each country and total KRU covers 1989–1991, not 1987–1991. Trade values inside parentheses are net imports and without parentheses are net exports. Grain production and trade exclude rice, buckwheat, sorghum and pulses. Meat covers beef, pork and poultry.

average annual KRU grain output increased 42%, from 100 Mt to 142 Mt. The growth in grain output created surpluses for export, such that Russia and Ukraine became large-sized grain exporters and Kazakhstan a medium-sized one. The KRU region moved from average annual net grain exports over 1996–2000 of 4.0 Mt to 40.5 Mt over 2006–2012.

One likely reason for the rebound in KRU agricultural production since 2000 has been favourable state policy. From 2000 to 2005, Russian annual budget subsidies to agriculture (from both federal and regional governments) declined in real terms by 26% (Russian Federal Service of State Statistics, 1987–2012). However, in 2005, the Russian federal government identified agriculture as a national priority area that would receive increased funding (along with health, education and housing). From 2005 to 2010, total state support to agriculture rose by 135% in real roubles (RUB). The Russian government stated that the main goal of agricultural policy was to revive the livestock sector, which received the bulk of the new subsidies (Interfax, 1987–2012).

In Ukraine and Kazakhstan, agricultural subsidies have been increasing since 2000. In 2010, Ukrainian agricultural subsidies were about three times higher in real terms than in 2000, while Kazakhstani agricultural subsidies were about seven times higher in real terms than in 2000 (Kazakhstan Republic Statistical

Agency, 1987–2012; Ukrainian State Statistics Service, 1987–2012).

In addition to budget subsidies, the Russian government also helped its livestock sector with trade protection. In 2003, Russia established restrictive tariff rate quotas (TRQ) for imports of beef and pork and a pure quota for poultry that converted to a TRQ in 2006. The annual quota for poultry was set at 1.05 Mt and the low-tariff quota for beef and pork at around 0.45 Mt (Interfax, 1987–2012). In comparison, poultry, beef and pork imports in 2002 equalled 1.37 Mt, 0.50 Mt and 0.60 Mt, respectively (USDA/PSD, 2013). The low in-quota tariff for beef and pork was kept at the previous rate of 15%, while the out-of-quota tariffs were fixed at 60% and 80%, respectively. Although the TRQ regime was liberalized in 2005/06, it was strengthened in 2009. Since 2000, Russia has imposed many sanitary-based restrictions, and often complete bans, on imports of meat (and especially poultry) and other livestock products.

After 2006, the KRU governments aided the livestock sector, and all grain consumers, by periodically restricting grain exports. In response to the surge in world food prices in 2006–2008, the Ukrainian government first banned wheat exports, then replaced the ban with an export quota. In 2008, both the Russian and Kazakhstani governments put a tax on wheat exports, and the disastrous grain harvest of 2010 motivated Russia in August of that year to impose a complete ban on all grain exports, which remained in effect until the end of June 2011. The ban required Russian grain traders to abrogate their existing supply contracts with foreign buyers.

Despite the rebound in KRU agricultural output after 2000, agrofood imports have grown substantially, especially in Russia (Table 1.2). From 2000 to 2008, Russian agrofood imports increased from US$7 billion to US$33 billion (nominal); they did not grow between 2008 and 2010, mainly because of the economic crisis that hit the country in 2008/09. The KRU import rise in nominal values overstates the growth in real (inflation adjusted) terms, especially the jump from 2006 to 2008, when world agrofood prices increased substantially. However, the largest price spikes were for bulk crops (e.g. wheat and rice), and Russia imports little of these products.

There were two main causes of the large increase in Russian agrofood imports after 2000. The first cause was high GDP growth that averaged 4.8% annually, which increased consumer income and agrofood demand. The second cause was a strong appreciation of the Russian currency (RUB) in real terms. Although the RUB was fairly stable during the decade vis-à-vis the US$ and other major currencies in nominal terms, Russia had higher price inflation than did its major trading partners. This appreciated the currency in real terms, or in other words, decreased the prices of imported goods relative to competing domestic output (Liefert *et al.*, 2009). The meat import TRQ regime created in 2003 and other protectionist measures were a response to the surging meat imports and to the negative agrofood trade balance in general.

Transition and KRU Farm-level Changes

By the late Soviet period, two types of large farms dominated agricultural production in the USSR: collective farms (*kolkhozy*) and state-owned farms (*sovkhozy*). Although their origins differed, by the late Soviet period, these farms were very similar in structure and behaviour. Most households on the USSR farms also independently operated a private plot (household plot), typically less than half a hectare in size. The households could either consume their plot output or sell it freely in farmers' markets (the only completely free markets in the Soviet system). The household plots specialized in high-value products such as livestock goods, fruit and vegetables, and in 1990, produced 26% of total Soviet agricultural output (Russian Federal Service of State Statistics, 1987–2012).

In the post-Soviet period, three major types of agricultural producers have existed: the former state and collective farms, household plots and new private/smallholder farms. (For surveys of KRU farm developments during the transition, see Shagaida and Lerman, 2008; OECD, 2004, 2013.) The dominant producer (if not always in total value of output, at least in institutional structure and influence) has been the former state and collective farms. During the 1990s in Russia and Kazakhstan, and beginning in 1999 in Ukraine, the former state and collective farms officially reorganized as corporate farms. Many became joint-stock companies, while others became some sort of cooperative or collective association. As joint-stock companies, the new corporate farms issued ownership vouchers to all their workers and managers, which gave the latter a share in the farms' land and other assets. In Russia and Ukraine, individuals could use these vouchers to obtain land and leave the farm to work as private/smallholder farmers or lease or sell their voucher to the farm management and remain on the farm as hired labour. Most workers chose the latter option. In Kazakhstan, farm members were given leasing, rather than ownership, rights to land (OECD, 2013).

Table 1.5 gives the share of agricultural enterprises, household plots and private/smallholder farms in agricultural output and landholdings in Russia during the transition. (In Table 1.5, the former state and collective farms are called agricultural enterprises, following usage in the KRU region.) By 2010, the share of agricultural enterprises in total agricultural output in Russia, Ukraine and Kazakhstan was 44%, 40% and 30%, respectively; the share of household plots in output in the three countries was 48%, 55% and 44%, respectively; and the share of private/smallholder farms was 7%, 5% and 26%, respectively (Kazakhstan Republic Statistical Agency, 1987–2012; Russian Federal Service of State Statistics, 1987–2012; Ukrainian State Statistics Service, 1987–2012).

With the collapse of central planning, farm managers were given the freedom and responsibility to make their own production decisions, obtain inputs and market their output. As a result, their position within the farms strengthened considerably. Farm management generally has been conservative, and on many farms little real change has occurred since the Soviet period concerning internal organization, administration and work incentives. Throughout the KRU region, the corporate farms remain large, with thousands of hectares (ha) – around 4000 ha on average in Russia and 2000 ha in Ukraine – and hold over one-half of all the farmland. They also produce most of the three countries' bulk crops.

Workers on the large corporate farms have retained control over their household plots, with the right either to consume or to sell their output

Table 1.5. Percentage share of Russian agricultural output and land by farm type, 1990–2010. (From Russian Federal Service of State Statistics, 1987–2012.)

	1990	1995	2000	2005	2010
Share of output					
Agricultural enterprises					
Total output	73.7	50.2	45.2	44.6	44.4
Crops	75.9	45.1	47.9	44.0	40.8
Livestock products	72.0	56.0	42.2	45.2	47.6
Household plots					
Total output	26.3	47.9	51.6	49.3	48.3
Crops	24.1	52.4	47.8	46.5	48.0
Livestock products	28.0	42.7	56.0	52.0	48.6
Private/smallholder farms					
Total output	**	1.9	3.2	6.1	7.2
Crops	**	2.5	4.4	9.6	11.2
Livestock products	**	1.4	1.8	2.9	3.9
Share of agricultural land					
Agricultural enterprises	98.1	89.4	86.1	78.4	69.5
Household plots	1.8	5.2	6.0	10.5	16.9
Private/smallholder farms	**	5.4	7.9	11.1	13.5

Note: ** = insignificant. Agricultural enterprises include farms that are part of agroholdings.

on free markets. The total amount of agricultural land used by the household plots has increased since 1991 in both Russia and Ukraine (Table 1.5). The expansion has occurred not from growth in the size of the household plots, but from the plot-holders' ability to lease land from the local government or simply use the free public meadows and pastures to graze their livestock. The rise in the household plots' share in agricultural land also includes the growth in the number of garden plots tended by the general population.

The share of the household plots (including garden plots) in total agricultural output grew substantially between 1990 and 2010 to around one-half. However, this apparently stellar performance requires some qualifications. One is that during this period, agricultural output by the former state and collective farms decreased. Also, corporate farms have the incentive to under-report their production (one motive being to avoid paying profit tax), which understates their output share in official statistics. Another point is that the household plots specialize in high-value products, such as meat and other livestock goods, fruit and vegetables. Last, the household plot-holders have a symbiotic relationship with their parent farms through which they obtain inputs (such as animal feed and fuel) inexpensively or for free.

The third type of agricultural producer is the private/smallholder farms. These have been created mainly by workers on the former state and collective farms using their ownership vouchers to obtain land and break away as private farmers. These farms typically range from 50 to 150 ha in size, but in Russia and Ukraine they produce less than one-tenth of total agricultural output. Like the agricultural enterprise farms, private/smallholder farms specialize in bulk crops, such as grain and sunflower seeds.

In Kazakhstan, private/smallholder farms account for a much higher share of output than in Russia and Ukraine (about one-quarter). This is because in southern Kazakhstan, dominated by ethnic Kazakhstanis, the large farms have been eliminated and the land has been distributed to families (OECD, 2013). These southern private/smallholder farms use a large amount of labour, which is why in Table 1.1 the share of agriculture in the total labour force is greater for Kazakhstan (28%) than for Russia and Ukraine.

Since 1991, the KRU countries have lacked clear and consistent laws and regulations of land ownership and land markets (especially of the latter). A plethora of parliamentary legislation, presidential decrees and special governmental resolutions have been passed on land issues, much of them contradictory (Shagaida and Lerman, 2008). KRU laws and bureaucracy make the leasing of agricultural land easier to achieve than the buying and selling of agricultural land. Although the official legislation exists for the market sale of land, there are major impediments for those who try to do so. The obstacles include weak administrative and technical infrastructure (such as the absence of public registries of plans and maps) and arbitrary hurdles created by local official agencies. As a result, well-functioning agricultural land markets do not exist in the KRU region, and transaction costs for market sales are high.

In Ukraine, the only way land can be obtained both *de jure* and *de facto* is through leasing. However, the land leasing market appears to work fairly well, and leasing as opposed to ownership does not require the large expenditure of initially purchasing land. This has encouraged the inflow of investment and quality management into agriculture, contributing to the growth in Ukrainian agricultural productivity (Deininger *et al.*, 2013).

Upstream and Downstream Operations and Services

As a legacy of the planned economy period, input supply, agricultural wholesaling and food processing are heavily concentrated. Consequently, enterprises in these activities can exert market power on farms as either buyers or sellers.

Since 1991, the KRU agrofood systems have suffered from weak service industries. These include equipment rental and repair, veterinary care, banking and finance and legal services. Finance has been particularly deficient. The banking industry has been reluctant to lend to farms, whether large or small. Major borrowing impediments faced by the large corporate farms are large existing debt and high default risk, while the private/smallholder farms suffer from a lack of collateral (which the absence of effective land markets exacerbates) and high transaction costs (especially relative to their small size) from the complex loan procedures (Yastrebova *et al.*, 2008).

Institutional and Physical Infrastructure

A market-driven agricultural economy also needs strong supporting public institutions. These include a commercial legal system that protects property and enforces contracts; a market information system; and programmes for agricultural education, extension and research. Throughout the reform period, all these public support services in KRU agriculture have been weak. KRU agriculture also suffers from inadequate physical infrastructure, especially involving transportation and storage, although in recent years the KRU governments and private sector have made improvements in this area through increased attention and investment.

Weak infrastructure increases transaction costs, segments domestic regional markets from each other and cuts them off from the world market. These negative effects particularly hurt small producers and help explain why private/smallholder farming has developed so modestly. The complicated and contradictory laws and procedures for owning and selling land make it difficult for an individual to secure a private/smallholder farm or expand it. Commercial services and public support institutions are weak or non-existent, and the private/smallholder farms face concentrated market power from input suppliers, wholesalers and processors. Private/smallholder farms also must often contend with unsympathetic, if not hostile, local and regional governments and organized crime. In the rough-and-tumble world of the KRU economy and agriculture, small private/smallholder farmers can be squeezed from all sides.

Rise of Agroholdings

Despite the problems just discussed, a growing number of KRU farms appear to be adapting successfully to their market environment by responding more strongly to price signals, cutting waste and other costs and becoming more profit-oriented and efficient. In particular, a progressive element has entered KRU agriculture in the form of large agroholdings (Gataulina *et al.*, 2005; Serova, 2007; Demyanenko, 2008; Rylko *et al.*, 2008; Wandel, 2009). The new agroholdings are vertically integrated enterprises that typically combine primary agriculture, processing and distribution. The agroholdings usually acquire a number of existing corporate farms and improve them, as well as bring investment, superior technology and better management practices into the entire agrofood system. These producers often introduce advanced technology through imports, such as higher-quality seeds, machinery and animal breeding stock. Although the data are not firm, agroholdings currently appear to control about one-fifth of the arable land in the KRU region.

Research has generated opposing views about agroholdings. On the one hand, Rylko *et al.* (2008) find that Russian agroholdings outperform other domestic agricultural producers, a conclusion supported by anecdotal evidence (Interfax, 1987–2012; FAO, 2009). On the other hand, Gataulina *et al.* (2005) and Hockmann *et al.* (2009) find that agroholdings are not more productive or profitable than other large Russian agricultural enterprises. Critics of agroholdings argue that independent of whether these enterprises outperform other types of agricultural producers, they have their own limitations, the main one being that they have become so large and unwieldy they suffer from diseconomies of scale (Gataulina *et al.*, 2005; Hockmann *et al.*, 2009; Wandel, 2011). The benefit of reducing transaction costs through vertical integration must be weighed against the cost of being too large.

Another factor behind the development of agroholdings is the influence of regional governments. Regional officials, particularly in Russia, are worried about the viability of many of their large farms, and especially the economic and social consequences if they are liquidated. Local officials have strongly encouraged agroholdings to take direct ownership and responsibility for the farms that supply their primary product, perhaps even making this a requirement for the agricultural enterprises to operate in local processing and distribution. In return, the governments might provide soft loans, tax relief and other assistance. Yet, the agroholdings are interested in the more lucrative and manageable agribusiness activities of processing, distribution and export, and might not want to be involved in the actual management of agricultural enterprises.

However, the limitations of agroholdings notwithstanding, a superior class of large farms has emerged in KRU agriculture that includes agroholdings that appear to be improving productivity and overall performance. Rylko *et al.* (2008) have a more general term of *new operators*, which covers a wider range of producers than *agroholdings* and might be more appropriate to describe these superior agricultural enterprises. The Deininger *et al.* (2013) study of Ukraine's agriculture supports and qualifies this point. For large commercial farms, they find no evidence of economies of scale (higher productivity from greater size). Rather, the main drivers of productivity growth are the quality of the farm management and the local supporting infrastructure.

The new operators/agroholdings can be viewed as a response to the serious problems and dysfunction in KRU agriculture that involve both farm operations and weak infrastructure (Demyanenko, 2008; Rylko *et al.*, 2008; Hockmann *et al.*, 2009). Since the new operators/agroholdings appear to outperform most of the domestic competition, they should continue to expand in numbers and influence. Given that these producers apparently represent current best KRU agricultural production practices, their expansion should boost production.

KRU Agricultural Policy Outlook

The main objective of Russian agricultural policy since 2000 has been to revive the livestock sector, and this priority will likely continue. In addition to favoured treatment with state budget subsidies and trade protection, the Russian government is trying to help the livestock sector by improving the quality of the animal breeding stock, mainly by investing in importing superior animals (Karlova *et al.*, 2006). Although Russia has become a big grain exporter, it continues as a large net agrofood importer in value terms, mainly because it imports meat and other high-value products. The government is displeased with the negative trade balance in agriculture and appears determined to reduce it. In particular, the government has set the goal of eliminating poultry imports (Interfax, 1987–2012).

Russia's favouring of the livestock sector is contrary to the country's apparent comparative advantage vis-à-vis the world across major agricultural commodities (Liefert, 2002). The policies diminish Russia's impact on world agricultural markets. They reduce imports of meat and other livestock products, and by increasing domestic demand for animal feed, decrease grain exports.

Ukraine and Kazakhstan also wish to revive their livestock sector, but unlike in Russia, their agricultural policies have not favoured that sector strongly compared to the crop production sector (OECD, 2007; Pomfret, 2008). This policy position reflects the larger share of the crop production sector in these countries' agricultural economies, relative to Russia, and the importance of crops (especially grain) in their export earnings.

Yet, during the surges in world agricultural and food prices that began in 2006, all three KRU governments restricted grain exports. These controls conflicted with the interests of grain producers, but aided livestock producers. In addition to reducing exports, the KRU grain export controls decreased the KRU countries' production of the crop by depressing domestic prices. Combined with variable weather conditions, the export controls also hurt the KRU region's reliability as a grain supplier to the world market. Yet, if the past is any guide, such restrictions likely will continue as a KRU policy option.

However, the agricultural policies of all three countries will be constrained strongly by membership of the World Trade Organization (WTO). Russia, Ukraine and Kazakhstan all officially began their WTO membership bids in the mid-1990s. Ukraine joined in 2008 and Russia in 2012, while Kazakhstan was still negotiating WTO membership in 2014.

Two key pillars of the Agreement on Agriculture of the Uruguay Round are market access and domestic support. Ukraine's terms of WTO accession set an average import tariff ceiling (the bound rate) for agricultural products of 10.66% that required a drop in average import tariff rates of about one-quarter (WTO, 2008). WTO accession also fixed the country's bound (maximum allowable) *trade-distorting* domestic support at US$613 million per annum.

In its accession terms, Russia agreed to bind its average import tariff for agricultural goods at 10.8%, a drop from its average tariff in 2011 of 13.2% (WTO, 2011). Russia also liberalized its meat import TRQ regime to some degree. The country was allowed out-of-quota tariff rates for beef, pork and poultry of 55%, 65% and 80%, respectively. However, Russia wholly eliminated the in-quota tariff for pork, and agreed that in 2020 the TRQ for pork would be replaced by a flat maximum tariff of 25%.

Regarding domestic support, Russia agreed to a bound annual level of trade-distorting subsidies of US$9 billion in 2012, to fall to US$4.4 billion by 2018. In comparison, Russian agricultural subsidies (from both the federal and regional governments) in 2010 equalled US$8.6 billion (Russian Federal Service of State Statistics, 1987–2012). However, the measure for agricultural support used by the WTO, called the Aggregate Measure of Support (AMS), does not include all the budget subsidies that Russia provides to agriculture (or that many WTO members also provide). Russia's pre-accession AMS category of support, therefore, is not as close to the US$9 billion bound level for 2012 that the 2010 US$8.6 billion subsidy figure suggests. None the less, Russia's commitment to reduce its AMS to US$4.4 billion by 2018 will entail either a drop in production- and trade-distorting support or limited potential to increase it. The Russian government has stated that it will move away from agricultural subsidies that directly affect (increase) production which are subject to the WTO support limit, and will move toward subsidies that are *decoupled* from production and thereby not constrained by the WTO bound support level, such as direct payments to producers (USDA, 2012).

In addition to the bounds placed on future agricultural import tariffs and market-distorting agricultural subsidies, Russia's agricultural trading partners will benefit from its WTO accession by gaining an official forum for challenging the country's sanitary and phytosanitary import restrictions. WTO accession could benefit the Russian economy and consumers by lowering prices for import-competing goods, motivating domestic producers to become more efficient (because of increased competition from imports) and increasing foreign investment and technology transfer. However, the gains to domestic producers from improved access to foreign markets could be limited. Although Russia exports some industrial and agricultural products for which market access gains are possible, such as steel, chemicals, fertilizer and grain, the bulk of Russia's export earnings come from energy (oil and natural gas). Most of the Russian economy is import competing. Major sectors of the economy lobbied strongly against accession, including aviation, furniture, financial services, telecommunications and agriculture (Cooper, 2008). This helps to explain the slow pace of Russia's accession effort.

Conclusions

During the transition decade of the 1990s, in addition to replacing state allocation of resources with markets, the KRU countries also privatized farms and other agrofood enterprises. At the same time, the large former state and collective farms stayed fairly intact, with most officially reorganized as corporate farms that remained largely unreformed in operational terms. Many of these corporate farms are weak and chronically unprofitable. Private/smallholder farming has not flourished, land reform is muddled and land markets function poorly. In contrast, the large new operators/agroholdings that arose in the KRU region around 2000 have brought investment, new technology and better management into the agrofood sector. KRU primary agriculture is a complicated blend of huge agroholdings, other large corporate farms (agricultural enterprises), struggling private/smallholder farms and small household plots and gardens.

All the KRU countries experienced a large drop in agricultural output during the 1990s, especially the overexpanded livestock sector. The substantial decline in domestic feed demand from the contraction of the livestock sector ended the large Soviet era imports of grain, soybean and soybean meal and decreased domestic grain area and production.

After the difficult decade of the 1990s, all three KRU countries have had greater agricultural success in the 21st century. On the one hand, the combination of rising consumer income from GDP growth and currency appreciation in real terms stimulated agrofood imports, especially by Russia, which became a large meat importer. On the other hand, grain production increased substantially, such that the KRU region became a major grain exporter and the livestock sector began to rebound. The agroholdings and other new and modern agrofood enterprises appear to have played a role in these positive developments, although the Russian livestock sector also has benefited from policy favouritism. A key outlook question is the degree to which the revival of the KRU livestock sectors will cut into the KRU countries' grain exports.

Even in the most optimistic future scenario for KRU agriculture, the industry faces major challenges. These include non-viable large farms, a glut of unskilled workers, a shortage of skilled labour and deficient commercial services and public institutions to support the sector. The new agroholdings appear to be a second-best response to the many dysfunctional aspects of KRU agriculture, with their own innate weaknesses. Some specialists even question whether these agricultural enterprises will last for the long run. Key outlook questions therefore are will the agroholdings and other new operators continue to expand; can they further improve their performance; and how long will they exist? For the short term at least, KRU agriculture is likely to remain a complicated mix of continuing improvement and success for some aspects and entities and continuing dysfunction and struggle for others.

References

CIA (2008–2012) *The World Factbook*. Central Intelligence Agency (CIA), Washington, DC. Available at: https://www.cia.gov/library/publications/the-world-factbook/ (accessed 5 June 2013).

Cooper, W.H. (2008) *Russia's Accession to the WTO*. CRS Report for Congress, Order Code RL31979, Congressional Research Service, Washington, DC.

Deininger, K., Nizalov, D. and Singh, S.K. (2013) *Are Mega-farms the Future of Global Agriculture? Exploring the Farm Size–Productivity Relationship for Large Commercial Farms in Ukraine*. Policy Research Working Paper 6544, World Bank, Washington, DC.

Demyanenko, I. (2008) *Agriholdings in Ukraine: Good or Bad?* Policy Paper Series AgPP No 21. German-Ukrainian Policy Dialogue in Agriculture, Kiev.

EBRD/FAO (2008) *Fighting Food Inflation through Sustainable Investment: Grain Production and Export Potential in CIS Countries*. European Bank for Reconstruction and Development (EBRD), London, and Food and Agriculture Organization (FAO), Rome.

FAO (2009) *Russian Federation: Analysis of the Agribusiness Sector in Southern Russia*. Report Series No 13. FAO Investment Centre/EBRD Cooperation Programme, Rome.

FAOSTAT (2013) Food and Agriculture Organization Statistics (FAOSTAT). Available at: http://faostat3.fao.org/home/E (accessed 2 December 2014).

Gataulina, E.A., Uzun, V.Y., Petrikov, A.V. and Yanbykh, R.G. (2005) Vertical integration in an agroindustrial complex: agrofirms and agroholdings in Russia. In: Swinnen, J.F.M. (ed.) *The Dynamics of Vertical Coordination in Agrifood Chains in Eastern Europe and Central Asia, Case Studies*. World Bank, Washington, DC, pp. 45–71.

Global Trade Information Services (2011) World Trade Atlas. Available at: http://www.gtis.com/english/GTIS_GTA.html (accessed 5 June 2013).

Godfray, H.C.J., Beddington, J.R., Crute, I.R., Haddad, L., Lawrence, D., *et al.* (2010) Food security: the challenge of feeding nine billion people. *Science* 327(5967), 812–818.

Hockmann, H., Bokusheva, R. and Bezlepkina, I. (2009) Agroholding membership: Does that make a difference in performance? *Quarterly Journal of International Agriculture* 48(1), 25–46.

Interfax (1987–2012) *Food and Agriculture Report*. Interfax, Moscow.

Karlova, N., Mokshina, P., Serova, E., Tikhonova, T. and Shick, O. (2006) Russian agrifood sector: basic trends in 2005. In: *Russian Economy in 2005: Trends and Outlook*. Institute for the Economy in Transition, Moscow, pp. 218–243.

Kazakhstan Republic Statistical Agency (1987–2012) *Kazakhstan statisticheskii ezhegodnik (Kazakhstani Statistical Yearbook)*. Kazakhstan Republic Statistical Agency, Astana.

Liefert, W.M. (2002) Comparative (dis?)advantage in Russian agriculture. *American Journal of Agricultural Economics* 84(3), 762–767.

Liefert, W.M. and Liefert, O. (2012) Russian agriculture during transition: performance, global impact, and outlook. *Applied Economic Perspectives and Policy* 34(1), 37–75.
Liefert, W.M. and Swinnen, J. (2002) *Changes in Agricultural Markets in Transition Economies*. Agricultural Economic Report No 806. Economic Research Service, USDA, Washington, DC. Available at: http://www.ers.usda.gov/media/890565/aer806_002.pdf (accessed 20 September 2013).
Liefert, W.M., Koopman, R. and Cook, E. (1993) Agricultural reform in the former USSR. *Comparative Economic Studies* 35(4), 49–68.
Liefert, W.M., Liefert, O. and Shane, M. (2009) *Russia's Growing Agricultural Imports: Causes and Outlook*. Outlook Report WRS-09-04, Economic Research Service, USDA, Washington, DC. Available at: http://www.ers.usda.gov/media/157740/wrs0904_1_.pdf (accessed 20 September 2013).
Liefert, W.M., Serova, E. and Liefert, O. (2010) The growing importance of the former USSR countries in world agricultural markets. *Agricultural Economics* 41(s1), 65–71.
OECD (1999) *Agricultural Policies in Emerging and Transition Economies*. Organisation for Economic Co-operation and Development (OECD), Paris.
OECD (2004) *Achieving Ukraine's Agricultural Potential*. OECD, Paris.
OECD (2007) *Agricultural Policies in Non-OECD Countries: Monitoring and Evaluation*. OECD, Paris.
OECD (2013) *OECD Review of Agricultural Policies: Kazakhstan 2013*. OECD, Paris.
Pomfret, R. (2008) Kazakhstan. In: Anderson, K. and Swinnen, J.F.M. (eds) *Distortions to Agricultural Incentives in Europe's Transition Economies*. World Bank, Washington, DC, pp. 219–263.
Russian Federal Service of State Statistics (1987–2012) *Rossiiskii statisticheskii ezhegodnik (Russian Statistical Yearbook)*. Russian Federal Service of State Statistics, Moscow.
Rylko, D., Khramova, I., Uzun, V. and Jolly, R. (2008) Agroholdings: Russia's new agricultural operators. In: Lerman, Z. (ed.) *Russia's Agriculture in Transition: Factor Markets and Constraints on Growth*. Lexington Books, Lanham, Maryland, pp. 95–133.
Sedik, D.J. (1993) A note on Soviet per capita meat consumption. *Comparative Economic Studies* 35(3), 39–48.
Serova, E. (2007) Vertical integration in Russian agriculture. In: Swinnen, J.F.M. (ed.) *Global Supply Chains, Standards and the Poor: How the Globalization of Food Systems and Standards Affects Rural Development and Poverty*. CAB International, Wallingford, UK, pp. 188–206.
Shagaida, N. and Lerman, Z. (2008) Land reform and development of land markets. In: Lerman, Z. (ed.) *Russia's Agriculture in Transition: Factor Markets and Constraints on Growth*. Lexington Books, Lanham, Maryland, pp. 137–183.
Ukrainian State Statistics Service (1987–2012) *Statystychnyj shhorichnyk Ukrainy (Statistical Compendium of Ukraine)*. Ukrainian State Statistics Service, Kiev.
USDA (2012) *Russian Federation: Agriculture Development Program 2013–2020*. GAIN Report No RS1270. US Department of Agriculture (USDA).
USDA/PSD (2013) FAS production, supply and distribution online. Available at: http://www.fas.usda.gov/psdonline/ (accessed 28 May 2013).
Wandel, J. (2009) *Agroholdings and Clusters in Kazakhstan's Agro-food Sector*. Discussion Paper 126. Leibniz Institute of Agricultural Development in Central and Eastern Europe, Halle, Germany.
Wandel, J. (2011) Business groups and competition in post-Soviet transition economies: the case of Russian agroholdings. *Review of Austrian Economics* 24, 403–450.
World Bank (2008–2012) Data. Available at: http://data.worldbank.org/ (accessed 5 June 2013).
WTO (2008) WTO welcomes Ukraine as a new member. *Accessions* (WTO Press Release). World Trade Organization (WTO), Geneva, Switzerland (5 February 2008).
WTO (2011) Working party seals the deal on Russia's membership negotiations. *Accessions* (WTO Press Release). WTO, Geneva, Switzerland (11 November 2011).
Yastrebova, O., Subottin, A. and Epshtein, D. (2008) Farm finances and access to credit. In: Lerman, Z. (ed.) *Russia's Agriculture in Transition: Factor Markets and Constraints on Growth*. Lexington Books, Lanham, Maryland, pp. 257–301.

2 Kazakhstan's Agricultural Development Constraints: Evidence from the Wheat, Beef and Dairy Sectors

Martin Petrick[1]* and Dauren Oshakbaev[2]
[1]*Leibniz Institute of Agricultural Development in Transition Economies (IAMO), Halle (Saale), Germany;* [2]*Independent Consultant, Astana, Republic of Kazakhstan*

Abstract

Kazakhstan is a key player in the world agricultural markets, with considerable export potential in the wheat, beef and dairy sectors. Based on unique farm-level data covering all production systems, this chapter offers new insights into the constraints that hamper Kazakhstan's further economic growth and its integration into the global food trade. While a frequently mentioned bottleneck is agriculture's lacking access to finance, most farm managers are afraid to take loans. Unavailability of land is now the most cited constraint to land expansion in the highly regulated land market of the northern grain region. Another set of constraints in wheat production is related to the market power of elevators, the vagaries of trading over long distances in an underdeveloped rail and seaport infrastructure and the intervention activities of state agencies. In the cattle sector, there are significant problems in year-round fodder supply. The value chains for beef and dairy are bifurcated into an import-dependent chain of industrially processed products serving urban consumers, and a local chain of raw products serving rural consumers and urban bazaars. A more favourable economic environment for private/individual livestock farms requires simultaneous improvements on many fronts.

Acknowledgements

The authors are grateful to Ihtiyor Bobojonov, Carol Fountain, H. Carole Schmitz, Linde Götz, Andreas Gramzow, Andrew Schmitz and Peter Voigt for providing helpful comments/edits on earlier drafts of the chapter. Research assistance by Katharina Vantomme is acknowledged.

Introduction

Based on a solid overall economic performance and relative political stability, Kazakhstan has gained an internationally recognized position among the post-Soviet countries. While much of the economic growth is driven by the Kazakhstani oil and gas sectors, the country has also become one of the top ten global exporters of wheat and flour. International observers see Kazakhstan's agribusiness sector as a key investment target deriving its attractiveness from the country's extensive arable land resources, positive demand prospects in neighbouring countries, growing domestic consumption and a relatively liberal trade regime (OECD, 2011). However, with rising incomes, many countries are shifting toward more protein-rich diets, and toward

*Corresponding author; e-mail: petrick@iamo.de

creating other opportunities in addition to the export of wheat. Kazakhstan's domestic beef and dairy production may have considerable development potential when opening up regional export perspectives.

Analysts have pointed out repeatedly the bottlenecks of further sector development (OECD, 2011, 2013): restricted access to finance for agribusinesses; low land productivity; land tenure restrictions; an underdeveloped grain transport infrastructure; livestock inventories that are too small; constraints in fodder supply; inadequate quality standards; and truncated value chains in beef and dairy production. The vast distances of the landlocked country to markets pose considerable logistic challenges for all land-dependent producers in the country. At the same time, after the collapse of most of the industrial livestock producers during the transition crisis of the 1990s, cattle and sheep now are spread across 2.5 million households, mostly private smallholder producers.

This chapter provides an empirical analysis of some of the farm-level constraints to the further development of Kazakhstan's wheat, beef and dairy sectors. To this end, farm survey data are evaluated. A hallmark of the survey data is that the data cover the entire range of farm types currently operating in Kazakhstan. According to Dudwick *et al.* (2007), OECD (2013) and Petrick *et al.* (2013), there are four important farm groupings:

1. *Household producers*. These producers were an integral part of the rural food supply during the Soviet period, particularly with regard to vegetables and livestock products. During transition, many of them took over some of the livestock from the collapsing former state farms.
2. *Private/smallholder farms*. These farms were created during the land reforms of the 1990s.
3. *Agricultural enterprises*. Many of these enterprises are former state farms.
4. *Agroholdings*. These are agricultural enterprises that belong to a horizontally/vertically integrated business group, established by domestic and outside investors.

The survey data in this chapter are particularly instructive with regard to farmers' access to the key production factors of capital, land and fodder. Furthermore, this chapter provides an overview of production and marketing structures of the industry and discusses developments based on additional information from domestic sources. This chapter analyses farm survey data; investigates constraints on credit and funding, land access among wheat producers, marketing among wheat producers, livestock feeding among dairy farmers and marketing among beef and dairy farmers; and derives policy recommendations.

Data Sources

The bulk of the data for this analysis comes from a farm survey conducted by the Leibniz Institute of Agricultural Development in Transition Economies (IAMO) in 2012. The data were collected in Akmola and Almaty provinces, two important agricultural regions of Kazakhstan. Akmola is part of the northern grain region and its agricultural output is mostly grain, whereas agricultural production in the southern foothills province of Almaty is more diversified and oriented toward livestock. In each of these two provinces, two counties were preselected, one close to and one distant from the provincial capitals. Within the given counties, farms were selected randomly on the basis of company registers provided by the local government administration. Enumerators then arranged standardized face-to-face interviews with the farm managers. Household producers were identified by visiting a sample of village residents at home. Data collection was carried out during summer and autumn, often before the crop was fully harvested and marketed, so all economic performance indicators referred to the 2011 crop year. The survey was a follow-up of a 2007 collection of data by The World Bank (Dudwick *et al.*, 2007).

The data included information about the legal status of the farms. A minimum quota was set at 150 entities registered as agricultural enterprises in the Akmola sample and 10 entities registered in the Almaty sample. Furthermore, the survey instrument questioned if each enterprise belonged to a parent organization, such as an agroholding. In this way, it was possible to distinguish four types: (i) agroholdings; (ii) agricultural enterprises; (iii) private/smallholder farms; and (iv) individual household producers. In the survey, the category of agroholding denoted a single enterprise location or branch, and was not the entire company (Petrick *et al.*, 2013).

Table 2.1 gives an overview of the sample structure and some key measures of the operational scale in the different subsamples. It demonstrates the enormous variation in both utilized area and herd sizes per farm across farm types. Most individual household producers maintain gardens on less than 0.5 hectares (ha) of land and keep one or two cows. The median private/smallholder farm area utilized for agriculture is 75 ha, and the median livestock number is 10–30 animals. The median agricultural enterprise area utilized for agricultural production is 12,800 ha of land. Agroholdings are much larger than are the individual agricultural enterprises. Cattle herds on agroholdings and agricultural enterprises are also much larger than they are on private/smallholder farms.

Farmers' Access to Credit

In 2011, total bank loans to the agricultural and food sectors in Kazakhstan were worth US$2.3 billion. Nominal interest rates in 2011 for loans to legal entities stood at about 10–13% for loans in national currency and 7–10% for loans in foreign currency. The majority of the loans to the agricultural and food sectors were made in the national currency, the Kazakhstani tenge (KZT). In late 2007 and 2008, with commercial lending to the agricultural sector in Kazakhstan strongly contracting due to the unfolding global financial crisis, the agricultural and food sectors suffered from high default rates. For example, in September 2011, 10.6% of loans to agriculture were non-performing and 33.7% were at risk (Issayeva, 2012). This situation was worse in other sectors of the economy. Due to a large share of non-performing loans and recent government bailouts, international observers regarded the Kazakhstani banking industry as weak (EBRD, 2012).

As a reaction, the Kazakhstani government promoted access to funding from the state-owned holding, KazAgro. Officially, the government declared that it feared negative consequences for domestic food security from the contraction of the private credit supply. At the same time, there was a widespread inclination to control economic outcomes from the political centre (Wandel, 2010). The KazAgro subsidiary, Agrarian Credit Corporation (ACC), is the key government agency providing farmers with subsidized credit (OECD, 2013, pp. 138–150). To this end, ACC is linked to a network of some 160 so-called Rural Credit Partnerships. These partnerships consist of 30–40 farms, the managers of which must make deposits to the ACC to become members eligible for funding. Based on available farm collateral, farmers submit their credit proposals via Rural Credit Partnerships to the ACC. Rural Credit Partnerships are the local branches of the centralized governmental subsidy programme. If the proposal is accepted, the ACC will grant a credit at a subsidized rate to the Rural Credit Partnership. The interest rate for loans to farmers from

Table 2.1. Operational scale of different farm types in the survey data, 2011. (From authors' calculations based on Kazakhstan farm survey data; IAMO, 2012.)

	Household producers	Private/ smallholders	Agricultural enterprises	Agroholdings
Number of farms in survey sample	300	245	47	8
Utilized agricultural area (ha)[a]	0.04	75	12,800	24,000
	(0.01; 0.07)	(20; 421)	(4,732; 18,136)	(17,152; 34,618)
Farms with cattle (%)	55	37	36	13
Size of cattle herd (head count)[a]	2	30	271	920
	(1; 4)	(10; 89)	(77; 408)	
Farms with dairy cows (%)	52	32	13	13
Size of cow herd (head count)[a]	1	10	225	480
	(1; 2)	(4; 30)	(32; 350)	

Note: [a]Median (first and third quartiles). All statistics are based on available data only.

the Rural Credit Partnership is twice that of the subsidized rate. Unlike traditional credit co-operatives in other countries, the Rural Credit Partnerships have no autonomy when making credit decisions, since they are not allowed to accumulate regular savings and have no control over the ACC deposits made by farmers (Gaisina, 2007). Rural Credit Partnership membership is limited to registered legal entities such as agricultural enterprises and private/smallholder farms. Private individuals cannot become members. While well-managed private/smallholder farms and agricultural enterprises can attract funding from the ACC (Petrick *et al.*, 2011), overall participation is low and default rates are high – even in the provinces with the highest penetration of Rural Credit Partnerships such as South Kazakhstan and Almaty. Less than 2% of all agricultural entities are ACC members (OECD, 2013, p. 142). In addition to the Rural Credit Partnerships, there are also direct channels through which the ACC provides funding for the agricultural sector, and there is a separate leasing programme for farm machinery, also funded by KazAgro.

The agricultural enterprises and agroholdings in the 2011 survey sample obtained a total of 39 loans in that year. Of these loans, 49% were extended by KazAgro and 31% were offered through commercial banks. The other 20% came from private moneylenders co-financing investment projects on smaller operations. These figures implied that 6% of all farms and enterprises in the sample obtained credit from KazAgro and 4% received credit from bank loans. Reported annual interest rates ranged from 1.5% to 13%, with a mean of 6.5%. KazAgro loans had a mean interest rate of 5.8% and mean bank loan rate of 7.8%. When the non-borrowers were asked about their reasons for not borrowing, 94% indicated it was because of the high interest rate, 79% indicated it was because of the default risk and 39% indicated it was because of the expensive loan costs and the involvement of bureaucracy.

Wheat Sector

While Kazakhstan has large land resources for wheat production, its lack of rainfall results in a low and highly fluctuating yield level. Even though average yields of rainfed wheat production varied between 6.8 decitonne (dt)/ha in 2010 and 16.4 dt/ha in 2011, there is a high potential for quality wheat at low production costs (Oshakbaev, 2010). There is also a transport cost advantage to regional export destinations (OECD, 2011, p. 111).

About 80% of the wheat in Kazakhstan is produced in the three northern provinces of Akmola, Kostanay and North Kazakhstan. These provinces produce most of the country's grain exports. Figure 2.1 shows the land used from 2000 to 2011 by farm type in the Kazkhstani grain region. Most farmland growth has occurred on agricultural enterprises of 10,000+ ha and private/smallholder farms of 500+ ha (Petrick *et al.*, 2011).

Constraints in land access

Even though the land code of 2003 (implemented in 2005) established the legal conditions for fully private landownership by rural citizens, most land is still state property and is leased to farmers for a 49-year term at a very low rate (OECD, 2013, pp. 151–154; Petrick, 2013). In the northern grain region, many beneficiaries of the initial land redistributions of the 1990s contributed their land share to the transformed agricultural enterprises. Because secondary rentals of land leased from the state are prohibited, short- and medium-term adjustments in land use outside the land sales market are difficult. Interviews with farmers and local experts revealed that land use outside the legal conditions of land sales occurred mostly when existing farms had changes of ownership due to liquidations or mergers and the land shares were transferred to the new owner (Petrick *et al.*, 2011). Land transactions are controlled largely by local land commissions, in which directors of existing farms and local officials are represented. Agricultural enterprises benefited from the new legislation more than did private farms, because the latter could not acquire land shares from the rural residents.

From the survey data, it is noted that land purchases are both very recent and very rare (Petrick, 2013). Land rentals occur more frequently.

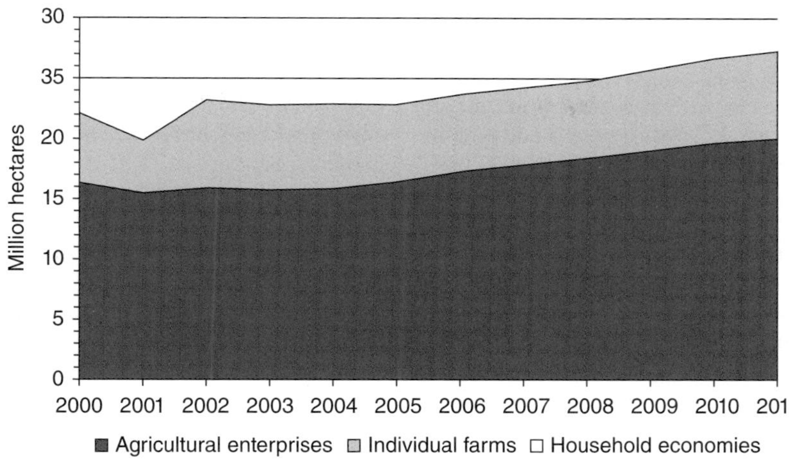

Fig. 2.1. Land use by farm type in Kazakhstan's grain region (million ha; Mha), 2000–2011. (From authors' calculations based on *Kazakhstan Statistical Yearbook of Agriculture, Forestry, and Fishery*, Agency of Statistics of the Republic of Kazakhstan, 2000–2011.)
Notes: Figure includes Akmola, Kostanay and North Kazakhstan provinces. Agricultural enterprises include agroholdings.

For example, in 2011, 52% of the private/smallholder farmers in the Akmola province rented additional land, as did 64% of the enterprises and 38% of the agroholdings. In addition, 64% of private/smallholder farms, 77% of enterprises, but only 44% of agroholdings stated that the unavailability of rental land was their main constraint.

Marketing constraints

Kazakhstan continues the Soviet era system of grain storage infrastructure, where grain is stored mainly in large, centralized storage and trading facilities called elevators. After privatization and a series of re-sales, ownership and management of the elevators has become the responsibility of the large grain producers, whose top priority is to store and handle their own grain. The provision of storage services to other producers and traders is an additional source of income to these large grain producers. In a bumper crop year, such as 2011, elevator managers exert considerable market power, which causes producers problems when trying to access the elevators. Because loan interest rates are high and the payback period for storage facilities is relatively long, many farmers cannot afford to set up their own on-farm storage facilities. Moreover, after the government introduced a grain warehouse receipt system in 2001, more than 200 elevators became licensed (OECD, 2013, p 145). Trading arrangements at the elevators are based on grain warehouse receipts issued for a certain elevator station. Farmers must deliver their wheat to the elevator by any means sooner or later to fulfil their contract obligations. Private/smallholder farmers do not sell directly to the elevator; instead, they use local traders or middlemen to market their grain. Household producers usually are not engaged in significant wheat production.

Farmers often complain that elevator laboratories try to underrate the gluten content in wheat and overrate humidity, impurities and admixtures. In the first case, the elevators may mix higher-quality wheat with lower-quality wheat, to improve the overall average quality of their yield. In the second case, the elevator obtains more wheat for less money, because an overrated content of impurities and admixtures allows the elevators to charge more for their cleaning services.

Some elevators arbitrage their grain trading gains by using grain stocks that are the property of their customers. The elevator management may sell a portion of stored grain on lucrative terms, and then may return the balance of the grain after the next harvest when the prices are relatively low, thus reaping a profit. This is illegal to do because the grain receipt system implies that holders can withdraw their grain at any time. As long as holders do not withdraw grain massively, the elevator's illicit grain deals will not be noticed. But once elevator stocks are depleted, the management may fail to find other sources quickly in case a farmer's grain must be returned. Grain owners may see themselves unable to withdraw their grain. Such hold-ups may delay farmers' trading activities, which can result in fines imposed by their business partners. Experienced farmers or traders thus keep a black list of elevators active in illegal trading practices. A recent report of the Kazakhstani Committee on State Inspection in Agriculture discovered unsupplied grain receipts totalling 180,000 tonnes (t) of grain between 2008 and 2010.

One of the few alternatives for farmers is to sell their wheat to a grain mill. With flour exports becoming more important, this practice has become a more frequently accessed marketing channel of Kazakhstani grain producers (OECD, 2013, p. 208). In the farm survey data for 2011, 34% of grain producers sold their grain directly to the elevator, 34% marketed their grain through intermediate traders, 8% sold their grain to grain mills, 14% sold no grain in the previous 12 months and 10% marketed their grain through other channels.

The transport of harvested grain is difficult because Kazakhstan is a landlocked country and the closest access to international ocean freight is via the Black Sea ports. In addition, the Russian railway system is fairly expensive because of high transport fees. The wheat price at elevators that deliver grain only by Russian railways is thus lower than it is in neighbouring elevators that have access to the Kazakhstani system. Because of the higher costs, exporting southward has become more attractive. For example, in August 2013, the wheat price at the Saryagash railway station (the main gateway to Uzbekistan) was US$22 higher than that at the Tobol railway station (the gateway to Russia). Figure 2.2 gives a snapshot of third-class wheat prices in different parts of Kazakhstan. Most contracts on the domestic market are on an elevator basis matching the ex works (EXW) rule of the International Commercial Terms. EXW means that the seller's delivery obligation is met when the seller places the goods at the disposal of the buyer at the seller's premises or at another named place. In contrast, export contracts usually follow delivered at place (DAP) rules, where the seller is responsible for delivering the product to a border station and has to clear customs there. Free on board (FOB) types of contracts, where the buyer is responsible for the freight costs, are used solely in Caspian Sea trades via Kazakhstan's only grain port in Aktau.

Although the Kazakhstani Aktau seaport grain terminal was the most used seaport for exporting grain to the Caspian Sea countries of Iran, Turkmenistan, Russia, Azerbaijan and Georgia, the capacity of the terminal at 0.5 million tonnes (Mt) was sufficient for only 30% of the exports to the Caspian Sea countries in 2009/10 (OECD, 2013, p. 211). Because of railway delivery problems in the Uzbekistan territory, Kazakhstan and Turkmenistan are constructing a direct

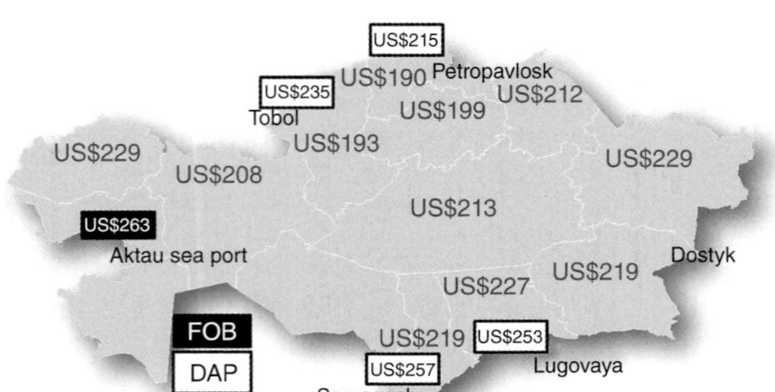

Fig. 2.2. Regional prices of third-class wheat in Kazakhstan, August 2013. (From authors' calculations based on data from Kazakh Zerno, 2013.) See text for abbreviations.

railway connection between their two countries that will also include Iran. Although Kazakhstan railways own 5200 rail cars, which is sufficient for operations in years with average harvests, this number is insufficient during peak periods when railcar shortages are compensated for by expensive rentals from Russian companies.

Key reasons for the low level of exports to China include high transportation costs and non-tariff barriers applied by China for grain deliveries. Supposedly for phytosanitary reasons, the Chinese government prohibits in-bulk transportation and requires grain to be packed in 50 kg bags. Packaging adds US$20 to each tonne of wheat, making it non-competitive on Chinese and South-east Asian markets.

Additional market risks for producers emerge from the activities of the Food Contract Corporation (FCC), a public agency managing the Kazakhstani grain reserves (OECD, 2013). Since 2003, FCC has also assumed the role of a market regulator through intervention purchases, and has a priority claim for storage and transport facilities. Beginning in 2009, the FCC pursued a countercyclical price stabilization strategy in the wheat market. In good harvest years, FCC prices may be higher than market prices, but transport to delivery points with FCC concessions may be particularly hard to get for farmers. In bad years, FCC prices may be lower than market prices, but mandatory deliveries may be established. The implied purchase licenses or obligations could lead to farmers offering bribes for FCC sales access in good years and bribes for FCC exemptions in bad years.

Beef and Dairy Sector

Before forced settlement under Russian rule, extensive pasture resources provided the basis for nomadic cattle production in Kazakhstan. Low production costs and geographical proximity to Russian markets suggested there could be a possible revival of pasture resources (OECD, 2011). Large-scale livestock production on former state and collective farms collapsed completely after the dissolution of the Soviet Union. With the disruption of mechanized supply channels of field forage and downstream processing and marketing in other Soviet republics, the socialist model of industrial cattle production was no longer viable (Pomfret, 2008). Some of the agricultural enterprises liquidated their cattle livestock. Only a small number of Kazakhstan's agricultural enterprises remained engaged in livestock production (Fig. 2.3).

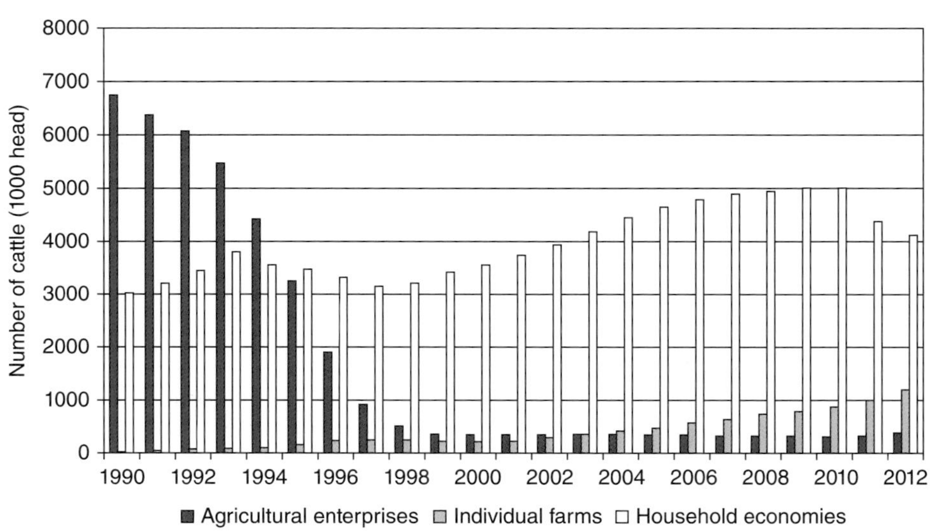

Fig. 2.3. Cattle number by farm type in Kazakhstan (1000 head). (From authors' calculations based on *Kazakhstan Statistical Yearbook of Agriculture, Forestry, and Fishery*, Agency of Statistics of the Republic of Kazakhstan, 1990–2013.)

Feeding and management constraints

Many analysts agree that improvements in feeding quality and in sanitary and management standards are necessary for further upgrading of the livestock sector in Kazakhstan (World Bank, 2004a; FAO, 2010a,b). Table 2.2 gives a snapshot of current feeding practices among different types of cattle producers in the 2012 Kazakhstan farm survey data. The large majority of household producers depend on communal grazing lands that are close to their villages. Private/smallholder farms and even agricultural enterprises also make use of this public land. Household producers have to purchase most of their supplementary fodder. Private/smallholder farms and agricultural enterprises engage in the production of hay in summer or use fodder stocks.

Compared to Western standards, commercial cattle live weight at market is lower and the fattening period is longer in Kazakhstan, due to insufficient fodder on private/smallholder farms and on agricultural enterprises (World Bank, 2004b). In addition, the prevalent use of communal grazing land is far from perfect, because of insufficient public grazing land, overstocked pastures, intermixing of livestock, spread of disease and poor management (Fig. 2.4). In contrast to the commercial producers, however, the cattle fattening period is much shorter for Kazakhstan's household producers, who probably sell their livestock early if winter feeding is scarce and if fodder purchases are expensive (FAO, 2010a,b).

Fodder shortages are critical in winter, where the main source of raw fodder for cattle is hay and silage. Household producers rely mostly on commercial sources of purchased raw fodder produced on agricultural enterprises that sell forage for cash. Compared to Western standards, the share of raw fodder purchases (not concentrate) appears to be high, which raises concerns about the economic and ecological sustainability of this sort of labour sharing in the rural economy. In contrast, cattle feeding on agricultural enterprises includes large feedlots that the government has promoted by a series of support measures. It is unclear as to what extent this system is economically viable (FAO, 2010a,b), since meat processors in Kazakhstan are widely dependent on the import of frozen meat (OECD, 2013).

Table 2.2. Performance indicators of cattle producers in Kazakhstan, 2011. (From authors' calculations based on Kazakhstan farm survey data; IAMO, 2012.)

	Household producers	Private/ smallholder farms	Agricultural enterprises, including agroholdings
Using communal grazing land (per cent of farms)	82	53	44
Strategies for winter feeding (per cent of farms, multiple answers possible):			
Complementary fodder purchases	95	47	28
Reduction of animal stock	13	10	6
Sending animals to distant pastures	4	0	0
Producing hay in summer	1	10	33
Using fodder stocks from previous years	1	44	33
Typical cattle live weight when sold for beef production (kg)[a]	N/A	280 (225; 350)	290 (250; 350)
Typical cattle age when sold for beef production (months)[a]	12 (10; 24)	24 (18; 24)	24 (24; 24)
Average daily gain (g)	N/A	403 (333; 657)	416 (361; 486)
Number of respondents in subgroup	166	90	18

Note: [a]Median (first and third quartiles). N/A = no or insufficient data. All statistics are based on available data only.

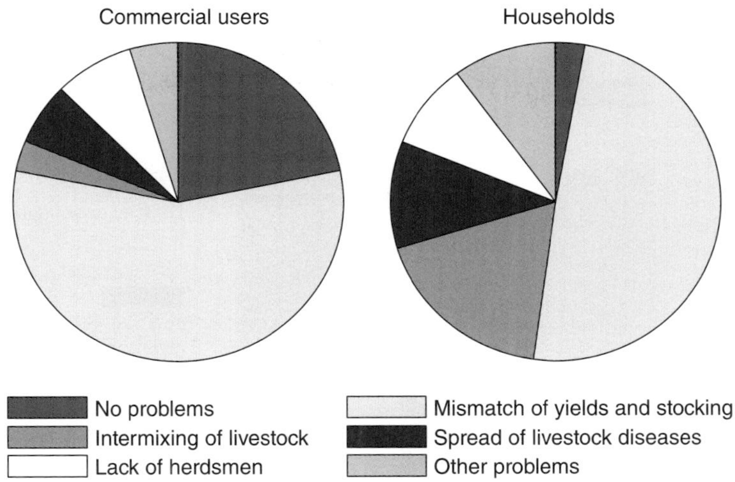

Fig. 2.4. Problems with communal grazing land by type of user in Kazakhstan, 2011. (From Kazakhstan farm survey data; IAMO, 2012.)

A crucial issue in the Kazakhstani context is whether feedlots based either on purchased or self-produced field forage can compete effectively with production systems that utilize vast communal grazing pastures on an extensive basis.

Compared to international standards, dairy yields of Kazakhstani farms are low. In 2011, median cow milk yields were 2090 kg/year for private/smallholder farms, 2100 kg/year for agricultural enterprises and 1350 kg/year for household producers.

Marketing constraints and the value chain

Cattle and dairy value chains are required to access high-value markets for processed high-quality Kazakhstani livestock products. Marketing channels differ considerably between the individual household producers, the private/smallholder farmers with small herds and the agricultural enterprises with large herds (Table 2.1). Household producers and private/smallholder farmers sell the majority of their cattle to traders (middlemen), who are responsible for acquiring slaughter venues and for acquiring the necessary veterinary certificates (OECD, 2013, pp. 231–234). Neither of these channels is subject to the strict quality and sanitary standards typical of high-value meat chains. Agricultural enterprises raising cattle sell directly to processors (50%), to marketing channels such as export markets (30%) and to traders/middlemen (20%).

The picture is similar in milk marketing (Fig. 2.5). Most household producers sell their cow milk directly to consumers in their vicinity. The typical herd size on private/smallholder farms is larger than that of household producers, so the majority of the milk from private/smallholder farms is sold to the commercial market through local intermediaries, who collect the milk and deliver it to the dairy processors for a certain monthly rate (OECD, 2013, pp. 217–219). Seasonality and milk quality are important issues for this value chain, since appropriate cooling and sanitary production conditions are not necessarily given. Also, transporting fresh milk over large distances increases the transaction costs for producers and processors. Agricultural enterprises with dairy cows often are linked directly to dairy processors, but many of Kazakhstan's dairy processors are undersupplied and depend on milk-powder imports or source their raw milk from neighbouring countries such as the Kyrgyz Republic.

Both the beef and dairy value chains are subject to a problematic bifurcation that prevents the integration of small-scale producers with high-value processing and outlets (Fig. 2.6). Industrial processors are dependent on imports for their raw material, since significant export

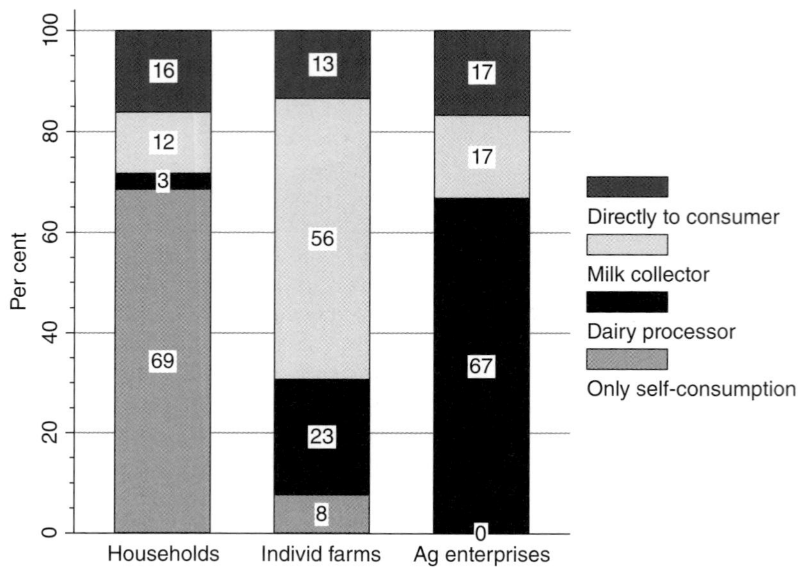

Fig. 2.5. Main marketing channels for fresh milk by type of producer in Kazakhstan, 2011. (From Kazakhstan farm survey data; IAMO, 2012.)

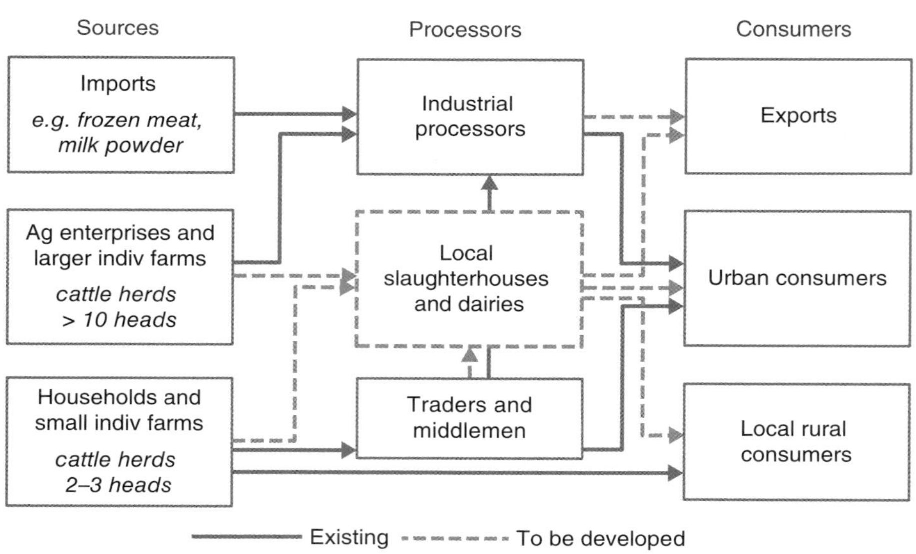

Fig. 2.6. Value chains of beef and dairy production in Kazakhstan. (From authors' illustration.)

channels of domestically produced livestock are not being developed. Small producers have limited connections to high-value markets through semi-professional intermediaries. Local slaughterhouses and dairies represent the missing link between these two branches of the value chain, but this link is yet to be established in Kazakhstan.

Conclusions and Future Direction

Kazakhstan is regarded widely as a key player on world agricultural markets, with considerable export potential in the wheat, beef and dairy sectors. While wheat cultivated in the vast northern cropland area is a main export commodity, beef and dairy products are not being generated in structures that connect easily to international value chains.

Some economists think that lack of access to value chains is because there is limited access to credit sources for agriculture in Kazakhstan. Despite low interest rates for government subsidized credit, loans to the farming sector have been very modest. This may be due to the fact that lending operations are very centralized and there is little active involvement in the banking system by farmers (Gaisina, 2007).

It has been suggested that the changes necessary to raise the sector's profitability are not dependent on more credit for Kazakhstani farmers. Rather, the evidence suggests that the management skills of the operators and of the local institutional environment should be adjusted to allow more domestic and foreign investment.

Wheat production is dominated by large and super-large farms operating in the northern rainfed grain region of Kazakhstan. Agricultural enterprises cultivating 10,000 ha/farm control about three-quarters of the total cropland in that region, and their share is increasing. Private/smallholder farmers with an average 500 ha each produce on the remaining land. The most cited constraint to land expansion is the lack of available land; most of which is rented from the government at a symbolically low price. Hence, there is little competition for land based on the economic performance of the land users. Rather, land is allocated through mechanisms in which local land commissions play key roles. Farmers receive annual payments for priority crops (OECD, 2013, p. 135). One strategy for raising productivity would be to develop land-tenure institutions that would allow more competition by transferring land to the most productive tenants. Higher revenue from state land rentals may generate financial resources to alleviate other agricultural development constraints.

Wheat producers are further restrained by the market power of elevator companies, the vagaries of trading over long distances in an underdeveloped rail and seaport infrastructure and the intervention activities of the state-mandated Food Contract Corporation. Better transport and storage facilities would make trading and shipping more reliable and would make financial gains at the farm gate more lucrative.

Kazakhstan's domestic beef and dairy value chains are currently less developed than is the wheat value chain. The value chains for beef and dairy are bifurcated into an import-dependent chain for industrially processed products serving urban consumers, and a local chain for raw products serving rural consumers. The beef and dairy value chains suffer from atomized production structures that have a weak resource base and a fragmented processing and marketing network. There are significant problems in year-round fodder supply for cattle and dairy producers. The large majority of households depend on communal grazing land, where problems of overstocking are prevalent. In winter, they have to rely on fodder purchases, mostly from agricultural enterprises. Daily weight gains and sales weight of livestock cattle are low for both household producers and private/smallholder farmers.

Improvements are needed for a more favourable economic environment for private/smallholder beef or dairy production farms, as well as for slaughterhouses and dairy processors. Upgraded local management, financial programmes for smaller entrepreneurs in rural areas and political will are likely to be more effective than are centralized funding programmes for feedlot operations. The profitability of such operations appears to be questionable and is subject to significant risks. Incentives could be provided to companies of the trade and retail sectors to establish medium-sized processing units similar to those in central and Eastern Europe (Dries *et al.*, 2009). In any case, an emerging stratum of medium-sized livestock producers is now clearly visible in official statistics and in survey data. The stratum is likely to expand further if the environment and the overall income development in Kazakhstan continue to improve.

References

Agency of Statistics of the Republic of Kazakhstan (1990–2013) *Kazakhstan Statistical Yearbook of Agriculture, Forestry, and Fishery*. Agency of Statistics of the Republic of Kazakhstan, Astana, Kazakhstan.

Dries, L., Germenji, E., Noev, N. and Swinnen, J.F.M. (2009) Farmers, vertical coordination, and the restructuring of dairy supply chains in Central and Eastern Europe. *World Development* 37(11), 1742–1758.

Dudwick, N., Fock, K. and Sedik, D.J. (2007) *Land Reform and Farm Restructuring in Transition Countries: The Experience of Bulgaria, Moldova, Azerbaijan, and Kazakhstan*. World Bank, Washington, DC.

EBRD (2012) *Integration Across Borders*. Transition Report. European Bank for Reconstruction and Development (EBRD), London.

FAO (2010a) *Sub-sectoral Cross-cutting Features and Issues*. FAO Investment Center, Food and Agriculture Organization (FAO), Rome (highlights on four livestock subsectors in Kazakhstan).

FAO (2010b) *The Meat Sub-sector*. FAO Investment Center, FAO, Rome (highlights on four livestock subsectors in Kazakhstan).

Gaisina, S. (2007) Rural credit partnerships and their role in the development of agriculture in Kazakhstan. In: Petrick, M. and Buchenrieder, G. (eds) *Sustainable Rural Development. What Is the Role of the Agri-food Sector?* IAMO, Halle, Germany, pp. 148–163.

IAMO (2012) *2011 Farm Data*. Leibniz Institute of Agricultural Development in Transition Economies (IAMO), Halle (Saale), Germany.

Issayeva, G. (2012) Agriculture in Kazakhstan. Presentation at the Global Forum for Food and Agriculture (GFFA), Berlin (20 January 2012).

Kazakh Zerno (2013) Kazakh Zerno. Available at: http://kazakh-zerno.kz (accessed 1 September 2013).

OECD (2011) *Competitiveness and Private Sector Development: Kazakhstan 2010*. Sector for competitiveness strategy. Organisation for Economic Co-operation and Development (OECD), Paris.

OECD (2013) *OECD Review of Agricultural Policies: Kazakhstan 2013*. Organisation for Economic Co-operation and Development (OECD), Paris.

Oshakbaev, D. (2010) Kazakhstan – agricultural land of opportunity. In: Agri Benchmark (ed.) *Cash Crop Report 2010*. Benchmarking Farming Systems Worldwide, Braunschweig, Germany, pp. 44–47.

Petrick, M. (2013) *Competition for Land and Labour among Individual Farms and Agricultural Enterprises: Evidence from Kazakhstan's Grain Region*. IAMO Discussion Paper 141. IAMO, Halle, Germany.

Petrick, M., Wandel, J. and Karsten, K. (2011) *Farm Restructuring and Agricultural Recovery in Kazakhstan's Grain Region: An Update*. IAMO Discussion Paper 137. IAMO, Halle, Germany.

Petrick, M., Wandel, J. and Karsten, K. (2013) Rediscovering the virgin lands: agricultural investment and rural livelihoods in a Eurasian frontier area. *World Development* 43, 164–179.

Pomfret, R. (2008) Kazakhstan. In: Anderson, K. and Swinnen, J.F.M. (eds) *Distortions to Agricultural Incentives in Europe's Transition Economies*. World Bank, Washington, DC, pp. 219–263.

Wandel, J. (2010) *The Cluster-based Development Strategy in Kazakhstan's Agro-food Sector: A Critical Assessment from an Austrian Perspective*. IAMO Discussion Paper 128. IAMO, Halle, Germany.

World Bank (2004a) *Kazakhstan's Livestock Sector – Supporting Its Revival*. World Bank Technical Paper, World Bank, Washington, DC.

World Bank (2004b) *Kazakhstan: Rangelands in Transition – the Resource, the Users, and Sustainable Use*. World Bank Technical Paper. World Bank, Washington, DC.

3 The Rise of the Former Soviet Union Region as a Major Grain Exporter

William M. Liefert* and Olga Liefert
Economic Research Service, US Department of Agriculture, Washington, DC, USA

Abstract

This chapter examines how and why the Kazakhstan, Russia and Ukraine (KRU) region has become a large grain exporter since 2000. Particular emphasis is on explaining the growth in KRU grain production, which provides domestic surpluses available for export. Rising input productivity has propelled output growth, with the large new operators being a likely driver of the productivity increase. Empirical evidence is presented showing productivity growth specifically for Russia's grain economy. This chapter also provides an outlook for the KRU region's grain production and exports over the short to medium term.

Acknowledgements

Disclaimer: The views expressed in this chapter are those of the authors and may not be attributed to the Economic Research Service or the US Department of Agriculture.

Introduction

In the 1990s, the former Soviet Union was a large grain importer. Following the country's dissolution in December 1991, the major grain producing countries of the former Soviet Union – Kazakhstan, Russia and Ukraine (the KRU region) – began their transition from a centrally planned economy to a market economy. During the 1990s, grain imports by the KRU region largely ended and the KRU countries collectively became a small net grain exporter.

Since 2000, the KRU region has emerged as a large grain exporter. Average annual (gross) KRU grain exports rose from 9 million tonnes (Mt) during the 1996–2000 period to 42 Mt during the 2006–2012 period. Over the 2006–2012 period, the KRU countries accounted for 14% of total world grain exports and 21% of world wheat exports (Fig. 3.1). The move by the KRU region from a large grain importer to an exporter has increased the volume of grain available on the world market by 60–70 Mt (Table 1.4 in Chapter 1, this volume, and Fig. 3.2). Wheat is the dominant KRU grain export, followed by maize and barley (Fig. 3.3).

This chapter examines why the KRU region has become a large grain exporter, and also provides an outlook for the region's production and exports over the short to medium term. Chapter 1 (this volume) discusses why this

*Corresponding author; e-mail: wliefert@ers.usda.gov

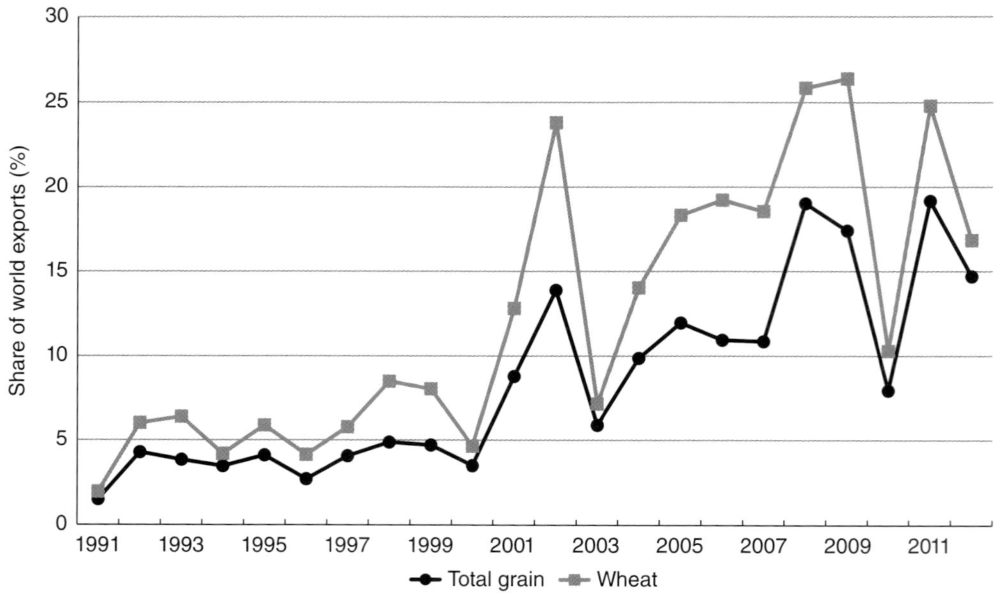

Fig. 3.1. KRU region combined gross exports in the world grain trade, 1991–2012. (From USDA/PSD, 1987–2012.)

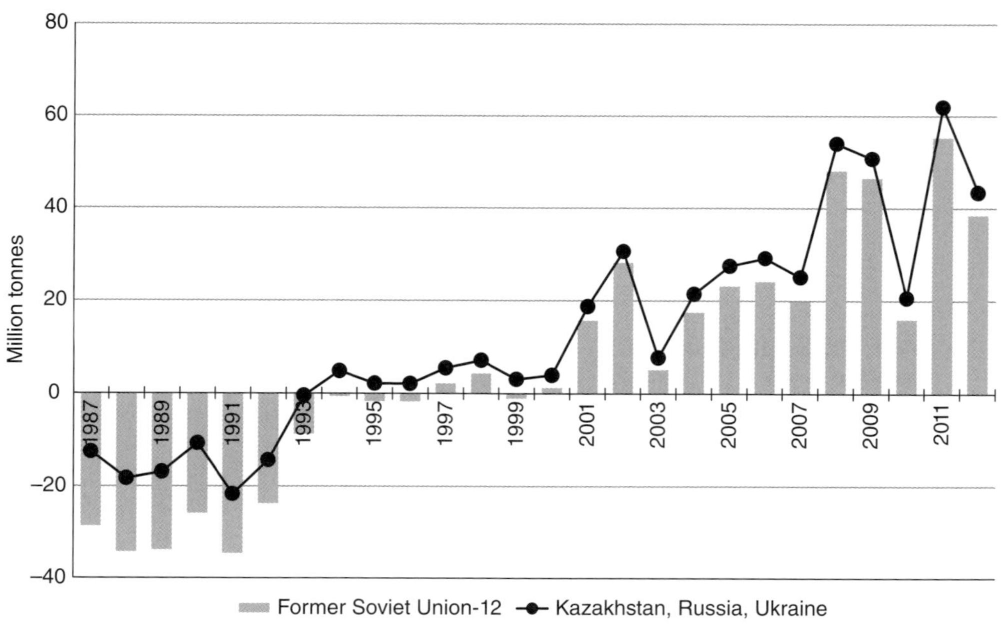

Fig. 3.2. Net grain exports (Mt) of KRU region combined and FSU-12, 1987–2012. (From USDA/PSD, 1987–2012.)

Note: Imports and exports are net of trade among the KRU countries and net vis-à-vis the rest of the world. The FSU-12 represents the republics of the former Soviet Union minus Lithuania, Latvia and Estonia.

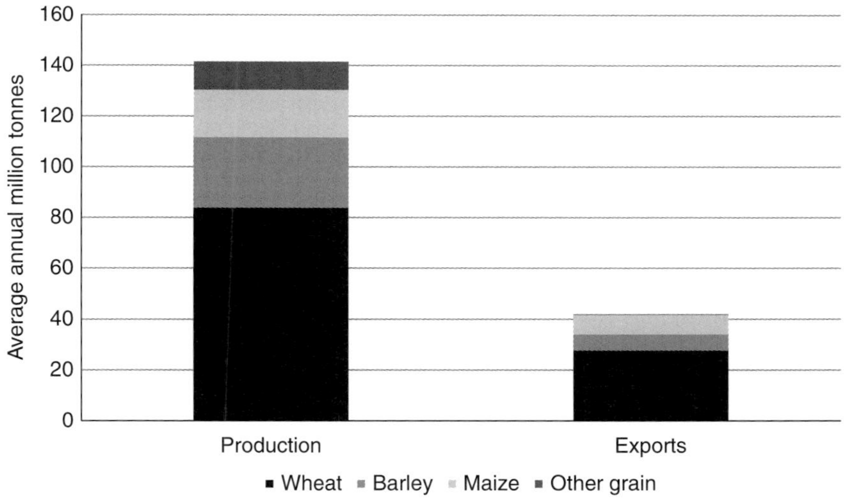

Fig. 3.3. Commodity breakdown of KRU combined grain production and exports (Mt), 2006–2012. (From USDA/PSD, 1987–2012.)

development is important for both world commercial agriculture and world food security. A key reason why the KRU region has become a large grain exporter since 2000 is because of increased grain output, which created surpluses for export. Consequently, a major issue examined is why grain production has risen since 2000, with special focus on whether productivity in the grain economy has increased. However, due to data limitations, empirical evidence for productivity growth in grain production is presented for Russia only.

Why KRU Grain Production has Increased Since 2000

There are two main reasons why the KRU region has become a large grain exporter: (i) an overall restructuring of the region's agricultural production, consumption and trade as these countries moved to (fairly) open market economies in the 1990s; and (ii) the large increase in the region's grain production since 2000.

Chapter 1 (this volume) explains how the KRU countries' transition from a planned economy to a market economy during the 1990s severely contracted the livestock sector; ended the large imports of grain, soybean and soybean meal to feed livestock; and reduced grain production. Downsizing the livestock sector was a key reason why the KRU region moved from being a grain importer to being a grain exporter. The sector's contraction substantially reduced domestic demand for animal feed, such that the large Soviet era imports of grain, soybean and soybean meal ended. Russia became a large net meat importer, Ukraine and Kazakhstan became smaller net meat importers, and all three countries started to export grain. The decrease in the need for livestock feed was substantial enough to turn the KRU region from a large grain importer during the Soviet time into a leading grain exporter, despite grain production being lower than during the late Soviet period. The KRU average annual grain output during the 2006–2012 period was 142 Mt, which was down from 160 Mt during the 1987–1991 period (Table 1.4 in Chapter 1, this volume).

After the large decline during the 1990s, grain production rose well beyond the level needed to meet domestic demand. KRU average annual grain output increased from 100 Mt during the 1996–2000 period to 142 Mt during the 2006–2012 period (Table 1.4 in Chapter 1, this volume). Production expanded primarily because of a rise in yield rather than an expansion in crop area.

KRU grain area

During the 1990s, the grain area declined in all three KRU countries (Table 3.1). To a large degree, this corrected the Soviet policy of pushing grain production on to marginal land, which had resulted in inefficient, high-cost production. Although grain area increased in Ukraine and Kazakhstan after 2000, in Russia it continued to fall slightly. The total KRU grain area during the 2006–2012 period was somewhat above that of the 1996–2000 period, but 29% below the level of the 1987–1991 Soviet period.

KRU grain yields

Given that the rise in aggregate KRU grain production since 2000 cannot be explained by growth in area, output had to increase because of the growth in yield. Table 3.1 shows that grain yields in all three KRU countries fell during the 1990s, but rebounded after 2000. Aggregate KRU grain yield increased from a yearly average of 1.52 t/ha during the 1996–2000 period to 2.10 t/ha during the 2006–2012 period.

Rising grain yields suggest that either the volume of input use increased or input productivity increased, or some combination of the two occurred. The only empirical works we could find on productivity growth in KRU agriculture since 2000 are two studies for Russia. Both indicated a substantial improvement in productivity. Bokusheva *et al.* (2012) calculated that total factor productivity (TFP) in Russian agriculture grew by about 25% between 2000 and 2008, while Swinnen *et al.* (2012) computed that agricultural TFP rose by 54% between 2000 and 2007. These studies provide strong empirical support for the argument that productivity-enhancing farm-level improvements occurring in Russian agriculture are contributing to rising output. Yet, for our purposes, this work is limited, because it covers the agricultural sector as a whole rather than focusing only on grain.

Table 3.2 provides evidence more specific to grain production that, since 2000, Russian grain yields have increased because of productivity growth rather than a rise in input use. Russia's average annual grain output during the 2006–2010 period was one-third greater than it was during the 1996–2000 period. Yet, during the 2006–2010 period, the volume of inputs used in grain production was lower compared to that of the 1996–2000 period: grain area was down 4%; agricultural labour was down 21%; grain combines decreased 37%; tractors decreased 40% (although labour and tractors cover all agriculture, not just grain production); and oil-based fuel (gasoline and diesel) was down 33%. However, it is notable that fertilizer input more than doubled.

So, the one key input for grain production that showed an increase in use during the 2001–2005 period was fertilizer. Table 3.2 shows that relative to the 1996–2000 period, fertilizer use (measured as tonnes of fertilizer per hectare of grain area, excluding maize) rose during the 2001–2005 period by about two-fifths. By the 2006–2010 period, fertilizer use was more than double that of the 1996–2000 period. However, calculations based on Russian fertilizer price data (Russian Federal Service of State Statistics, 2006) showed that the share of fertilizer in the total value of Russian grain production in 2005 was only about 5%. Given this relatively

Table 3.1. KRU grain yield (Mt/ha) and area (million ha), 1987–2012.[a] (From FAS Production, Supply and Distribution Online, USDA/PSD, 1987–2012.)

	Yield (Tonnes per hectare)			Area (Million hectares)			
	Russia	Ukraine	Kazakhstan	Russia	Ukraine	Kazakhstan	KRU
1987–1991	1.64	3.27	0.89	58.5	13.3	23.3	95.1
1992–1995	1.62	2.91	0.90	51.3	12.2	20.6	84.0
1996–2000	1.54	2.18	0.84	40.6	11.9	13.3	65.7
2001–2005	1.91	2.62	1.06	40.0	13.0	13.6	66.6
2006–2012	2.10	2.87	1.16	38.5	14.1	15.1	67.7

Note: [a]Figures are average annual values during the period identified. Area refers to harvested area only.

Table 3.2. Russian grain output and input use, 1990–2010.[a] (From USDA/PSD, 1987–2012; Russian Federal Service of State Statistics, 1987–2012.)

	1990/91	1992–1995	1996–2000	2001–2005	2006–2010
Grain output	100	88	66	80	87
Input use					
Area (harvested hectares)	100	93	74	73	71
Agricultural labour	100	103	89	80	70
Grain combines	100	85	73	66	46
Tractors	100	96	75	59	45
Fertilizer	100	46	24	34	50
Oil-based fuel	100	54	27	21	18

Note: [a]The table gives indices with the average annual value of the variable during 1900/91 = 100, except for grain output and area, where the average annual value of the variable during 1987–1991 = 100. Combines and tractors are deliveries of units to farms, per thousand hectares of sown area. Fertilizer use is tonnes per hectare of grain area. Area, combines and fertilizer are specific to grain production, while agricultural labour, tractors and oil-based fuel apply to all agriculture.

low share, and the fact that all the other inputs in Table 3.2 show a decline in use since 2000, we conclude that the overall volume of inputs used in Russian grain production has fallen since 2000 by a non-trivial degree. When combined with the large increase in grain output since 2000, the input decline indicates that input productivity has risen substantially.

There is a second (and complementary) way to assess whether input productivity has grown in the Russian grain economy since 2000 and why input use has generally fallen. The approach focuses on analysing changes in the Russian grain market in which the volume of output is determined mainly by four key factors or variables: (i) the price of grain; (ii) the prices of inputs used in production; (iii) government subsidies (for output or input use); and (iv) production technology improvements that increase input productivity. We define production technology to include farm organization and managerial and work practices as well as pure technology, such that it covers any farm-level changes that raise input productivity. If we can determine the direction and general magnitude of change in the first three variables, productivity-raising technological change becomes a residual, and we can gain insight into its likely direction of change, order of magnitude and impact on output.

We applied the above analysis to the Russian grain market for the 2001–2005 period, which was a time of relative stability for agricultural output and input prices, as well as for the macroeconomy (for Russia and the world in general). In contrast, the 2006–2010 period was less amenable to analysis because of the volatility in world grain output and input prices and because of the unstable macroeconomic conditions. As opposed to the turbulent 2006–2010 period, the relative stability of the 2001–2005 period allowed us to determine reasonably what the isolated effects of changes in key variables were on Russian grain production. If the results indicated that productivity-raising technological change occurred during that time, then the farm-level improvements responsible for the enhanced performance likely involved changes in farm management and behaviour that were preserved beyond 2005, and perhaps were even strengthened.

Table 3.3 presents the changes in the key variables that are the major drivers of the Russian grain market. The figures give the average annual value in the variables during the 2001–2005 period as a percentage change in the average annual value during the 1996–2000 period. The table figures are specific to grain production, except for agricultural input prices (including fertilizer prices) and government subsidies that cover all agriculture. The percentage changes given for prices and subsidies are based on values in real terms.

During the 2001–2005 period, Russian grain output prices, agricultural input prices in the aggregate that covers material inputs, such as fertilizer, machinery, fuel and certain agricultural services, and government subsidies all moved in directions that decreased rather than increased grain output. Grain prices fell 5%, input prices rose 13% and agricultural subsidies dropped 26%. The increase in input prices contributed to

Table 3.3. Key variables driving changes in the Russian grain market, 2001–2005 compared to 1996–2000. (From USDA/PSD, 1987–2012; OECD, 2007.)

Variable	Change (%)[a]
Output	21
Area (harvested)	–1
Fertilizer use	42
Yield	24
Output prices	–5
Agricultural input prices	13
Fertilizer prices	–6
Government subsidies to agriculture	–26
Input subsidies affecting grain production	–39
Fertilizer subsidies (within subsidies to all agriculture)	–11

Note: [a]Figures give the per cent change in the average annual value of the variable during the 2001–2005 period compared with the average annual value during the 1996–2000 period. Figures are specific to grain production, except for agricultural input prices, fertilizer prices and government subsidies, which cover all agriculture. Input subsidies affecting grain production involve such subsidies as those for fertilizer and seed use, crop insurance and soil improvement. Fertilizer use is tonnes per hectare for all grain. Prices and subsidies are in real terms.

the decline in input use (Table 3.2). Yet, during the 2001–2005 period, Russian average annual grain production was 21% higher than it was during the 1996–2000 period (Table 3.3; see Appendix Fig. A.3.1). Our analysis therefore indicated that grain output growth occurred because of a rise in input.

Productivity and yields were influenced by weather conditions. The continental climate of the KRU region has volatile weather conditions for grain production. Figure 3.4 gives KRU annual grain production over the 1995–2012 period. Grain output was low every year over the 1995–2000 period, except for an upward spike in 1997. In contrast, grain production over the 2001–2012 period was high every year except for 2003, 2010 and 2012. The weather indicators (mainly rain and temperature) showed that, in every year during the second half of the 1990s, the KRU region had unfavourable weather for grain production *except for 1997*, while in every year between 2000 and 2012, it had good weather *except for 2003, 2010 and 2012*. This information suggests that favourable weather has played some role in the growth of KRU grain yields and production since 2000.

A partial way to separate out and isolate the effect on grain yield and output growth of a rise in input productivity versus weather is by calculating grain yield based on harvested area rather than on sown area (Tables 3.1, 3.2 and 3.3). This approach partially corrected for bad weather in that it eliminated from the yield calculation area that was affected so severely by adverse weather that it produced no output. Yield comparisons between periods were therefore more meaningful because adverse weather was partially adjusted for in both periods. Based on sown area, the average annual Russian grain yield during the 2001–2005 period was 35% higher than it was during the 1996–2000 period, while based on harvested area, the average grain yield was only 24% higher. This reflects the fact that better weather also reduces the gap between planted and harvested area.

Table 3.1 shows that from the 1996–2000 period to the 2006–2012 period, yield based on harvested area increased in all three KRU countries by a significant degree. This indicates that, weather notwithstanding, productivity improvements in the KRU grain economies most likely played a key role in the growth of yields and output.

What KRU farm-level improvements could account for growing input productivity and grain output since 2000? Chapter 1 (this volume) discussed the apparent contribution to farm-level improvement in KRU agriculture that was made by new operators, and especially by agroholdings, that combined primary agriculture, processing and distribution into large vertically integrated enterprises. The new operators/agroholdings appear to be a force for productivity-enhancing technological change in KRU agriculture through such advances as the use of higher-quality seeds, improved fertilizers (and techniques such as soil testing for fertilizer rates), modern machinery and better agricultural animal breeding stock. Also, these enterprises can be linked to our previous market analysis in that their superior management and concern for cost-cutting and profitability motivate them to use all inputs more efficiently.

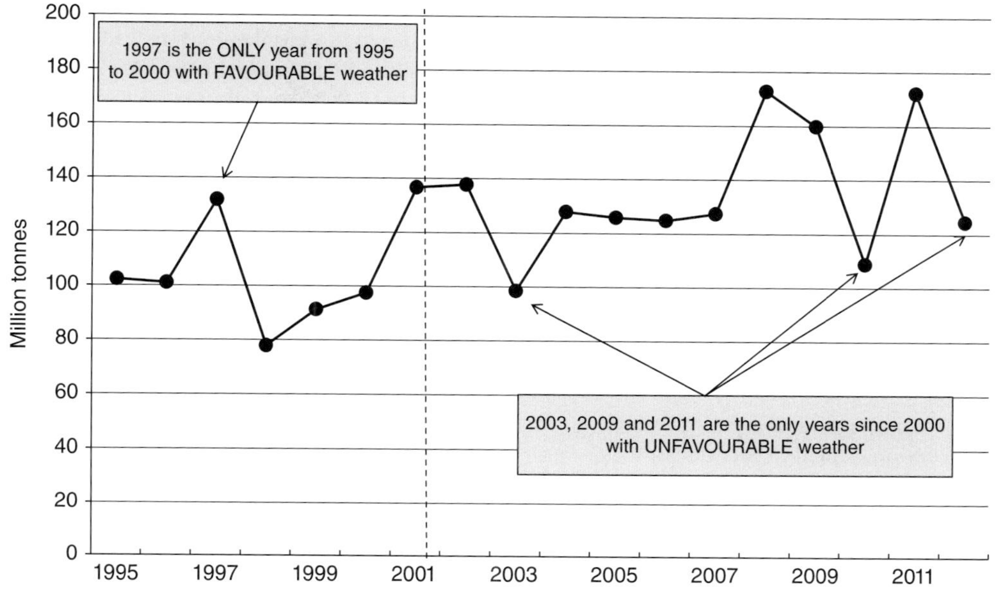

Fig. 3.4. KRU (countries combined) grain production (Mt) and weather, 1995–2012. (From USDA/PSD, 1987–2012.)

Outlook for KRU Grain Production and Exports

The future of KRU grain exports depends mainly on whether productivity and yield in grain production will continue to grow and whether domestic feed use remains relatively low, thereby increasing surplus output for export. That, in turn, depends largely on whether agroholdings and other new operators will keep growing in number and influence, and whether improvements in the overall KRU grain economy that coincide with the emergence of these new enterprises will continue.

Future KRU grain output will depend on both yield and crop area. As discussed earlier, Russia's grain area decreased substantially during the 1990s, and continued to do so after 2000 (Table 3.1). Grain area also decreased during the 1990s in Ukraine and especially in Kazakhstan, but began to increase in both countries after 2000. Over the 2006–2012 period, the grain area in Ukraine surpassed the level of the 1987–1991 period, but in Kazakhstan the grain area in the 2006–2012 period was still far below that of the late Soviet period. The surge in world agricultural and food prices in the 2006–2008 period and in 2011/12 kindled interest, both in the KRU region and the world, in returning idled KRU grain area to production. Some observers have argued that the area expansion might substantially increase world grain supplies, especially if combined with yield growth on existing KRU grain area. This could improve overall world food security, by putting downward pressure on world grain prices, mitigating any future price jumps, and benefiting the consuming poor of the world.

Given that Russia is the largest KRU grain producer and has experienced the largest drop in grain area (in absolute terms), most of the KRU grain area expansion would need to occur in Russia. As mentioned earlier, much of the area lost to grain production during the transition was probably due to low productivity/yield and high production costs. For this area to return to production, world prices will have to increase for a sustained period to cover the large costs incurred. Returning fallowed land into production again will require a substantial fixed cost of clearing the land to make it suitable for farming. In addition, the KRU region will have to invest heavily in building the physical and commercial infrastructures for storing and transporting grain, especially for production that will be expanding into remote areas.

Because the KRU ports are geared to the importation of grain rather than the exportation of grain, the storage infrastructure must be improved. This is due to the fact that import-oriented grain storage requires large capacity at the consumption sites, whereas export-oriented storage requires large capacity at the export sites (FAO, 2009). The respective KRU governments have made the improvement of infrastructure for their grain economies a priority, and infrastructure investment has increased substantially.

Since 2000, all three KRU countries have established some sort of state or parastatal grain company. In Kazakhstan, Russia and Ukraine, the companies are called the State Food Contract Corporation, United Grain Company and Agrarian Fund, respectively. The officially identified functions of these companies are to increase each government's involvement in the domestic grain market, increase grain exports and improve the physical infrastructure for the grain sector (Interfax, 2000–2013).

Since 2000, both the KRU grain and livestock sectors have grown substantially. Table 1.4 in Chapter 1 (this volume) shows that during the 2006–2012 period, average annual KRU meat production was 35% higher than it was during the 1996–2000 period, rising from 5.7 Mt to 7.7 Mt. By increasing domestic demand for agricultural animal feed, livestock sector growth has had a positive effect on grain production, but a negative effect on exports. KRU grain exports are mainly food wheat, although of low quality, and strong substitutability exists between using low-quality wheat as either food or feed.

A mitigating point concerning the trade-off between domestic feed use and exports, however, is that the KRU region appears to be improving the efficiency of its feed use, such that more meat and other livestock product output can be produced per unit of feed. (During the Soviet period, feed efficiency was very poor by Western standards.) For example, although total Russian meat production rose from the 1996–2000 period to the 2006–2010 period by 33%, grain used as animal feed increased by only 10% (USDA/PSD, 1987–2012).

The US Department of Agriculture (USDA, 2013) projected that KRU grain production and exports would grow steadily through 2022. Russian and Ukrainian grain production were projected to increase from the 2006–2012 period (average annual output) to 2022 by 21% and 60%, to 98 Mt and 64 Mt, respectively. The growth in KRU grain production would increase surplus output available for export. The USDA projection was that Russian and Ukrainian grain exports would rise by 75% and 113%, to 30 Mt and 36 Mt, respectively. By 2022, total KRU grain exports are projected at 78 Mt (Fig. 3.5). Wheat will continue as the dominant KRU grain export product, although maize's share in KRU grain exports will rise to 27%. By 2022, the KRU region's share in total world grain and wheat exports will be 20% and 30%, respectively. KRU wheat exports are projected to exceed those of the USA by 92%, with Russia's exports of wheat just about equalling the US volume. USDA also projects that from the 2006–2012 period to 2022, Russian and Ukrainian grain used as animal feed will rise by 28% and 26%, respectively. Grain exports by both countries are projected to increase substantially, despite growth in domestic grain output consumed by their livestock sectors.

We do not provide projections for Kazakhstan. The USDA models that generate the projections do not break Kazakhstan out as an individual country; rather, Kazakhstan is grouped with nine other former USSR countries (Armenia, Azerbaijan, Belarus, Georgia, Kirghizstan, Moldova, Tajikistan, Turkmenistan and Uzbekistan). However, Fig. 3.5 does include grain export projections for Kazakhstan that are the authors' own estimates based on the USDA model-generated projections for the region encompassing the country.

Conclusions

During the first decade of the 21st century, the KRU region collectively became a large grain exporting region, especially of wheat. Over the 2006–2012 period, the KRU region provided 14% of total world grain exports and 21% of total world wheat exports. The two main reasons for the KRU region's grain export growth were: (i) restructuring of the KRU countries' agricultural production and trade during the transition decade of the 1990s, in particular the downsizing of the livestock sector, which reduced domestic demand for feed grain; and (ii) growth in

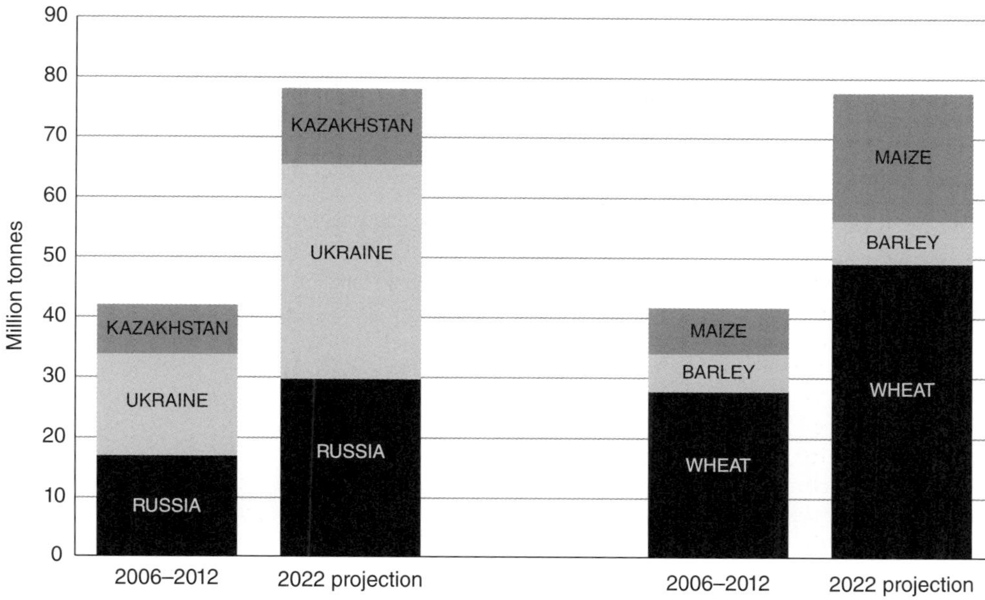

Fig. 3.5. KRU grain exports (Mt), 2006–2012 and 2022 projection. (From USDA, 2013.)
Note: The bars for 2006–2012 give average annual gross exports over the period.

grain production since 2000. Grain output rose because of apparent improvement in farm-level management and technology, which increased productivity and yields, although favourable weather during most of the decade was probably also a contributing factor.

KRU wheat and overall grain exports will likely continue to grow. USDA projects that by 2022 the KRU region's total grain and wheat exports will rise by 86% and 77%, respectively, relative to average annual volumes during the 2006–2012 period, boosting exports to 78 Mt and 49 Mt, respectively. By 2022, the KRU region is projected to supply 20% of the world's total grain exports and 30% of the world's wheat exports. Russia's wheat exports alone are projected to equal roughly those of the USA, and total KRU wheat exports will be 92% larger than the US volume.

The main reason for the projected growth in KRU grain production and exports is further farm-level improvements that increase input productivity and yields. The improvements appear to be led by large new operators that have upgraded KRU agricultural management and technology. However, KRU governments must improve the physical, commercial and institutional infrastructure for the grain economy, especially if production is to expand into more remote regions.

References

Bokusheva, R., Hockmann, H. and Kumbhakar, S.C. (2012) Dynamics of productivity and technical efficiency in Russian agriculture. *European Review of Agricultural Economics* 39(4), 611–637.

FAO (2009) *Russian Federation: Analysis of the Agribusiness Sector in Southern Russia*. Report Series No 13. FAO Investment Centre/EBRD Cooperation Programme, Food and Agriculture Organization, Rome.

Interfax (2000–2013) *Food and Agriculture Report*. Moscow (weekly).

Liefert, W.M., Gardner, B. and Serova, E. (2003) Allocative efficiency in Russian agriculture: the case of fertilizer and grain. *American Journal of Agricultural Economics* 85(5), 1228–1233.

OECD (2007) *Agricultural Policies in non-OECD Countries: Monitoring and Evaluation*. Organisation for Economic Co-operation and Development (OECD), Paris.

Russian Federal Service of State Statistics (1987–2012) *Rossiiskii Statisticheskii Ezhegodnik (Russian Statistical Yearbook)*. Russian Federal Service of State Statistics, Moscow.

Swinnen, J.F.M., Van Herck, K. and Vranken, L. (2012) Agricultural productivity paths in central and eastern European countries and the former Soviet Union: The role of reforms, initial conditions and induced technological change. In: Fuglie, K.O., Wang, S.L. and Ball, V.E. (eds) *Productivity Growth in Agriculture: An International Perspective*. CAB International, Wallingford, UK, pp. 127–144.

USDA (2013) *USDA Agricultural Projections to 2022*. Long-term Projections Report OCE-2013-1. US Department of Agriculture, Washington, DC.

USDA/PSD (1987–2012) FAS production, supply, and distribution (PSD) online. Available at: http://www.fas.usda.gov/psdonline/ (accessed 5 September 2013).

Varian, H.R. (1992) *Microeconomic Analysis*. W.W. Norton, New York.

Appendix

Assessing Productivity Growth in the Russian Grain Economy Using Output Market Analysis

Figure A.3.1 can be used to analyse the effect of changes in the key market variables on Russian grain production during the 2001–2005 period (relative to the 1996–2000 period). For simplicity, we assume that grain is one generic product. *D* is the domestic demand curve for Russian grain while S^1 is the initial domestic supply curve. We assume that P^2 is the world price for grain. This price sets the Russian domestic price, such that Q^3 and Q^4 are the quantities of grain supplied and demanded domestically, and $Q^4 - Q^3$ the quantity of grain imported (during the 1996–2000 period, Russia was a net grain importer of 3 Mt/year on average).

Our focus is on the nature of the supply curve for a firm. From a firm's profit function, we can derive its supply curve or output function which is $q = f\ (p, W)$, where q is output, p is the output price and W is a vector of input prices (Varian, 1992). Government subsidies can change the real price that producers receive for their output or pay for their inputs, and thereby also impact production. The three key price-related variables that affect grain supply therefore are the grain output price, input prices and any possible subsidies to grain producers.

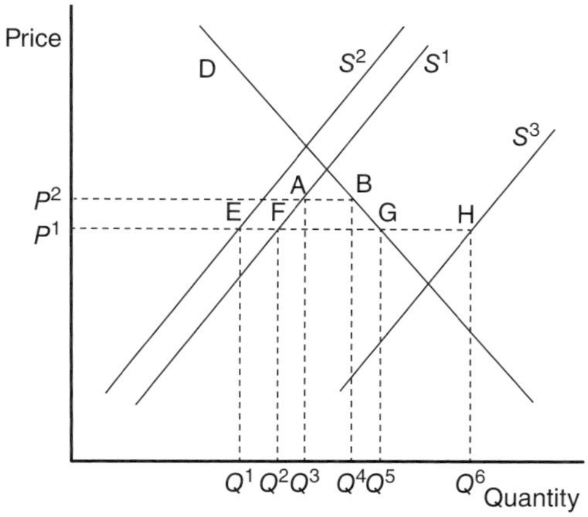

Fig. A.3.1. Changes in the market for Russian grain, 1996–2005.

Table 3.3 shows that in the 2001–2005 period, the Russian grain output price fell by 5% relative to the 1996–2000 period. In Fig. A.3.1, this drops the price from P^2 to P^1, with the isolated effect that production falls from Q^3 to Q^2. Table 3.3 also shows that in the 2001–2005 period, Russian agricultural input prices rose in the aggregate by 13%. In Fig. A.3.1, higher input prices shift the supply curve from S^1 to S^2, such that output falls further to Q^1. Higher input prices reduce input demand and use, which in turn lowers output. (The changes in output price and quantity and shifts in the supply curve as given in the figure do not necessarily reflect the actual magnitudes that occurred. Rather, the priority in drawing Fig. A.3.1 is conceptual and visual clarity.)

However, Table 3.3 also identifies that the price of fertilizer decreased by 6%. This price drop helped motivate an increase in fertilizer use (per hectare of grain area) of over 40%, which had the isolated effect of shifting the supply curve for grain right rather than left, thereby increasing production. Yet, we mentioned earlier that in 2005, the share of fertilizer in the total value of Russian grain production was only about 5%. Input prices for Russian grain production in the aggregate (including for fertilizer) increased during the 2001–2005 period, with the total input price effect being a decrease in production (leftward shift in the grain supply curve).

Russian fertilizer use fell sharply during the 1990s (Table 3.2) because domestic fertilizer prices increased substantially and fertilizer producers exported the bulk of their output (around 80%). To counter the difficulties in obtaining fertilizer, Russian regional governments during the 2001–2005 period often pressured fertilizer producers to increase deliveries to farms, and at reduced prices (Liefert *et al.*, 2003). Without such government help, fertilizer use by grain producers during that time would not have increased as much and prices might have increased rather than decreased.

Table 3.3 shows that in the 2001–2005 period, Russian government subsidies for all agriculture fell in real terms by 26% (compared to the 1996–2000 period). For producers receiving subsidies, this also shifted the supply curve leftward and reduced output. The negative effect on grain producers probably was not large, given that the bulk of subsidies went to the livestock sector. However, grain producers did benefit from subsidies on the interest paid for loans and for fertilizer purchases.

During the 2001–2005 period as compared to the 1996–2000 period, the key price-related variables that affected gain production – the grain output price, input prices and subsidies – all moved in directions that decreased rather than increased output. The total negative output effect probably was not large, but what was most important was that the effect was not positive. Hence, all the increase in output had to come from input productivity growth (resulting from either technological change/farm-level improvements or favourable weather conditions). In Fig. A.3.1, these two effects combined are represented by a rightward shift in the supply curve from S^2 to S^3. The shift increases output from Q^1 to Q^6, and in the 2001–2005 period, was large enough to move Russia from a net import to export position in grain (of $Q^6 - Q^5$).

Table 3.3 identified that in the 2001–2005 period, as compared to the 1996–2000 period, Russian grain output rose by 21%. In Fig. A.3.1, this increase equals $(Q^6 - Q^3)/Q^3$. Consequently, our results indicated that during the period in question, farm-level improvements that increased input productivity, combined with favourable weather, raised Russian grain production by a minimum of 21%.

4 Policy Issues and Prospects for Ukraine's Grain Exports

Kateryna G. Schroeder[1]* and William H. Meyers[2]
[1]The World Bank Group, Washington, DC, USA; [2]Food and Agricultural Policy Research Institute, University of Missouri, Columbia, Missouri, USA

Abstract

The Ukrainian agricultural sector has been plagued by erratic and often unfavourable government policies. Both policy uncertainty and the actual nature of policies have the effect of increasing producer risk and reducing incentives to invest and improve management and inputs in agricultural production. Progress has also been hampered by the poor marketing infrastructure, by the moratorium on land sales and by the costly transportation systems. The farm-to-port costs in Ukraine are estimated to be more than 50% higher than they are in comparable markets in the European Union (EU) and the USA. Moreover, there is very little attention drawn to research and development in the agricultural research system and the very slow and costly certification procedures for the improvement of seed technologies. This chapter focuses on the major contextual issues that impact the development and implementation of land and agricultural policies in Ukraine and discusses their implications for the integration of Ukraine with world markets. Also, this chapter provides suggestions on how to improve the policy environment in the growing Ukranian grain sector.

Acknowledgements

Disclaimer: The findings, interpretations and conclusions expressed in this chapter are entirely those of the authors and do not involve their respective organizations.

Introduction

In the 2012/13 marketing year, Ukraine was among the top ten global exporters of maize (13.5 million tonnes (Mt)), wheat (7.2 Mt) and barley (2.2 Mt) (USDA, 2013a). These numbers are quite impressive, especially because Ukraine was a net importer of the same grains in 1991. In spite of improvements, Ukrainian grain production is well below its potential, due to a number of institutional and technical inefficiencies. According to APK-Inform, an agribusiness analytical agency in Ukraine (2013), about 20 million hectares (Mha) of land in Ukraine are farmed at a subsistence level (Bond, 2013). They estimate that if yields are improved on this land, production volumes could grow from 50 Mt to 80 Mt, and grain exports have a potential to increase to 50 Mt. This may be an optimistic scenario, but there clearly is a potential for growth.

This chapter highlights major contextual issues that impact the development of the grain sector in Ukraine and points out the key barriers

*Corresponding author; e-mail: Kateryna.schroeder@gmail.com

to achieving the grain production potential within the country. Specifically, we focus on Ukraine's land and agricultural policies, infrastructure, marketing systems and financial support for agricultural research and development. This chapter discusses the policy implications these factors have for the development of Ukraine's grain sector.

Land Reform

After the collapse of the Soviet Union in 1991, land reform in Ukraine became a controversial subject due to the importance of its agricultural sector and rich land resources. Beginning with its first governmental decree, *On Land Reform* (Verkhovna Rada, 1990b), Ukraine entered an era of frequent debates and legislative changes that continued to shape its land reform (Table 4.1).

The three major stages of agricultural land privatization that have occurred in Ukraine between the dissolution of the Soviet Union in 1991 and 2014 include denationalization (transfer of land from state to collective property); allocation of land resources (distribution of collective agricultural enterprise land and assets among members, along with the issuance of landownership certificates); and land transfer (land demarcation process and acknowledgement that the land is private property) (Meyers *et al.*, 2005). Despite the accomplishment of these three stages and transfers of land to private property, the government has not ensured more preconditions for a functioning land market in the country. The freedom to buy and sell land is on hold until at least

Table 4.1. Chronology of major events of the Ukrainian land-reform process, 1990–2016. (From Verkhovna Rada (various years); Csaki and Lerman, 1997; Melnychuk *et al.*, 2005; APK-Inform, 2013.)

Year	Summary
1990/91	Laws *On Land Code* and *On Land Reform* were enacted by the Ukrainian Parliament. All lands were declared to be subject to land reform.
1992	Laws *On Forms of Land Ownership* and *On Collective Agricultural Enterprises*, and Decrees *On Acceleration of Land Reform and Privatization of Land* and *On Privatization of Land Plots* were ratified. All farmland was transferred into collective and individual ownership, except for a limited number of categories of land that stayed in state ownership. All former Soviet state farms (*sovkhozy* and *kolkhozy*) were restructured into collective agricultural enterprises, with all members entitled to a share of land in private property. In 1992, the *On Land Code* was further amended, with a 6-year moratorium on land sales.
1993–1995	Government Resolutions *On Priority Measures for Preparation and Implementation of Land Reform* and *On Immediate Measures for Acceleration of Land Reform in the Sphere of Agricultural Production*, and Decree *On Regulations for Division into Shares of Land Transferred to Collective Ownership of Agricultural Enterprises and Organizations* marked a new stage of land reform. During this time, the government introduced ownership certificates that were distributed to the landowners, allowing them to leave the collective agricultural enterprises. However, no actual exit mechanisms were introduced.
1998/99	In 1998, Law *On Land Lease* was introduced that allowed rental of land plots. Additionally, Presidential Decree *On Immediate Measures to Accelerate the Reform of the Agrarian Sector of the Economy* abolished collective ownership and ordered the reorganization of collective agricultural enterprises into enterprises of other ownership forms, such as private enterprises, associations with limited liability, joint-stock companies and private farms. This decree was critical for privatizing agrarian land and restructuring collective enterprises (Meyers *et al.*, 2005).
2001	Law *On Agreements about Alienation of Land Shares* was enacted. It prohibited selling or transferring agricultural land in any way except as an inheritance, gift or exchange. *On Land Code* introduced that same year extended the moratorium on land sales until January 2005 (Fedorenko, 2013).
2004	The Parliament of Ukraine further extended the moratorium on land sales until January 2007, with amendments to *On Land Code*.
2006–2016	The moratorium on land sales was extended until January 2016.

2016, when the moratorium on land sales is set to expire. (Note: the moratorium could be extended further.)

The moratorium on buying and selling land has delayed the development of the agricultural sector because it has prevented the movement of land plots from less to more efficient landowners. It has also limited small- and medium-sized farmer access to credit because their land plots cannot be used as collateral. Finally, the general atmosphere of land policy uncertainty slows down the development of market infrastructure in the sector. Despite all the arguments raised by numerous experts, the moratorium on land sales continues, due primarily to the fear that land sales would lead to low transparency and unfair distribution of land resources in Ukraine. Policy makers also have expressed many concerns about foreign ownership, purchases by large enterprises and lack of appropriate land market institutions. However, the longer the moratorium has been extended, the more complicated it has become to open up the land market.

Meanwhile, lack of legal ways to transfer landownership among market players has resulted in the expansion of land rental markets in the country. According to the Ukrainian Center for Land Reform Policy (2013), about 17.5 Mha (42%) of total agricultural land in Ukraine was rented out in 2013. Retirees constituted 52% of all the lessors. In 2011, the average land rental price was equal to 303 Ukrainian hryvnia (UAH)/ha, or US dollar (US$) 37/ha. After the revaluation of the nominal value of land in different regions in Ukraine in 2012, the land rental price increased to an average of 533 UAH/ha (or US$66.6/ha). Even though the Land Code of Ukraine has allowed the duration of rent to be up to 50 years, according to the Center for Land Reform Policy in Ukraine (2013), more than 81% of all rental agreements are for less than 10 years.

At times, the moratorium on land sales has also led to the development of black market operations directed at changing land ownership. As anecdotal evidence suggests, the illegal instruments include preliminary land sales agreements and land leases that allow the lessee to become the landowner once the moratorium ends (Fedorenko, 2013). Policy makers are concerned about how these arrangements could be abrogated or nullified once the moratorium ends.

Farm Structure

Among the eastern European countries that share a common Soviet heritage, Ukraine has the second weakest land market performance after Belarus (Goychuk and Meyers, 2013a). This has influenced strongly the development of types of farm structure and the types of agricultural production systems in the country. Most farm structure in Ukraine is characterized by two categories of farms: (i) agricultural enterprises and (ii) private/smallholder farms or household plots. Agricultural enterprises in Ukraine differ from private/smallholder farms in their organizational, legal and ownership origins and include state-owned enterprises, private enterprises, economic partnerships and production cooperatives, among others. Agricultural enterprises account for 63% of agricultural land use in Ukraine, while private/smallholder farms account for 27% of landownership (Meyers and Goychuk, 2012).

Between 2000 and 2013, a rapid increase in the number and size of agricultural enterprises of 3000 ha or more, and the emergence of agroholdings, occurred in Ukraine. Agroholdings are megafarms (sometimes larger than 100,000 ha) that are often vertically integrated with processors and/or exporters. Legally, such farms can be registered as a business enterprise that is a joint-stock or limited liability company owned by a group of stakeholders, or they can be registered as a private enterprise that is owned by a private individual or individuals (van Leeuwen *et al.*, 2012). The emergence of agroholdings dates back to 1999, when the Presidential Decree *On Immediate Measures to Accelerate the Reform of the Agrarian Sector of the Economy* abolished the Soviet form of collective agricultural ownership (Verkhovna Rada, 1999). Instead, it ordered the reorganization of collective agricultural enterprises into enterprises of other ownership forms such as private enterprises, associations with limited liability, joint-stock companies and private farms. The first accumulation of land by agroholdings took place between 1999, when the Decree came into effect, and 2001, when the

Government of Ukraine announced the moratorium on land sales (Fedorenko, 2013). However, it was not until 2006, when the Law *On Holding Companies in Ukraine* was enacted, that the vertical integration process took place in agroholdings, primarily in the grain, oil, sugar, meat and dairy sectors (Verkhovna Rada, 2006).

The next stage in the development of agroholdings took place in 2008/09 after the financial crisis, when smaller companies were bought out by the larger agroholdings. This led to even further concentration of land in the hands of fewer enterprises. Lissitsa (2010) estimated that Ukraine's largest agroholdings controlled 14% of the cultivated agricultural land.

Such a trend of land accumulation can be observed in Table 4.2. From 2005 to 2011, there was a 35% increase in the number of farms of more than 3000 ha and a 59% increase in the sown area of Ukraine. About 85% of this shift came from the decline in the number of 500–2000 ha farms (Meyers and Goychuk, 2013). Anecdotal evidence is that agroholdings often offer medium-sized farms a substantial price to transfer lease contracts for land these farmers originally leased from the landowners.

In terms of grain production, farms of 2000 ha or more accounted for about 15% of the total wheat production in Ukraine in 2011, about 8% of barley production and more than 30% of maize production in the country (State Statistics Service of Ukraine, 2011). The majority of this wheat production was used to feed livestock and poultry at the same farms that controlled about 28% of the total animal production in Ukraine (Kobuta *et al.*, 2012).

The reasons for the interest of investors in agroholdings include the vast availability of relatively cheap fertile land, sufficient infrastructure development, world market access, productive and relatively cheap labour, preferential terms of agricultural enterprise taxation and increasing commodity prices that convey a promise of higher profits. However, a number of experts who study the emergence and development of agroholdings point out that there are also reasons for such megafarms to be a post-USSR, rather than a Western, phenomenon. These reasons include underdeveloped institutional conditions and political economic forces pertaining to the transition economies that allow for such large capital accumulations (Demyanenko, 2008; Epshtein *et al.*, 2013). A study by Deininger *et al.* (2013) supports this further by showing that the growth of large farms in Ukraine cannot be attributed to economies of scale. When Deininger *et al.* (2013) accounted for differences of location (rayon-fixed effects) and differences of management skills (farm-fixed effects), the data showed constant returns to scale. Such results indicate that management skill, location, policies, vertical integration benefits and access to credit are the more likely reasons for the expansion of these large farm structures.

Policy Environment

Policies in the Ukrainian grain sector can be subdivided into three major categories: (i) domestic support of agricultural producers via budget transfers; (ii) fiscal measures in support of grain producers; and (iii) trade policies. Budget transfers include per hectare payments, partial reimbursements for the cost of domestically produced machinery and mineral fertilizers, reductions in the cost of insurance premiums, compensation for wheat crop losses and partial reimbursement of interest rates for loans from the commercial banks (Kobuta *et al.*, 2012). Even though Ukraine spends about 3.8% of its governmental budget on agriculture, the overall effect is that budget transfers for the improvement of grain production have been rather negligible (FAO/EBRD, 2010; OECD, 2011). For example, using the percentage of farm gate value of production (PSE) gives Ukraine a protection level that is comparable with those of Brazil and the USA, but is much lower than that of the EU (Table 4.3). The other two measures, nominal protection coefficient (NPC) and nominal assistance coefficient (NAC), also show Ukraine's protection levels to be comparable with the similar estimates of Brazil and the USA, but are lower than the protection levels of the EU. However, total producer support as a percentage of gross domestic product (GDP) is substantially higher for Ukraine than it is for Brazil, the USA and the EU. In detailed commodity analyses, the Organisation for Economic Co-operation and Development (OECD) indicates that the meat, dairy and sugar industries receive positive support, while the grain industry receives negative support (OECD, 2011).

Fiscal measures in the Ukrainian grain sector are, in turn, comprised of two main elements:

Table 4.2. Distribution of the Ukrainian agricultural enterprises by size, 2005–2011. (From authors' compilation of data from State Statistics Service of Ukraine, 2011.)

	Number of enterprises				Sown area			
	Number	Per cent	Change (%)		1000 ha	Per cent	Change (%)	
Total land area (Ha)	2011	2011	From 2008	From 2005	2011	2011	From 2008	From 2005
Total land area	44,919	100.0	–6.0	–11.03	19,494	100.0	0.17	5.85
<50	24,464	54.5	–8.0	–14.56	537	2.8	0.39	–3.57
50–100	4,236	9.4	1.0	7.68	310	1.6	0.98	6.61
101–250	4,582	10.2	–6.0	–5.78	749	3.8	–6.65	–6.36
251–500	3,199	7.1	–7.0	–10.92	1,154	5.9	–6.58	–10.96
501–1000	2,901	6.5	–6.0	–20.54	2,010	10.7	–5.67	–20.51
1001–2000	2,777	6.2	–7.0	–18.28	3,977	20.4	–6.04	–17.99
2001–3000	1,322	2.9	–4.0	–2.29	3,216	16.5	–4.26	–2.36
>3000	1,438	3.2	3.0	35.02	7,459	38.3	10.15	58.78

Table 4.3. Comparison of estimates of support to agriculture in Ukraine, Brazil, the USA and the EU, 2008–2010 average. (From OECD, 2011.)

	Ukraine	Brazil	USA	EU
Producer support estimate (PSE), %	7	5	9	22
Nominal protection coefficient (NPC), ratio	1.01	1.03	1.01	1.07
Nominal assistance coefficient (NAC), ratio	1.07	1.05	1.09	1.28
PSE as per cent of gross domestic product (GDP)	1.45	0.41	0.20	0.69

(i) fixed agricultural tax (FAT) and (ii) a system of value-added tax (VAT) exemptions. FAT was introduced in 1998 by substituting it for 12 other taxes and duties that existed at that time. To qualify for this tax, 75% of the gross revenue of agricultural enterprises must come from sales of self-produced agricultural products. In 2010, the FAT rate was equal to an average of 6 UAH/ha (or about US$0.75/ha) of arable land for all eligible enterprises, regardless of their size (BE Berlin Economics GmbH, 2011). While application of the FAT rate resulted in simple and more transparent tax procedures for agricultural producers, it also resulted in rather low contributions of the agricultural sector to Ukraine's budget.

Comparable with international practices up to 2011, Ukrainian exporters were entitled to a refund of VAT they paid while procuring their export grains. However, the implementation of such refunds has always been troublesome for the government, and in 2011, the amount of the VAT non-returnable debt to grain companies accounted for up to US$315 million (Kobuta *et al.*, 2012). Additionally, it became a common practice for the grain producers to use part of their expected VAT refund to bribe an official in charge of such refunds. To deal with this problem, the government banned VAT export refunds until 1 January 2014, by amending the tax code of Ukraine. To account for the lost VAT refunds, grain traders passed their expenses on to the producers, which resulted in the farm gate price decreasing by as much as 20% (BE Berlin Economics GmbH, 2011; Kobuta *et al.*, 2012). In effect, such a policy on the non-refund of VAT is the same as an export tax that discriminates against grain exports relative to other exports.

When it comes to grain export policy, the Ukrainian government is prone to intervene in the grain export markets in a rather *ad hoc* manner, with quotas and tariffs used as instruments for such intervention. For example, in October 2006, in response to rising global grain prices, Ukraine introduced a wheat export quota that ranged from 3000 t to 1.2 Mt between 2006 and 2008. In August 2010, following the Russian ban on wheat exports, Ukraine implemented a new export quota in the amount of 500,000 t, which was increased to 1 Mt in December 2010. In May 2011, export quotas for wheat were substituted with export tariffs that remained in place until October 2011.

Apart from the inconsistency in the announcement and actual implementation of the quotas, the decision to grant the licence to export was not done in a transparent way. For example, 43 traders applied for such a licence to export barley in 2006, but only five companies (Serna, Suntrade, Reider-Trade, Louis Dreyfus Ukraine and Barge) were granted permission to export more than two-thirds of their barley exports (UkrAgro-Consult, 2013).

In 2011, however, under pressure from grain traders, the Ukrainian government signed a Memorandum of Understanding (MOU) that established voluntary restrictions on exported grains for the 2011/12 marketing year. In that year, the MOU allowed for 24.8 Mt of grain exports to leave the country, while the actual grain exports totalled 21.8 Mt, of which 5.2 Mt of exports were of wheat, 13.9 Mt were of maize and 2.6 Mt were of barley (A. Kaliberda, Kiev, 2013, personal communication). Similar memoranda were signed in 2012 and 2013. Even though it is an improvement over the *ad hoc* approach to the export interventions, this MOU is far from ideal – it still does not allow market forces to allocate recourses efficiently, and it is prone to generate rent-seeking behaviour. As anecdotal evidence suggests, the decision on who exports what amounts of grain is still subject to decisions that are not transparent.

As expected, Ukrainian farmers seem to be the losers from the interventionist policy of their government. The average difference between French free on board (FOB) and Ukrainian ex-warehouse (EXW) prices increases twofold (in absolute terms) from about US$34/t to US$73/t in periods when export restrictions are in place as opposed to more tranquil policy periods. As an example, Fig. 4.1 displays the dynamics of the French wheat FOB to Ukrainian wheat farm price ratios. The solid vertical lines in the figure represent the introduction of wheat export quotas and the dotted vertical lines represent the abolishment of the export quota periods. Except for high ratios in late 2008 that may have been attributed to Ukrainian wheat imports by the Egyptian government, it can be seen that during times when quotas are in place, the ratio of French to Ukrainian wheat prices is higher compared to similar periods without government intervention. According to World Bank (2008) estimates, the imposition of grain export restrictions in the 2007/08 marketing year resulted in losses to agricultural producers equal to US$1.8 billion for wheat and barley combined. Apart from losses to farmers, research has found that various export restrictions in the Ukrainian market decrease the efficiency of the wheat market, which may lead to further welfare losses in the economy (Goychuk and Meyers, 2013b).

Despite the negative effects of export interventions, the Ukrainian government implemented restrictions to prevent volatile world market prices in its domestic market. However, research by Götz *et al.* (2013) revealed that the export restrictions did not result in decreased domestic wheat price volatility. On the contrary, the market uncertainty created by the government coincided with a much higher price volatility of Ukrainian wheat compared to world wheat price volatility.

Infrastructure and Marketing Systems

A study by Goychuk (2013) analysed short-run dynamics between French wheat export prices and Ukrainian farm wheat prices. The results reveal that it takes about 9 months for Ukrainian farm prices to correct 90% of the disequilibrium after a change in the French export price. Compared to the literature on short-run price adjustments in the grain market (Dawson *et al.*, 2006; Ghoshray, 2007), this is a slow rate of adjustment, signifying the presence of inefficiencies in the market and resulting in Ukrainian farmers being unable to take full advantage of increasing world wheat prices.

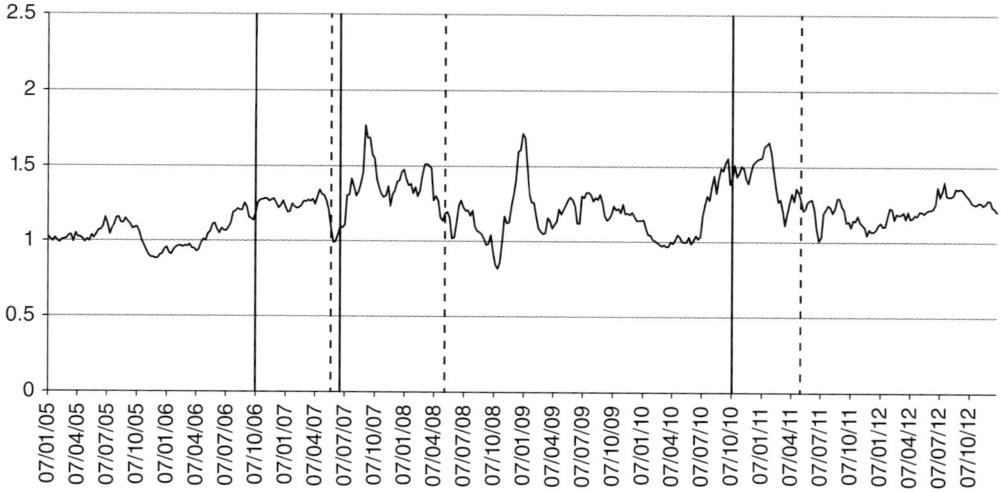

Fig. 4.1. Dynamics of French FOB to Ukrainian farm wheat price ratios, 2005–2012. (From authors' calculations.)
Notes: Solid vertical lines represent introduction of export quota period. Dotted vertical lines represent abolishment of export quota period. Dates are in day/month/year format.

Frequent policy intervention in grain exports is one such inefficiency. The Goychuk (2013) study also concluded that the long-run price transmission from world to Ukrainian farm prices decreased by 0.34 in 2007/08 and decreased by 0.14 in 2010/11, when there was higher government involvement in the grain market.

However, there existed other inefficiencies in the Ukrainian grain market that prevented Ukraine from reaching its full potential. Between 2009 and 2011, the average difference between FOB and EXW wheat (class 3) prices accounted for US$50/Mt. This suggested that Ukrainian grain producers received about 80% of the export price equal to the 2009–2011 average price of US$245. This was 10% (US$25/t) less when compared to what wheat producers received in France (BE Berlin Economics GmbH, 2011). Due to the inefficiency in the Ukrainian supply chain, their producers have lost US$500 (in absolute terms) annually since 2009. These estimations do not include adjustments for the elimination of VAT refunds or the implementation of export tariffs. The major inefficiencies in the grain marketing chain that cause major cost increases are related primarily to cumbersome grain certification processes (CIDA, 2007; BE Berlin Economics GmbH, 2011).

The overall condition of the Ukrainian physical infrastructure is satisfactory. Most of the roads in the country are paved, the country has a wide network of rail lines and grain storage capacity is sufficient to meet demand (Table 4.4).

There are, however, a number of traffic and storage bottlenecks that occur primarily during the peak months of August–October and can squeeze profit margins of the grain market players at times. Anecdotal evidence suggests that one of the major bottlenecks experienced by the grain exporters year after year is the shortage of railcars. The importance of this bottleneck is not surprising, given that 70% of grain is transported by rail, 27% by truck and 3% by river transportation (CIDA, 2007).

Lack of capacity in ports is another problem that constrains the movement of grain out of Ukraine. Existing storage capacity is being improved by agroholdings building their own storage facilities (FAO/EBRD, 2010). However, only 40% of all the grain is stored in elevators; the rest of the grain is stored at outdoor warehouses (CIDA, 2007).

Agricultural Research and Development

Research and development (R&D) in agriculture is important. Studies have shown the positive impact of public and private agricultural R&D on the growth of productivity in the agricultural sector (Alston, 2010) and the importance of agricultural growth for overall economic development and poverty reduction (Johnston and Mellor, 1961).

Between 2006 and 2010, Ukraine's average spending on agricultural R&D was US$36 million. The average research intensity ratio (percentage of agricultural R&D in agricultural GDP) for

Table 4.4. State of Ukrainian infrastructure as an average from 2010 to 2013. (From authors' compilation of data from FAO/EBRD, 2010; Deloitte, 2012; World Bank, 2013.)

Railways	Roads	Ports	Storage facilities (both grain and oilseed silos)
21,705 km of rail lines	97.9% of roads are paved	Capacity: 30 Mt annually	Capacity: 36 Mt annually
6 local railway networks	Many roads are in unsatisfactory condition	10 large and mid-size ports on the Black Sea	15% of Ukraine's silos are state owned
All rail lines are owned by the state company, *Ukrzaliznytsya*		3 ports on Sea of Azov 3 Dnieper ports 3 other river ports Several smaller ports	Storage facilities are not distributed proportionally across Ukraine

Ukraine was 0.33 during this period. Compared to ASTI's (Agricultural Science and Technology Indicators, 2013) global assessment of public agricultural R&D spending (Beintema *et al.*, 2012), this is a low share. For example, in 2008, Ukraine's average research intensity ratio was equal to 0.31, as compared to the corresponding average of 0.51 for Eastern Europe and the former USSR countries. For the lower middle-income countries, a category to which Ukraine belongs under the World Bank rankings, the average ratio in the same year was equal to 0.39. This again puts Ukraine on the lower spectrum when it comes to agricultural R&D spending.

The importance of research spending in agriculture can be shown when comparing Ukraine with Brazil. Both Ukraine and Brazil are rich in agricultural resources. Yet, in 2010, while Ukraine and Brazil had similar shares of agricultural value added in their GDP (8% for Ukraine and 6% for Brazil), they had very different research intensity ratios (0.31 for Ukraine and 1.52 for Brazil). Beintema *et al.* (2006) showed that a twofold increase in the research intensity ratio in Brazil coincided with doubled yields for its major export crops.

In 2013, Ukrainian maize yields were about 60% of those in the EU, and maize yields were less than 50% of those in the USA (Fig. 4.2). The study by Tothova *et al.* (2013) showed that the 2000–2012 Ukrainian maize yields were 24% lower than were the world averages for that period. The average barley yields in Ukraine were about 50% of those in the EU and about 66% of those in the USA. A similar situation exists with wheat yields, although some of this difference is likely due to the market and to the policy risks faced by farmers. This suggests a strong potential for the increased production of barley, maize and wheat if Ukrainian yields are to close the gap even partially with the EU and US averages. As the Brazilian example shows, increases in agricultural R&D spending could be one of the ways to achieve this.

Policy Implications

We have elaborated on various barriers that are preventing or slowing the progress of Ukraine's grain economy toward achieving the potential based on its natural resource endowments and on its existing human and physical assets. Some of these barriers may be reduced or removed through government action, and others have been created by past government action. Finding the right balance between the *invisible hand* of the market forces and the *visible hand* of the government may be difficult, but there is a widespread perception that Ukrainian policy makers have been very reluctant to let market forces play a larger role in their grain economy.

The likely market context in which Ukraine formulates and implements future policies is important. The commodity price surges of 2007/08 and 2011/12 have led to a growing consensus

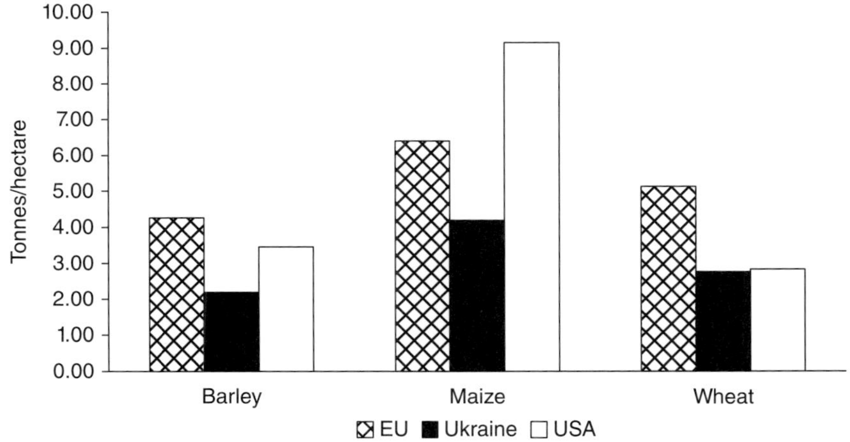

Fig. 4.2. Comparison of grain yields in the EU, Ukraine and the USA, 2000/01–2012/13. (From USDA, 2013a.)

among analysts that future prices will be higher and more volatile (FAPRI, 2013; OECD/FAO, 2013; USDA, 2013b). Two new factors influencing the direction and variance of commodity prices are higher energy prices and global climate events (FAPRI, 2013).

Because higher prices have been part of the stimulus for growth in Ukrainian grain production and exports, it is a good omen for the future. The next step is for Ukraine to build its reputation as a reliable supplier. The first priority is to reduce uncertainty through stable and predictable policies to achieve the growth and sustainability of Ukraine's grain economy.

The land market issues must be solved. Instead of more government restrictions, barriers to entry and exit should be removed. There needs to be fewer government interventions within the trade and domestic markets. Improving the infrastructure and regulatory systems will reduce farm-to-market costs and will benefit both farmers and traders. The apparent constraint on railcars suggests that more capacity and increased competition are needed. More public investment in agricultural R&D will improve the development and adoption of better management practices for crops that are higher yielding and are drought tolerant.

Increased grain production does not necessarily mean that exports will increase by the same amount. Poultry production has grown rapidly, and pig meat production has the potential to grow as well. The government needs to create a better business environment for foreign and domestic investment in meat production. A turnaround in domestic meat and dairy production may increase domestic grain use and may reduce grain exports. In either case, farmers will benefit from an expanded grain market. Creating conditions for a more efficient agricultural sector through minimum government intervention has proven to be the best formula for a dynamic and productive grain economy. That is the future Ukraine should strive to achieve.

References

Alston, J.M. (2010) *The Benefits from Agricultural Research and Development, Innovation, and Productivity Growth*. OECD Food, Agriculture, and Fisheries Papers 31. Organisation for Economic and Co-operative Development, Paris.

APK-Inform (2013) Press releases (various issues).

ASTI (2013) Agricultural Science and Technology Indicators website. Available at: http://www.asti.cgiar.org/ (accessed 30 August 2013).

BE Berlin Economics GmbH (2011) Turning Ukrainian agriculture into an engine of growth: a strategy for the development of the grains and oilseeds sector. Available at: http://www.berlin-economics.com/download/policypapers/Strategy%20Agriculture%20Ukraine_ACC-EBA_Berlin%20Economics_23%20Jan%202012.pdf (accessed 29 August 2013).

Beintema, N., Pardey, P. and Avila, F. (2006) Brazil: maintaining the momentum. In: Pardey, P., Alston, J. and Piggott, R. (eds) *Agricultural R&D in the Developing World: Too Little, Too Late?* International Food Policy Research Institute (IFPRI), Washington, DC.

Beintema, N., Stads, G., Fuglie, K. and Heisey, P. (2012) *ASTI Global Assessment of Agricultural R&D Spending: Developing Countries Accelerate Investment*. International Food Policy Research Institute, Washington, DC.

Bond, J. (2013) Investing in global farm productivity. Paper presented at the Federal Reserve Bank of Kansas City, Kansas City, Missouri.

Center for Land Reform Policy in Ukraine (2013) Internet database. Available at: http://myland.org.ua/index.php?id=3490&lang=uk&razd= (accessed 30 August 2013).

CIDA (2007) *Supply Channels for Grain and Its Products in the Ukrainian Market*. Report prepared for the Canada–Ukraine Grain Project. CanEd International, Inc, Canadian International Development Agency (CIDA), Ottawa, Canada.

Csaki, C. and Lerman, Z. (1997) *Land Reform in Ukraine: The First Five Years*. World Bank Discussion Paper No 371. The World Bank, Washington, DC.

Dawson, P., Sanjuan, A. and White, B. (2006) Structural breaks and the relationship between barley and wheat futures prices on the London International Financial Futures Exchange. *Review of Agricultural Economics* 28(4), 585–594.

Deininger, K., Nizalov, D. and Singh, S.K. (2013) *Are Mega-farms the Future of Global Agriculture? Exploring the Farm Size Productivity Relationship for Large Commercial Farms in Ukraine*. Policy Research Working Paper 6544. The World Bank, Washington, DC.

Deloitte (2012) Agricultural industry. Available at: http://investukraine.com/wp-content/uploads/2012/10/Agri_business.pdf (accessed 30 August 2013).

Demyanenko, S. (2008) *Agriholdings in Ukraine: Good or Bad?* The Institute for Economic Research and Policy Consulting, Kiev.

Epshtein, D., Hahlbrock, K. and Wandel, J. (2013) Why are agroholdings so pervasive in Russia's Belgorod oblast? Evidence from case studies and farm-level data. *Post-Communist Economies* 25(1), 59–81.

FAO/EBRD (2010) *Ukraine: Grain Sector Review and Public Private Policy Dialogue*. Report Series 15. Food and Agriculture Organization/European Bank for Reconstruction and Development (FAO/EBRD), Rome.

FAPRI (2013) *FAPRI US Baseline Briefing Book*. FAPRI-MU Report No 01-13. Food and Agricultural Policy Research Institute (FAPRI), University of Missouri, Columbia, Missouri.

Fedorenko, J. (2013) Development of agricultural holdings in Ukraine in the process of reorganization of agriculture of the XXI century. *Istoriya* 2(122), 107–111.

Ghoshray, A. (2007) An examination of the relationship between US and Canadian durum wheat. *Canadian Journal of Agricultural Economics* 55(1), 49–62.

Götz, L., Goychuk, K., Glauben, T. and Meyers, W.H. (2013) The dynamics of the Ukrainian farm wheat price volatility: evidence from a dynamic conditional correlation GARCH model. In: *Proceedings of the NCCC-134 Conference on Applied Commodity Price Analysis, Forecasting, and Market Risk Management*, St Louis, Missouri. Available at: http://www.farmdoc.illinois.edu/nccc134/conf_2013/pdf/Gotz_Goychuk_Glauben_Meyers_NCCC-134_2013.pdf (accessed 23 December 2014).

Goychuk, K. (2013) *Analysis of Asymmetric Price Transmission in the Ukrainian Wheat Supply Chain*. FAPRI-MU Report No 05-13. Food and Agricultural Policy Research Institute (FAPRI), University of Missouri, Columbia, Missouri.

Goychuk, K. and Meyers, W.H. (2013a) East European agriculture. In: Thompson, P. and Kaplan, D. (eds) *Encyclopedia of Food and Agricultural Ethics*. Springer, New York, pp. 467–475.

Goychuk, K. and Meyers, W.H. (2013b) Black Sea and world wheat market price integration analysis. *Canadian Journal of Agricultural Economics* 62(2), 245–261.

Johnston, B. and Mellor, J. (1961) The role of agriculture in economic development. *American Economic Review* 51(4), 566–593.

Kobuta, I., Sikachyna, O. and Zhygadlo, V. (2012) *Wheat Export Economy in Ukraine*. FAO Regional Office for Europe and Central Asia, Policy Studies on Rural Transition 4. Food and Agriculture Organization, Rome.

Lissitsa, A. (2010) The emergence of large-scale agricultural production in Ukraine: lessons and perspectives. Presentation, Ukrainian Agribusiness Club, Kiev, Ukraine. Available at: http://siteresources.worldbank.org/EXTARD/Resources/336681-1236436879081/5893311-1271205116054/lissitsa.pdf(accessed15 August 2013).

Melnychuk, V., Parkhomenko, S. and Lissitsa, A. (2005) Creation of Agricultural Land Market in Ukraine: Current State of Development. Available at: http://www.iamo.de/dok/dp86.pdf (accessed 15 August 2013).

Meyers, W.H. and Goychuk, K. (2012) The U in KRU: prospects for the Ukrainian grain economy. *Wies i Rolnictwo (Village and Agriculture)* 4(157), 75–89.

Meyers, W.H. and Goychuk, K. (2013) After 20 years of transition in Ukraine, will the market find a way? In: Kimhi, A. and Lerman, Z. (eds) *Agricultural Transition in Post-Soviet Europe and Central Asia After 20 Years*. Studies on the Agricultural and Food Sector in Central and Eastern Europe. Leibniz Institute of Agricultural Development in Transition Economies (IAMO), Halle (Saale), Germany, pp. 75–91.

Meyers, W.H., Demyanenko, S.I., Johnson, T.G. and Zorya, S.I. (2005) *Refocusing Agricultural and Rural Development Policies in Ukraine: Action Plan for the Road Ahead*. US Agency for International Development (USAID), Washington, DC.

OECD (2011) Agricultural policy monitoring and evaluation: OECD countries and emerging economies. Available at:http://www.oecd.org/tad/agricultural-policies/ukraine-agriculturalpolicymonitoringandevaluation2011.htm (accessed 2 September 2013).

OECD/FAO (2013) Agricultural outlook 2013–2022. Available at: http://www.oecd.org/site/oecd-faoagriculturaloutlook/ (accessed 2 September 2013).

State Statistics Service of Ukraine (2011) Agriculture of Ukraine: Statistical Yearbook. Available at: http://www.ukrstat.gov.ua/ (accessed 24 May 2013).

Tothova, M., Meyers, W.H. and Goychuk, K. (2013) Productivity growth and variability in KRU: evidence and prospects. Paper presented at the *International Agricultural Trade Research Consortium Symposium on Productivity and Its Impact on Global Trade*, Seville, Spain.

UkrAgroConsult (2013, various issues) Press releases.

USDA (2013a) Internet database: Production, supply and distribution online. Available at: http://www.fas.usda.gov/psdonline/ (accessed 30 August 2013).

USDA (2013b) *USDA Agricultural Projections to 2022*. Office of the Chief Economist, World Agricultural Outlook Board, US Department of Agriculture (USDA), Washington, DC.

van Leeuwen, M., Salamon, P., Fellmann, T., Banse, M., von Ledebur, O., *et al.* (2012) Extension of the AGMEMOD model towards Ukraine. In: Fellmann, T., Nekhay, O. and M'barek, R. (eds) *The Agri-food Sector in Ukraine: Current Situation and Market Outlook until 2025.* JRC Scientific and Policy Reports. European Commission, Seville, Spain.

Verkhovna Rada (1990a) *On Land Code*. Verkhovna Rada, Kiev.

Verkhovna Rada (1990b) *On Land Reform*. Verkhovna Rada, Kiev.

Verkhovna Rada (1992a) *On Acceleration of Land Reform and Privatization of Land*. Verkhovna Rada, Kiev.

Verkhovna Rada (1992b) *On Collective Agricultural Enterprise*. Verkhovna Rada, Kiev.

Verkhovna Rada (1992c) *On Forms of Land Ownership*. Verkhovna Rada, Kiev.

Verkhovna Rada (1992d) *On Privatization of Land Plots*. Verkhovna Rada, Kiev.

Verkhovna Rada (1993) *On Priority Measures for Preparation and Implementation of Land Reform*. Verkhovna Rada, Kiev.

Verkhovna Rada (1994) *On Immediate Measures for Acceleration of Land Reform in the Sphere of Agricultural Production*. Verkhovna Rada, Kiev.

Verkhovna Rada (1995) *On Regulations for Division into Shares of Land Transferred to Collective Ownership of Agricultural Enterprises and Organizations*. Verkhovna Rada, Kiev.

Verkhovna Rada (1998) *On Land Lease*. Verkhovna Rada, Kiev.

Verkhovna Rada (1999) *On Immediate Measures to Accelerate the Reform of the Agrarian Sector of the Economy*. Verkhovna Rada, Kiev.

Verkhovna Rada (2001) *On Agreements for Alienation of Land Shares*. Verkhovna Rada, Kiev.

Verkhovna Rada (2006) *On Holding Companies in Ukraine*. Verkhovna Rada, Kiev.

World Bank (2008) *Competitive Agriculture or State Control: Ukraine's Response to the Global Food Crisis*. Europe and Central Asia Region Sustainable Development Unit, Washington, DC.

World Bank (2013) Online database. Available at: http://data.worldbank.org/indicator (accessed 30 August 2013).

5 Wheat Export Development in Ukraine

Iryna Kobuta*
FAO Regional Office for Europe and Central Asia, Budapest, Hungary

Abstract

Wheat exports are very important for Ukraine's economy and agricultural development. This chapter examines the key factors determining the increase of wheat exports and assesses the consequences of the state regulation of wheat exports. In addition, the losses of wheat producers in Ukraine are calculated based on governmental restrictions on grain exports. Results suggest that Ukraine's export restriction policy heavily destabilizes agricultural markets and increases the risks to grain producers and restricts production.

Ukrainian Wheat Exports, 1990–2010

Ukraine is a country with well-developed agricultural traditions, where the aggregate of agroclimatic factors and the high quality of land resources favour the cultivation of cereal crops, wheat in particular. Wheat is the most popular agricultural crop, with more than one-half of all Ukrainian agrobusinesses engaged in its production. Wheat accounts for 40–50% of Ukraine's cereal crop production. Despite the quality potential of spring varieties, it was winter wheat production that prevailed in Ukraine, with a gross yield share of around 95% in the 2005–2010 period. Spring wheat crops were less than 0.1% in Ukraine until the early 1990s.

The quality structure of wheat production is dominated by its forage class. In accordance with a newly adopted standard (e.g. DSTU 3768:2009 wheat specification), forage wheat accounted for 67% of total production in 2009/10. Before adopting the new wheat standards, there was some confusion about export operations. For example, the quality and class of wheat are determined by gluten content in the domestic market and by protein content in the export market. As a result, second-class and third-class food wheat for the domestic market could be classified as feed wheat on the world market.

High internal production of wheat, coupled with a stable external demand, has established Ukraine as one of the key suppliers of this crop to the world market (State Statistics Service of Ukraine, 1995–2010). Export external deliveries are the most advantageous method to utilize the surplus, adjust the wheat supply and balance demand.

During the Soviet era, the grain market functioned under a government contract system. The organizations affiliated with state institutions or major companies that carried out barter

*E-mail: iryna.kobuta@fao.org

deliveries had access to public finances or to bank loans. At that time, grain exports were controlled by the government, so it was not surprising that external grain deliveries were few and irregular.

After the 1991 Soviet dissolution, more than 400 cereal product manufacturers were allowed to be privatized and independent private companies started to export cereals. These private exporters squeezed out many of the state operators that were replaced by large transnational companies, such as Toepfer, Cargill, WJ Ukraine and Serna. Access to low-interest, long-term lending resources gave transnational companies a competitive global market advantage. Trading companies owned by Ukrainians were also present in the cereal market, but their function was to act as an intermediary between the producers and the large grain traders. In 2007, this situation changed when exporters began investing in agricultural production and began forming their own grain resources through agroholdings. This caused the redistribution of market shares of the largest grain traders, such as Nibulon, Kernel, Serna, Mironivskyi Hliboprodukt and Rise, and the consolidation of grain elevators and reception centres by private companies.

The leading exporters have been increasing port capacities for cereals storage and trans-shipment. Developing their own infrastructure has allowed them to be independent from public ports and from intermediaries. It also has allowed them to benefit from grain trans-shipments and has enhanced the competitiveness of their own grain, having the advantage of US$10/tonne (t) to US$20/t (including trans-shipment costs). Nibulon, an agroholding company that is actively expanding its own port infrastructure, is the leader in this regard.

Wheat plays a key role in total cereal crop supplies to external markets. Its share in the 1990–2010 period averaged 47% (Fig. 5.1). In addition to wheat, Ukrainian grain exports have been dominated traditionally by barley and maize. During the same period, the share of barley in total grain exports was 35% and maize was 13%. These three crops accounted for 96% of the grain exports.

Since 2000, Ukraine has become one of the largest suppliers of feed wheat to the world market (FAS/USDA, 2013). Low- and medium-quality wheat averaged about 50% in the total structure of Ukrainian wheat exports to the world markets for the 1995–2010 period, reaching as high as 72–97% in some years (Fig. 5.1).

Traditionally, Near Eastern, North African and European Union (EU) countries have been the main consumers of Ukrainian wheat. This is due to lower transportation costs and need for these types of cereals. Asian and African countries demand Ukrainian milling wheat, whereas EU countries give preference to importing feed wheat.

Ukrainian wheat exports have grown significantly relative to Asian and African countries, especially for the 2006–2010 period, due to both increased demand for medium-quality food wheat in these countries and significantly reduced freight costs. Over the long term, exports

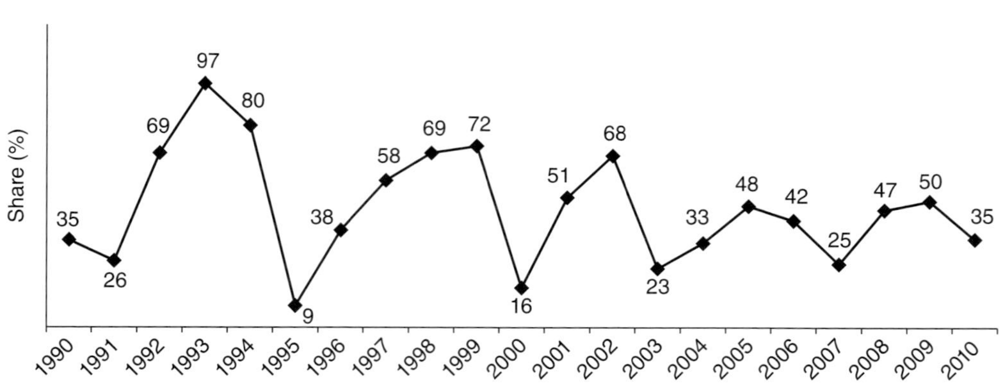

Fig. 5.1. Share of wheat (%) in total grain exports, 1990–2010. (From State Statistics Service of Ukraine, 1990–2010.)

to South-east Asia and to Africa's sub-Sahara will probably increase the most because of rapid population growth and because of accompanying food needs in these regions (Fig. 5.2). Over the same period, exports to the EU decreased, due to export quotas for low- and medium-quality wheat that have been evident in the EU since 2002. Exports to the former Soviet countries have been insignificant.

Grain quality presents a serious problem in Ukrainian cereal export operations because of factors that include bug infestations, dockage in grain lots and broken grain, which may threaten future wheat sales. For instance, Egypt, the largest importer of wheat in the world, had to suspend purchases of Ukrainian wheat in 2009/10 due to stricter requirements for the quality of imported wheat. This resulted in a 35% reduction of Ukrainian wheat deliveries to that country compared to the previous season. There are also problems with the export infrastructure and excess transport costs.

Key Factors Determining the Increase of Wheat Exports in Ukraine

During the 1990–2000 period, wheat production decreased and wheat exports increased. A major reason was the entry of large transnational companies into the Ukrainian market and high revenues obtained from Ukraine's export activities. In addition, Ukraine's internal operations were not as lucrative, due to considerable price fluctuations, administrative interference and limited market capacity.

However, beginning in 2000, wheat exports began trending upwards (Fig. 5.3). The correlation coefficient between wheat production and the export volume over this period was 0.67, which revealed a rather close relationship between these two indicators. It may be concluded that wheat production was the principal factor determining the volume of export deliveries of wheat from Ukraine in the 2000–2010 period.

Several factors put pressure on Ukraine's wheat market between 1990 and 2010 that reduced demand on its internal market and

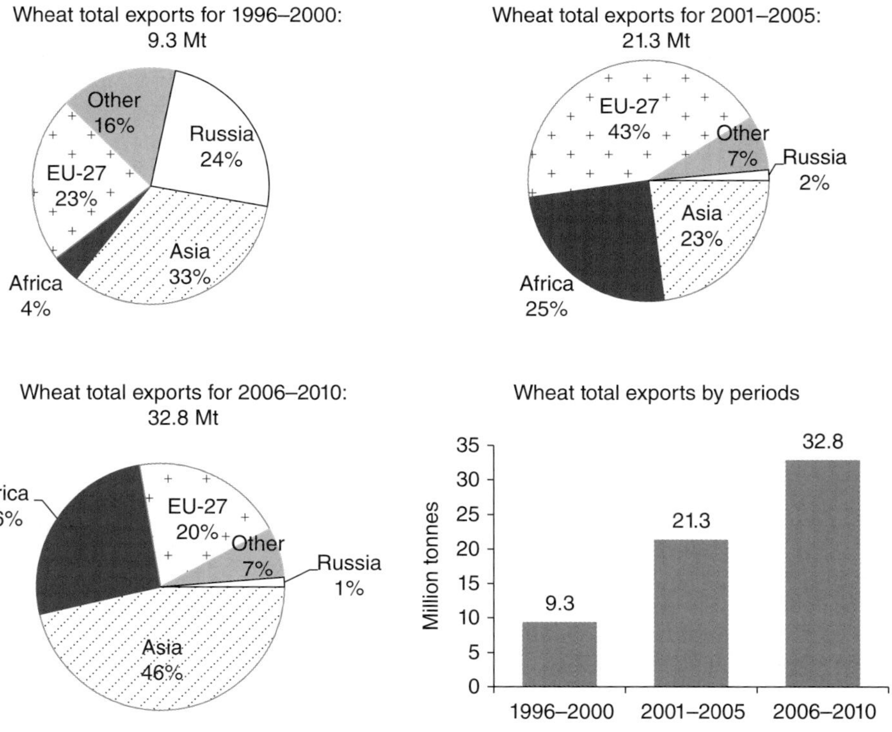

Fig. 5.2. Geography of wheat exports from Ukraine, 1996–2010. (From State Statistics Service of Ukraine, 1996–2010.)

Fig. 5.3. Changes in wheat production and exports, 1990–2010. (From State Statistics Service of Ukraine, 1990–2010).

increased its volumes of wheat exports. The first group of factors was due to decreased production in Ukraine. For example, annual per capita consumption of bread products decreased from 141 kg in 1990 to 111.7 kg in 2009. In addition, reduced feedstock caused stagnation of the domestic feed mill industry.

The second group of factors was related to the improvement and development of the Ukrainian export infrastructure, as well as its capacities for grain storage and processing. Low exports were caused by insufficient elevator facilities and low trans-shipment capacity at the ports. Due to the lack of public financing, infrastructure development funding came mostly from private investors. Investments were made in all elements of the logistic chain, which included elevator construction, port development infrastructure and equipment and technology acquisition. As a result, trans-shipment of exported wheat was about 36 million t (Mt) in 2010, compared to 6 Mt in 2003.

The third group of factors was the high level of government and administrative pressure put on the wheat market, such as export restrictions. While the first two groups of factors had a positive impact on export trends, the third group of factors had a negative impact on export trends.

Implications of Wheat Production Growth for Export

If wheat production growth continues in Ukraine, it will depend directly on export regulation. Because external markets are facing keen demand for wheat and are expecting favourable price trends, unrestricted exports will undoubtedly result in production growth, particularly for the short- and medium-term outlook.

Should Ukraine retain the tendency toward tightening both the export regime and the unjustified administrative interference on the grain crop market, agricultural producers will have to switch to growing more profitable crops such as maize or soybean. Even then, wheat most likely will remain a priority crop for export in Ukraine due to its limited domestic demand.

Given the upward trend of wheat demand since 2000, it may be assumed that Ukrainian wheat demand for domestic food consumption will be in the range of 5–5.5 Mt/year. Domestic wheat demand for feed purposes will not exceed 4–4.5 Mt annually. In the short- and medium-term perspective, aggregate domestic demand (or use) of wheat will be 11–12.5 Mt per annum (i.e. should production overstep that threshold, the difference will be exported, needless to say, without export restrictions). Given the absence of grain market destabilizing factors, Ukrainian government policy with regard to the regulation of grain exports may have a negative effect on domestic production, to the extent that sown area could be reduced.

Market operators intend to cover expenditures for export restrictions at the cost of the margin between domestic and world prices. For the time being, the world market is facing a big demand for inexpensive fodder wheat due to high prices for maize and its limited reserves.

Such a situation spurs mixed fodder producers on to use more fodder wheat.

It is necessary to note that Ukrainian wheat exports were the only effective mechanisms for regulating the wheat balance in the 1995–2010 period. Grain export deliveries from Ukraine could have been more significant and stable if the state policy regarding the export regulation of the grain market had been more well thought out and appropriate.

The situation on the world food market also contributed to the increase in export deliveries of wheat from Ukraine. World population growth requires large volumes of grain crops, both for food and feed purposes. In this regard, the role of wheat as a food crop has increased, especially in the Near Eastern, African and Latin American countries. Wheat consumption has also risen in traditional rice growing nations, especially in Indonesia and Bangladesh, due to urbanization and a significant shift to the Western dietary pattern.

Therefore, the situation existing within the world wheat markets, as well as the country's export potential, contributed to the establishment of Ukraine as a major world wheat supplier. The role of Ukraine strengthened after 2001, when world wheat reserves decreased and wheat prices increased.

Implicit Taxes for Producers through Trade Policies

Average annual prices for wheat in Ukraine (as well as annual monthly prices for the last years) follow the world wheat market, except the years of drastic changes in gross yield of wheat; for example, in 2003. The gap between the 2000 and 2010 world prices for wheat and domestic prices in Ukraine is shown in Table 5.1. For the purpose of comparison, two indices are given to demonstrate world prices: (i) the Chicago Board of Trade (CBOT) USA price; and (ii) the border reference price (export price for 1 t of wheat free on board (fob)), adjusted for transportation, storage and trans-shipment for Ukraine.

As shown in Table 5.1 and Fig. 5.4, the market price differential for the 2006–2008 period in Ukraine was higher because of the influence of export quantitative restrictions on wheat producer prices in Ukraine for those years.

Export restrictions

One of the main World Trade Organization (WTO) requirements applies to quantitative restrictions on exports. As a member of the WTO, Ukraine undertook not to apply quantitative restrictions on export unless it was called for by a critical shortage of food products in the country. Ukraine also assumed obligations to lift restrictions for the export of grain that were in force at the time of its accession to the WTO. Before entry into the WTO, export quotas for grain and oil crops were imposed to mitigate the effect of crop failure on internal prices of main foodstuffs. For example, within the 2006–2008 period, export quotas for wheat were imposed

Table 5.1. Wheat prices (t) and market price (UAH) differential in Ukraine, 2000–2010. (From author's calculations based on OECD database for Ukraine and US CBOT data, 2000–2010.)

		2000	2002	2004	2006	2008	2010
Producer price (farm gate)	UAH/t	487	310	493	527	753	1086
Reference price (farm gate)	UAH/t	410	343	510	620	1226	1176
Border reference price (fob/cif)[a]	US$/t	94	82	113	148	279	180
Handling and processing costs	UAH/t	131	107	146	126	244	253
Quality adjustment	Ratio	1.05	1.03	1.09	1.00	1.00	1.00
Official exchange rate	UAH/US$	5.44	5.33	5.32	5.05	5.27	7.94
Market price differential	UAH/t	77	–33	–17	–93	–473	–90
	US$/t	14	–6	–3	–18	–90	–11
CBOT (SRW[b] and EXW[c] wheat)	US$/t	95	119	127	148	293	213
Differential between Ukrainian producer price and CBOT	US$/t	–5	–61	–35	–43	–150	–77

Notes: [a]Fob/cif = free on board/cost, insurance, freight, whereby fob is paid by the seller and cif is paid by the buyer; [b]SRW = soft red wheat; [c]EXW = ex works (named place of delivery); CBOT = Chicago Board of Trade.

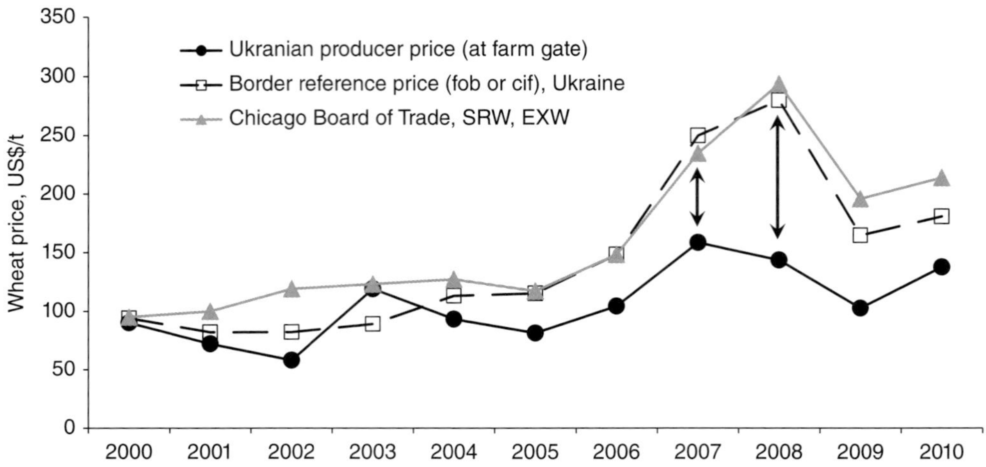

Fig. 5.4. Comparison of Ukrainian wheat producer price, border price and price on the Chicago Board of Trade (CBOT), USA, 2000–2010. (From author's calculations based on data from the State Statistics Service of Ukraine, the State Customs Service of Ukraine and CBOT of the USA information, 2000–2010.)

five times. The grain quota imposed from July 2007 to October 2007 made up only 12,000 t (3000 t wheat, 3000 t barley, 3000 t rye and 3000 t maize), essentially prohibiting exports. Immediately after WTO accession in 2008, an export quota (300,000 t) for sunflower seed oil and a prohibitive quota (1000 t) for sunflower seeds were imposed by the WTO for a short period. It was abrogated by the President of Ukraine as conflicting with Ukraine's commitments to the WTO in the Decree of the President of Ukraine No 481 (Kobuta *et al.*, 2012). According to World Bank estimations, the imposition of export restrictions for cereals and sunflower seeds in 2007/08 resulted in losses to agricultural producers totalling US$1.8 billion for wheat and barley and US$0.66 billion for sunflower seeds (World Bank, 2008).

In 2009, the Ukrainian government refrained from imposing quantitative restrictions on grain and oilseed exports. However, it reverted to setting quotas for grain exports in 2010. Specific quotas were imposed on five grain crops (wheat, maize, barley, rye and buckwheat). The quotas resulted in losses of grain exporters and producers, and placed its compliance with respective WTO agreements into question.

In 2011, Ukraine's Verkhovna Rada adopted Draft Law No 8324 on the sales of quotas by auctions, which the Ukrainian president vetoed (Kobuta *et al.*, 2012). One of the reasons for the veto was that the sales of quotas by auctions conflicted with Ukraine's obligations to the WTO, as specified in Article III of the General Agreement on Tariffs and Trade in 1994 (GATT, 1994). In addition, the cost of each lot was an additional duty placed on exports.

Figure 5.5 illustrates Ukrainian wheat quotas and changes in food wheat prices by domestic millers and bakeries between 2006 and 2010. Restrictions on exports were short term on preventing food wheat prices from increasing. There was a clear trend of increasing prices for food wheat purchases for domestic purposes immediately after the revocation of export quotas for wheat in Ukraine (Fig. 5.5).

The Ukrainian state policy of export restrictions on highly competitive agricultural products had a very short-term effect of restraining foodstuff price increases between 2006 and 2010 (Fig. 5.6). There was a clear and constant trend of growth in foodstuff prices. For example, the consumer price index for foodstuffs was 136.7% in 2008, but only 110.9% in 2010.

Export duties

For the 2000–2010 period, no export duties were applied in Ukraine for cereal crops (including wheat). Effective as of July 2011, Ukraine imposed the following export duties (shown in

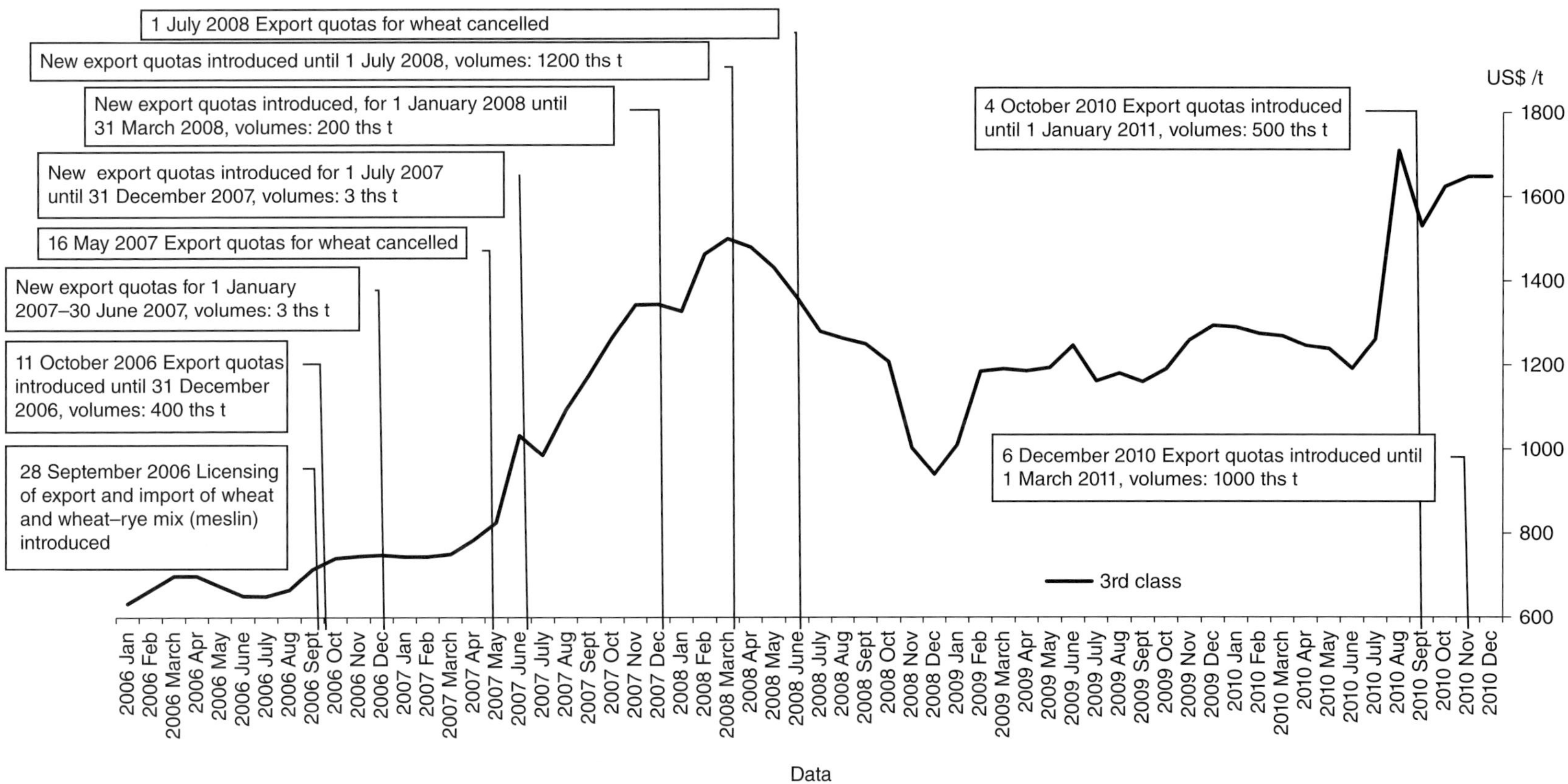

Fig. 5.5. Chronology of export quotas for EXW food wheat in Ukraine, 2006–2010. (From State Statistics Service of Ukraine, 2006–2010.)
Note: ths t = thousand tonnes.

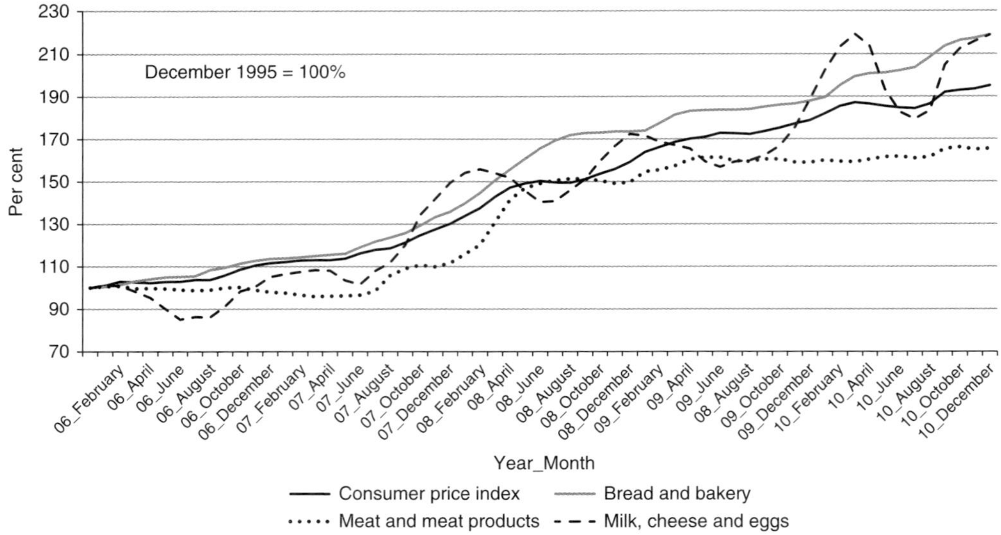

Fig. 5.6. Behaviour of consumer prices on the whole and by commodity groups, 2006–2010. (From State Statistics Service of Ukraine, 2006–2010.)

euros, EU currency): wheat (9%, but not less than €17/t); barley (14%, but not less than €23/t); maize (12%, but not less than €20/t). This measure was introduced for the same reasons as the export quotas in 2010 (Kobuta *et al.*, 2012). Export duties were levied to protect the domestic market and to prevent grain deficits and price fluctuations during periods of increased demand for grain on external markets. As with export quotas, this measure did not give the desired effect. Moreover, grain exports declined significantly, leading to an increase in grain domestic reserves, which in turn led to a loss in grain quality due to the lack of modern storage facilities. As a result, export duties for wheat and maize were abolished in October 2011 and for barley in January 2012. Losses to agricultural producers caused by the imposition of export duties on grain crops are estimated at from US$440 million to US$680 million as a result of selling price reductions (author's calculations).

Evaluation of the Effects of Government Intervention in the Ukraine Wheat Market

An evaluation of the government's intervention in the Ukrainian wheat market, based on Organisation for Economic Co-operation and Development (OECD) methodology, demonstrates that wheat is successively taxed (Table 5.2). Due to a series of political measures, agricultural producers receive less income from wheat sales because of understated domestic wheat prices compared to the world price level. For example, for the 2008–2010 period, negative market price support (MPS) totalled UAH 18 billion, and the percentage of product single commodity transfer (PSCT) was −33%. This meant that all governmental transfers (budget subsidy payments and negative market price supports) on average for the 2008–2010 period made up 33% of the total value of wheat production in Ukraine (Fig. 5.7).

For the 2006–2008 period, the difference in the domestic price of wheat producers and the world price was significant compared to other years between 2000 and 2010, due to the use of quantitative restrictions on wheat export from Ukraine. The largest negative price support (taxation) for wheat producers in Ukraine was in 2008 because of the lack of efficient state financial interventions on the grain market, export restrictions, financial crisis and crediting both producers and sellers of wheat. As a result, the percentage of PSCT for wheat was −59%, and the negative market price support totalled UAH 12 billion in 2008 (Table 5.2). At the same time, budget subsidies of UAH 414 million (or US$80 million) were negligible and wasted state resources

Table 5.2. Subsidies (taxes) to producers for wheat production cost in Ukraine, 2000–2010. (From author's calculations.)

	2000	2004	2008	2010
Total subsidies/taxes (%) to producers for wheat as a portion of value of production (after subsidies (taxes) (%)) [$\%SCT_{WT}$][a]	16	–3	–59	–8
Level of wheat production [Q_{WT}] (1,000 t)	10,197	17,520	25,885	16,845
Price of wheat production (farm gate) [P_{WT}] (UAH/t)	487	493	753	1,086
Value of production (farm gate) [$P_{WT}*Q_{WT}$] (million UAH)[b]	4,966	8,634	19,491	18,290
Total subsidies (taxes) for wheat producers [S_{WT} from Table 5.1] (million UAH)	783	–279	–11,842	–1,524
Value of production (farm gate) after subsidies (taxes) to producers [$(P_{WT}*Q_{WT}) + SCT_{WT}$] (million UAH)	5,749	8,355	7,649	16,766

Notes: [a]$\%SCT_{WT} = (100* SCT_{WT})/[(PP_{WT}*QP_{WT}) + SCT_{WT}]$; [b]million UAH (UAH = hryvnia, Ukrainian currency).

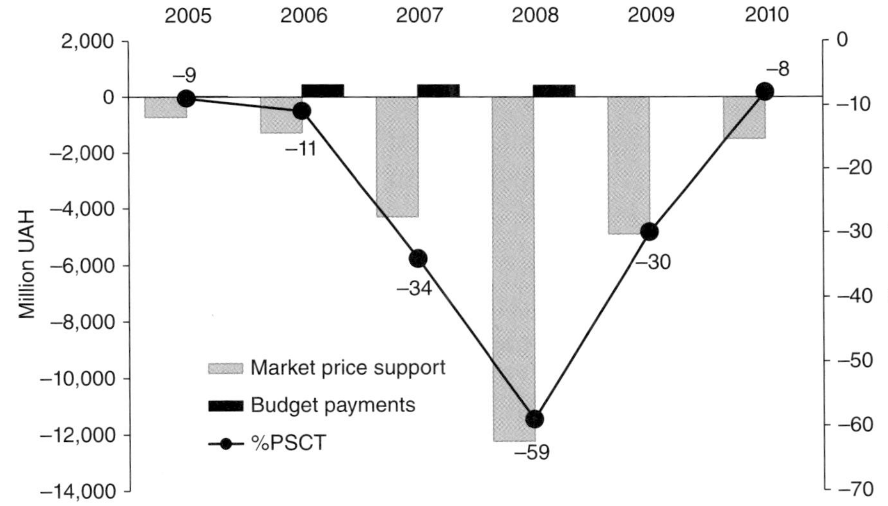

Fig. 5.7. Ukrainian wheat: percentage of PSCT = product single commodity transfer. (From OECD database for Ukraine, 2005–2010.)

against the large negative percentage of PSCT. Such governmental policy heavily destabilized agricultural product markets in Ukraine, thus increasing risks to grain producers and, consequently, restricting their production. In Table 5.2, we show the amount of subsidies (taxes) against the annual value of wheat production in Ukraine during the 2000–2010 period.

Conclusions

Export restrictions in Ukraine have had a direct effect on wheat production. The decrease in wheat profitability due to export restrictions and export contract registration charges has contributed to the reduction of wheat produced by agricultural producers. In addition, the taxation policy has benefited livestock producers in Ukraine because it has focused on rent by individual business entities, rather than on real political objectives.

Export restrictions in Ukraine were initiated by the government because of the inefficient work of state trading operators on the state price stabilization policy of the Ukrainian wheat market. The Agrarian Fund was inefficient when playing the role of price regulator for the domestic market. Actual volumes of state purchasing transactions on the Ukrainian wheat market were insignificant after 2000. They did not exceed 4% of the gross yield of wheat or 6% of the marketable wheat. Efficient functioning of the state purchasing system was not possible because of insufficient data on total grain stocks in the country, and because of the lack of official supply and demand balances for individual cereal crops, the unavailability of a state market price monitoring system and an uncoordinated purchasing activity of state trading operators (e.g. the Agrarian Fund and State Reserve Agency).

Minimum intervention prices for price regulation objects are set lower than actual market prices to ensure observance of Ukraine's commitment to the WTO and to avoid exceeding the total bound aggregate measurement of support (Total Bound AMS) of the base period. At the same time, among the positive factors of the state forward purchases of wheat is the absence of a sharp seasonal reduction of producers' prices at the point of sales of wheat during the harvesting period.

Efficient state policy of targeted aid to low-income population groups can be more effective than are export restrictions for solving the problem of the availability of important foodstuffs, especially during periods of price spikes.

References

CBOT (2000–2010) Data statistics. Chicago Board of Trade (CBOT), Chicago, Illinois.

FAS/USDA (2013) Wheat statistics. Foreign Agricultural Service, US Department of Agriculture (FAS/USDA), Washington, DC. Available at: http://www.fas.usda.gov/regions/europe-and-eurasia/ukraine (accessed 15 March 2014).

GATT (1994) Article III, General Agreement on Tariffs and Trade. Available at: http://www.wto.org/english/res_e/booksp_e/analytic_index_e/gatt1994_02_e.htm#article3 (accessed 15 March 2014).

Kobuta, I., Zhygadlo, V. and Sikachyna, O. (2012) Wheat export economy in Ukraine/FAO Regional Office for Europe and Central Asia. In: *Policy Studies on Rural Transition*. No 2012-4. Food and Agricultural Organization (FAO), Rome, pp. 56.

OECD (2000–2010) Data statistics. Organisation for Economic Co-operation and Development, Paris.

State Customs Service of Ukraine (2000–2010) State Customs Service of Ukraine, Kiev.

State Statistics Service of Ukraine (1990–2010) State Statistics Service of Ukraine, Kiev.

World Bank (2008) Competitive agriculture and state control: Ukraine's response to the food crisis. Available at: http://siteresources.worldbank.org/INTUKRAINE/Resources/WorldFoodCrisisandRoleofUkraine.pdf (accessed 4 April 2014).

6 Effect of Relative Export Price Changes of the Top Principal Crops in Russia

Ekaterina Vorotnikova,[1]* Serhat Asci[2] and James L. Seale, Jr[2]
[1]University of Idaho, Moscow, Idaho, USA;
[2]University of Florida, Gainesville, Florida, USA

Abstract

Following the collapse of the Soviet Union in 1991, the Russian economy was on the way to becoming more market based. The broadening of the market forces in Russian agriculture supports the hypothesis that the land allocation decision for grains and oilseeds is in large part determined by output prices. In this chapter, a land allocation model developed by Vorotnikova *et al.* (2013) uses post-Soviet data to determine whether or not output prices for grains and oilseeds affect land allocation significantly among crops and by what magnitude. We look at the effect of relative export price changes on land allocation among the top eight crops produced in Russia from 1992 to 2011. We find that the most price-responsive acreages are those of soybean, maize, sunflower, wheat, rye, barley and other crops, and conclude that post-Soviet Russian agriculture has become price responsive in cropland allocation decisions.

Introduction

While under the Soviet Union regime, Russian agricultural production and land allocation was based on a command economy. Following the collapse of the Soviet Union in 1991, the Russian economy was on its way to becoming more market based. Furthermore, the economy became more integrated into the global economy and in August 2012, after 18 years of negotiations, Russia became the 156th member of the World Trade Organization (WTO, 2012).

Land allocation decisions for grains and oilseeds are determined in large part by the output prices of crops. In this chapter, a land allocation model developed by Vorotnikova *et al.* (2013) uses post-Soviet data to determine if output prices for grains and oilseeds significantly affect land allocation for these crops and by what magnitude. The model postulates that land is allocated among different crops based on total land in production and based on the output prices of the crops. For example, land allocated to a particular crop is expected to increase if the output price of that crop increases. Changes in total land in agricultural production can also affect the amount of land allocated to the different crops. Unlike the effects of output prices, the effects of changes in total agricultural land are not necessarily predictable based on economic theory, but instead are estimated and measured based on model parameters.

First, post-Soviet Russian data are presented and described in terms of grain and oilseed land use, production, exports and prices. This is followed by a methodological section in

*Corresponding author; e-mail: evorotnikova@uidaho.edu

which the empirical model is presented and described. Data sources are then reported, followed by the presentation and discussion of parameter estimates. Based on the estimated parameters, price and land elasticities are calculated and discussed. Finally, interpretations and conclusions are drawn.

Post-Soviet Russia's Grain and Oilseed Sector

While total land in grains and oilseeds declined steadily from 1992 to 1998, it stabilized thereafter, fluctuating between 40 million and 50 million hectares (Mha) (Fig. 6.1). The largest proportion of land allocated to wheat was relatively stable throughout the 1992–2011 period, fluctuating between 20 Mha and 30 Mha. This was not the case for the other grains and oilseeds. For example, land allocated to barley decreased by around 10 Mha.

As shown in Fig. 6.2, significant decreases in acreage occurred for rye, while acreage seeded to sunflower, rapeseed and maize expanded, although at relatively low levels.

Wheat production was the largest among these crops (remaining steady at about 40 million tonnes (Mt) over the 1992–2011 period, followed by barley (Fig. 6.3). Maize production increased fivefold over the period, while rye, rice and oats production declined.

The most descriptive aspect of post-Soviet grains and oilseeds was the increase in their exports between 2001 and 2011. Wheat increased sevenfold over this period (Fig. 6.4). While starting from relatively low levels in 1992, maize exports increased 461 times and rye exports

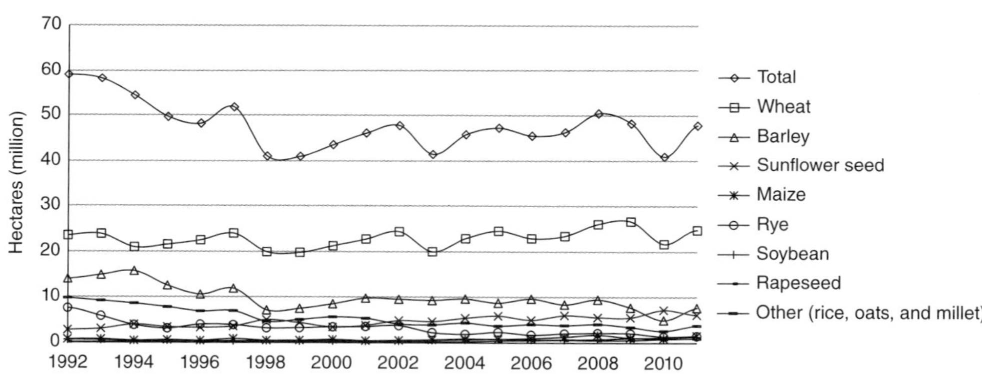

Fig. 6.1. Land use for selected crops in Russia, 1992–2011. (From data obtained from FAS/PSD, 2013.)

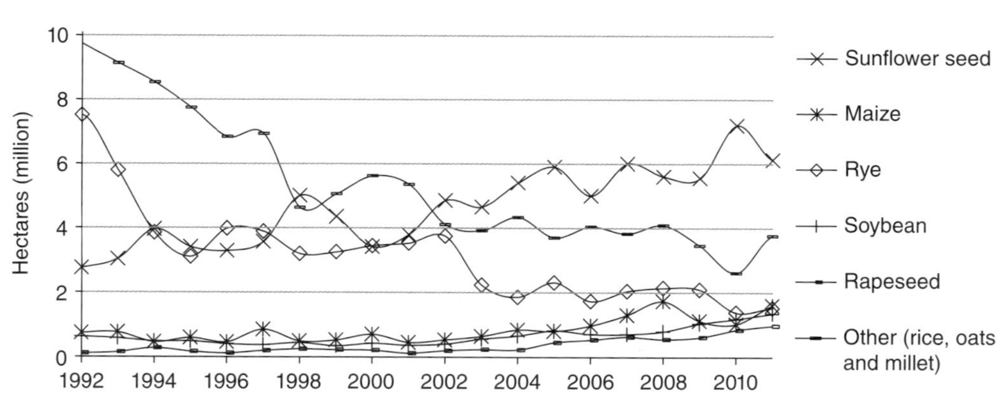

Fig. 6.2. Land use for crops of smaller production in Russia, 1992–2011. (From data obtained from FAS/PSD, 2013.)

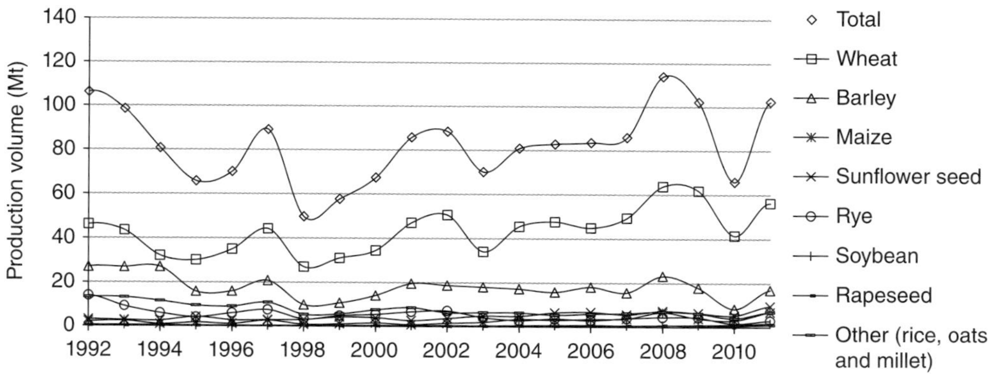

Fig. 6.3. Grain and oilseed production in Russia, 1992–2011. (From data obtained from FAS/PSD, 2013.)

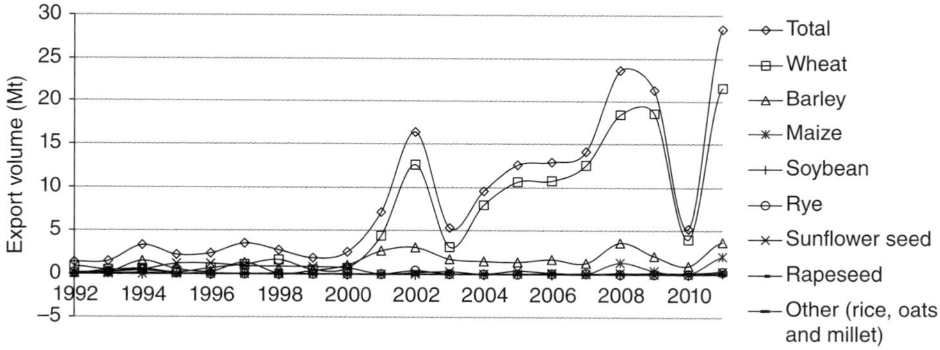

Fig. 6.4. Grain and oilseed exports of Russia, 1992–2011. (From data obtained from FAS/PSD, 2013.)

increased 29 times from 2001 to 2011 (FAS/PSD, 2013).

All crop prices increased steadily between 1998 and 2008. After 2008, however, prices of rapeseed, soybean and sunflower seeds continued to increase, while prices of maize, wheat and rye decreased before recovering (Fig. 6.5).

In Fig. 6.6, we analysed the price ratio of all crops from 1992 to 2011 relative to the price of wheat found in Fig. 6.5. The most volatile price ratio was the rapeseed to wheat price ratio, which increased significantly from 2000 to 2005 before decreasing steadily between 2005 and 2011. The price ratios of soybean to wheat and sunflower to wheat both showed a slight upward trend after 2005. However, the price ratios of maize to wheat declined before 2005 and then remained relatively stable. The price ratio of rye to wheat declined slowly after 2005. Overall, after 2005 in the two groups of crops, oilseeds and grains, the price ratios of crops in each group to wheat set a trajectory of convergence. By 2011, the price ratios of rapeseed, soybean and sunflower seed to wheat converged among the crops in the oilseeds group; the price ratio of maize, rye and barley to wheat also converged among the three grains.

The combination of increased agricultural exports and decreased production and acreage is surprising and interesting, especially in lieu of the US Department of Agriculture (USDA) forecast that Russian grain exports and production will continue to increase until at least 2021 (Liefert *et al.*, 2013). Relative to 2010, the USDA projections for 2021 indicate that the Russian grain area will increase 5%, grain production will increase 22% and grain exports will increase 82% (Liefert *et al.*, 2013). According to USDA forecasts, increases in production and technology will be the main drivers for Russia to reach its potential in agriculture. While USDA forecasts predict that Russian grain yields will increase 17% by 2021, only maize yield has increased thus far (Fig. 6.7).

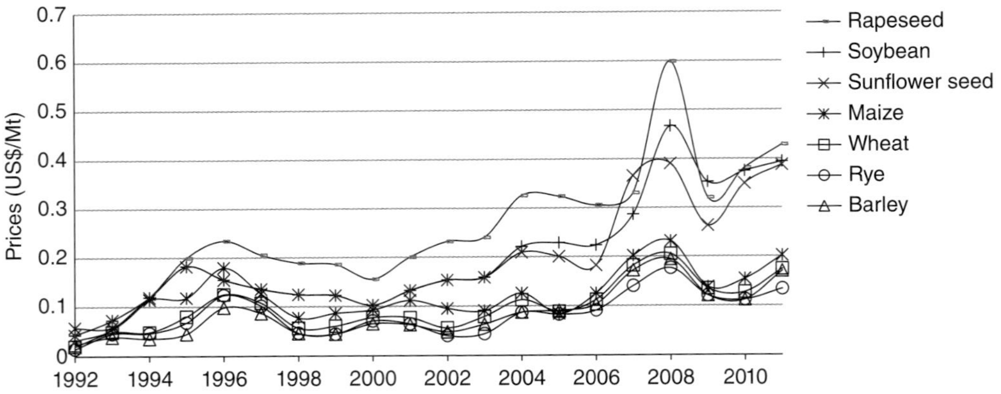

Fig. 6.5. Price changes of selected crops in Russia, 1992–2011. (From data obtained from FAOSTAT, 2013, and FAS/PSD, 2013.)

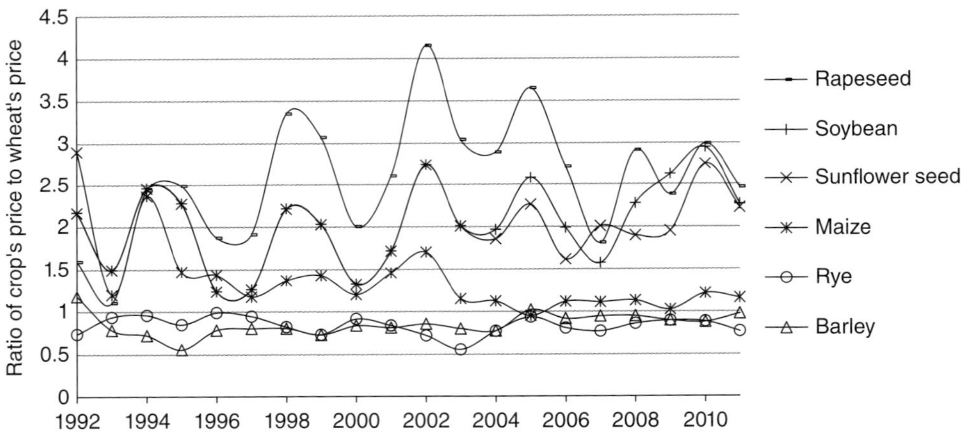

Fig. 6.6. Selected crop to wheat price ratios in Russia, 1992–2011. (From data obtained from FAOSTAT, 2013.)

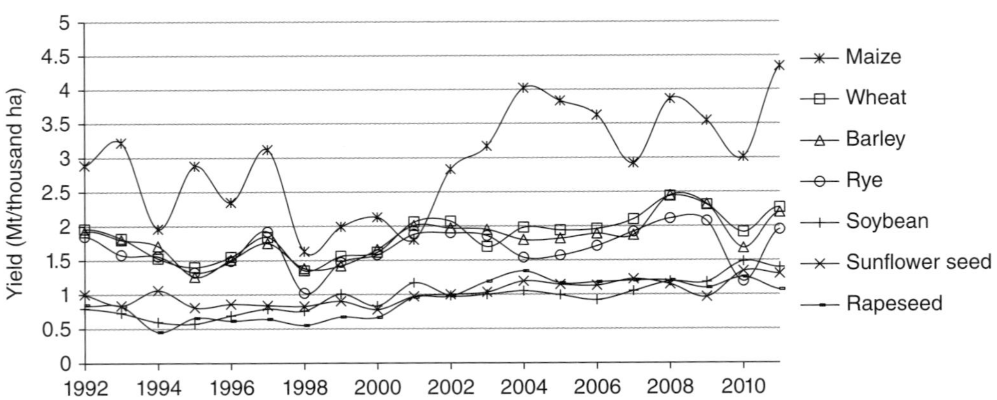

Fig. 6.7. Yield as a measure of productivity for selected crops in Russia, 1992–2011. (From data obtained from FAS/PSD, 2013.)

While there is some disconnect between the trends of decreased total land acreage and increased crop production among the top grains, there is anticipated expansion in production and exports. Thus, given such a paradigm of Russian agriculture in world trade, this chapter looks at the relative price change effects in land allocation dynamics among the top seven crops of Russia for 1992 to 2011. Also, since production in some crops has expanded while total land among the top crops has contracted, we suspect that the expansion might be happening at the expense of other grains. In particular, we are interested in whether output prices have significantly determined land allocation.

Methodology

Differential models developed by Barten (1964) and Theil (1965) and later modified by Theil (1977) for production theory can be used for cropland allocation decision making. This approach has been extended to a perfectly competitive multiproduct firm (Laitenen and Theil, 1978). It was modified further under the assumptions of homotheticity and input independence of a cost-minimizing firm (Theil, 1979) and improved for the supply response and input demand of a multiproduct firm, where an input was quasi fixed (Livanis and Moss, 2006). More recently, Vorotnikova *et al.* (2013) has developed a land allocation model based on the differential approach to producer theory for the multiproduct firm. We fit a variant of this model to post-Soviet data so as to examine the allocation of the quasi-fixed input factor of land in production, given the impacts of fluctuating crop prices.

The theoretical model is

$$f_i d(\ln L_i) = \theta_i d(\ln L) + \sum_{j=1}^{n} \pi_{ij} d(\ln P_{ij}), \qquad (6.1)$$

where L_i is the quantity of land devoted to crop i; f_i is the share of the total land allocated to crop i; P_i is the output price of crop j; θ_i is the marginal share of land allocated to i; π_{ij} is the (i,j)th Slutsky price coefficient of the firm; and $d(\ln L) = \sum f_i d \ln L$ is a Divisia land (quantity) index. To operationalize the model, let $\bar{f}_{it} = (f_{i,t} + f_{i,t-1})/2$ and $d(\ln X_t) = \ln X_t - \ln X_{t-1}$, where X represents L and P, and ε_t is an error term. The empirical model is then

$$\bar{f}_{it} d(\ln L_i)_t = \theta_i d(\ln L)_t + \sum_{j=1}^{n} \pi_{ij} d(\ln P_{ij})_t + \varepsilon_t. \qquad (6.2)$$

which we refer to as the Rotterdam version of the land allocation model. Note that the adding-up conditions are $\sum_i \theta_i = 1$ and $\sum_i \pi_{ij} = 0$. The homogeneity condition is $\sum_j \pi_{ij} = 0$, and the symmetry condition is $\pi_{ij} = \pi_{ji}$. The land volume elasticity and price elasticities (η_{ij}) of the land allocation equations are calculated by $\eta_i = \theta_i / f_i$ and $\eta_{ij} = \pi_{ij} / f_i$, respectively.

It is not necessary that θ_i is constant. One can hypothesize that θ_i, the marginal share, is equal to f_i, the average share, plus a constant parameter β_i. If one replaces $\theta_i = f_i + \beta_i$ in Eqn 6.2, we obtain a land allocation model similar to the Central Bureau of Statistics (CBS) model developed by Keller and van Driel (1985) and by Clements (1980), that is

$$\bar{f}_{it} d(\ln L_i - \ln L)_t = \beta_i d(\ln L)_t + \sum_{j=1}^{n} \pi_{ij} (d \ln P_{ij})_t + \varepsilon_t. \qquad (6.3)$$

We refer to this model as the CBS version of the land allocation model. The adding-up conditions for the CBS model are $\sum_i \beta_i = 0$ and $\sum_i \pi_{ij} = 0$.

Data

The data span the years 1992–2011. Acreage data are collected from the Production Supply and Distribution (PSD) data set provided by Foreign Agriculture Services (FAS/PSD, 2013). Price data at the producer level are collected from the Food and Agriculture Organization of the United Nations (FAOSTAT, 2013) and the FAS/PSD data set. The data consist of the annual quantity and unit prices of the main grains grown in Russia (barley, maize, wheat, soybean, sunflower seeds, rye, rapeseed and other crops). Other crops contain the total land area for rice, oats and millet, with its price calculated by dividing summed production value by summed production amount.

The model is estimated by dropping the other crops' land equation to avoid the singularity problem, as described by Barten (1964). We estimate all seven grain equations with iterative seemingly unrelated regressions (SUR) to obtain maximum likelihood estimators. When the estimation results are checked by replacing the dropped equation with another equation, the same results are obtained. This is accomplished by using the least squares command (LSQ) (Hall and Cummins, 2005) in Time Series Processing (TSP) version 5.0.

Results

Parameter estimation

We estimate the unrestricted parameterizations of the differential land allocation model using the Rotterdam model, the CBS model and a combined general model that includes both the Rotterdam and CBS models (Barten, 1993). We impose homogeneity and symmetry conditions. Table 6.1 shows the log-likelihood values obtained from these estimations. The parentheses contain the number of free parameters for each of the estimations, and the log-likelihood ratio test (LRT) statistics are provided below the estimation values. LRT is calculated by $-2[L(\theta^*) - L(\theta)]$, where $L(\theta^*)$ refers to the log value of the likelihood function when restriction is imposed, while $L(\theta)$ is used for the unrestricted estimation. LRT is compared with the critical value from the $\chi^2(q)$ distribution, where q is the difference in the number of free parameters between restricted and unrestricted estimations.

We compare three restrictions for our estimations: (i) the homogeneity-imposed model with the unrestricted model; (ii) the symmetry-imposed model with the homogeneity-imposed model for both the Rotterdam and CBS model; and (iii) the symmetry-imposed models of the Rotterdam and CBS systems with a general model that combines the two models (Barten, 1993). We reject homogeneity for the Rotterdam and CBS models, since the LRT for the Rotterdam (16.88) and CBS (15.33) models is greater than is the critical value of 14.07 at the 5% significance level. We do not reject homogeneity at the 10% significance level for either model. Laitinen (1978) shows that asymptotic tests reject homogeneity more often than they should when the number of equations increases in the model. Laitinen developed an exact test (Hotelling's T^2 test) multivariate hypothesis testing for samples with few degrees of freedom. Hotelling's T^2 test is calculated by $(n-1)(N-n-1)/(N-2n+1)\ F\{(N-2n+1), (n-1)\}$, where n is the number of commodities in the equations and N is the number of observations. F values are found for homogeneity as $F_{5,7} = 4.88$ at the 5% significance level. Based on the aforementioned formulation, the Hotelling's T^2 test statistic is 81.98, which is higher than the LRT statistics for both the Rotterdam and CBS models, so homogeneity should not be rejected. Meisner (1979) also shows the probability of rejecting symmetry when the number of goods or equations in the estimation increases. The low degree of freedom decreases the power of the asymptotic test. Since we have a high number of goods in the estimations, we increase the probability of rejecting the symmetry hypothesis, although it should not be rejected. Unlike for homogeneity, an exact test for small samples has not been developed, due to the complications from cross-equation restriction in symmetry tests.

Table 6.1. Test results for the log likelihood within models. (From authors' calculations.)

	Rotterdam	$-2[L(0^*) - L(0)]$	CBS	$-2[L(0^*) - L(0)]$
Unrestricted model (63)[a]	524.288		527.643	
Homogeneity imposed (56)[b]	515.849	16.88	519.978	15.33
$\chi^2(7)$ (95%)		14.07		14.07
$\chi^2(5,7)$ (95%)		81.98		81.98
Symmetry imposed (35)[a]	499.413	32.87	504.161	31.63
$\chi^2(21)$ (95%)		32.67		32.67
General model (36)[c]	474.707	9.55	474.707	0.06
$\chi^2(1)$ (95%)		3.84		3.84

Notes: [a] Number of free parameters for each estimation; [b] both symmetry and homogeneity are jointly imposed; [c] homogeneity and symmetry imposed log-likelihood value is provided for general model.

Finally, based on the LRT test results, we do not reject the CBS model but we do reject the Rotterdam model by comparing both of them to the combined general model. The LRT for the Rotterdam model is 9.55 and for the CBS model is 0.06, which are, respectively, greater and less than the chi-square critical value of 3.84 at the 5% significance level. This confirms that the CBS model fits the data better than does the Rotterdam model.

Table 6.2 gives the coefficients of the CBS model for the eight top grains and oilseeds produced in Russia, based on the price and acreage data from 1992 to 2011. (The parentheses below the coefficients include their asymptotic standard errors.) It is important to note that, due to the structure of the CBS model, the sign of its land coefficient β_i indicates whether the land elasticity will be greater than or less than unity. If the parameter is statistically zero, it indicates that the land elasticity will be unitary. Land elasticities are presented and discussed further below.

Marginal land shares may also be calculated from β_i, such that $\theta_i = f_i + \beta_i$ and they indicate the change in the quantity of land allocated to a crop when total land increases by one unit. The $\theta_{i.}$ for wheat, barley, other crops, maize and soybean are positive and significant at the 5% level. These results suggest that for a one unit of increase/decrease in total land, the hectares of wheat, barley, other crops, maize and soybean increase/decrease by 0.40, 0.33, 0.14, 0.02 and 0.01, respectively. Note that wheat hectares will increase the most, followed by barley hectares.

All own-price coefficients are positive, as expected, and those for wheat, sunflower, soybean and maize are statistically significant at the 5% level. These results indicate that land allocation decisions are partly determined by the output crop prices for these crops. Of the 28 cross-price coefficients, eight are significant statistically. Wheat/sunflower, rye/maize, soybean/rapeseed and soybean/maize are negative and significant at the 5% level, which indicates these crops behave as substitutes. Wheat/maize, soybean/other crops, rapeseed/other crops and sunflower/rapeseed coefficients are positive and significant at the 5% level.

Elasticity estimation

Land and price elasticities are provided in Table 6.3. Land elasticities are a measure of how much the quantity of land changes for each crop

Table 6.2. Coefficients of the Rotterdam and CBS models in Russia, 1992–2011. (From authors' calculations.)

	Barley	Maize	Wheat	Soybean	Sunflower	Rye	Rapeseed	Other crops
Price coef. (π_{ij})								
Barley	0.002	–0.001	–0.032	0.000	0.000	0.023	–0.002	0.010
	(0.028)[a]	(0.005)	(0.027)	(0.003)	(0.010)	(0.014)	(0.003)	(0.014)
Maize		0.010[c]	0.0143[c]	–0.004[c]	–0.002	–0.013[c]	0.000	–0.004
		(0.002)	(0.006)	(0.001)	(0.003)	(0.004)	(0.001)	(0.004)
Wheat			0.082[c]	–0.005	–0.042[c]	–0.012	–0.004	–0.003
			(0.036)	(0.003)	(0.011)	(0.017)	(0.003)	(0.017)
Soybean				0.012[c]	–0.003	–0.002	–0.008[c]	0.009[c]
				(0.003)	(0.002)	(0.002)	(0.002)	(0.002)
Sunflower					0.035[c]	0.010	0.006[c]	–0.004
					(0.007)	(0.007)	(0.002)	(0.007)
Rye						0.006	–0.001	–0.009
						(0.013)	(0.002)	(0.009)
Rapeseed							0.002	0.007[c]
							(0.002)	(0.003)
Other crops								0.007
								(0.013)
Land coef. (β_i)	0.131[c]	0.006	–0.090[b]	–0.005	–0.077[c]	0.008	–0.009[c]	0.036
	(0.046)	(0.010)	(0.052)	(0.004)	(0.022)	(0.029)	(0.004)	(0.024)
Land mar.	0.334[c]	0.024[c]	0.402[c]	0.007[c]	0.023	0.069[c]	0.002	0.144[c]
shares (θ_i)	(0.046)	(0.010)	(0.052)	(0.004)	(0.022)	(0.029)	(0.004)	(0.024)

Notes: [a]Figures in parentheses are standard deviations; [b]significant at 10% level; [c]significant at 5% level. $\theta_i = \beta_i + \bar{w}_i$.

Table 6.3. Output price and land elasticities of the CBS model in Russia, 1992–2011. (From authors' calculations.)

	Barley	Maize	Wheat	Soybean	Sunflower	Rye	Rapeseed	Other crops
Crop prices								
Barley	0.01	–0.01	–0.16	0.00	0.00	0.11	–0.01	0.05
	(0.14)[a]	(0.02)	(0.13)	(0.01)	(0.05)	(0.07)	(0.01)	(0.07)
Maize	–0.08	0.56[b]	0.81[b]	–0.24[b]	–0.10	–0.74[b]	–0.01	–0.21
	(0.28)	(0.14)	(0.34)	(0.06)	(0.18)	(0.23)	(0.07)	(0.22)
Wheat	–0.06	0.03[b]	0.17[b]	–0.01	–0.08[b]	–0.02	–0.01	–0.01
	(0.05)	(0.01)	(0.07)	(0.01)	(0.02)	(0.03)	(0.01)	(0.03)
Soybean	0.01	–0.34[b]	–0.36	0.96[b]	–0.24	–0.17	–0.61[b]	0.75[b]
	(0.20)	(0.08)	(0.22)	(0.21)	(0.15)	(0.16)	(0.13)	(0.18)
Sunflower	0.00	–0.02	–0.42[b]	–0.03	0.35[b]	0.10	0.06[b]	–0.04
	(0.10)	(0.03)	(0.12)	(0.02)	(0.07)	(0.07)	(0.02)	(0.07)
Rye	0.38	–0.21[b]	–0.20	–0.03	0.16	0.09	–0.02	–0.15
	(0.24)	(0.07)	(0.28)	(0.03)	(0.12)	(0.22)	(0.04)	(0.15)
Rapeseed	–0.33	–0.03	–0.57	–1.14[b]	0.94[b]	–0.22	0.30	1.05[b]
	(0.41)	(0.18)	(0.47)	(0.23)	(0.30)	(0.35)	(0.23)	(0.37)
Other crops	0.09	–0.03	–0.02	0.09[b]	–0.04	–0.09	0.07[b]	0.06
	(0.13)	(0.04)	(0.16)	(0.02)	(0.06)	(0.09)	(0.02)	(0.12)
Land	1.65[b]	1.36[b]	0.82[b]	0.57[b]	0.23	1.13[b]	–0.36	1.34[b]
	(0.23)	(0.56)	(0.11)	(0.29)	(0.22)	(0.47)	(0.57)	(0.22)

Notes: [a]Figures in parentheses are standard deviations; [b]significant at 5% level.

when total land increases by 1%. All elasticities are computed from their respective coefficients at the sample mean for the entire sample. First, the land elasticities for six out of eight crops (i.e. barley, maize, other crops, rye, wheat and soybean) are statistically significant at the 5% level. The CBS results indicate that if land expands (contracts) by 1%, then the land quantity for barley, maize, other crops, rye, wheat and soybean will go up/down by 1.65%, 1.36%, 1.34%, 1.13%, 0.82% and 0.57%, respectively. In summary, the results show that the quantity of land allocated to maize, barley, other crops and rye is most responsive to changes in total land. The quantity of land allocated to wheat and soybean is less responsive than those same crops are to a change in total land allocated to grains and oilseeds.

First, maize and barley follow the same trend in total land quantity, but wheat does not follow as closely as in Figs 6.1 and 6.2. Rapeseed goes in the opposite direction, gaining land share when total land decreases, which warrants the negative (but insignificant) sign of the land elasticity in the CBS model.

Second, own-price elasticities are displayed along the diagonals in Table 6.3. They provide a measure of the responsiveness of land quantity to changes in the own price of a particular crop. The own-price elasticities are significant for soybean, maize, sunflower and wheat at the 5% significance level. The results indicate that if the price of soybean, maize, sunflower and wheat goes up/down by 1%, the land quantity for these crops goes up/down by 0.96%, 0.56%, 0.35% and 0.17%, respectively. For the remaining crops, their elasticities are small, positive and insignificant, indicating the quantity of land allocated to these crops is not own-price responsive. Thus, maize and soybean hectares are most responsive their own crops' prices.

Third, cross-price elasticities measure land quantity responsiveness to price changes in competing crops. The cross-price elasticities for soybean/rapeseed, rye/maize, wheat/sunflower and maize/soybean combinations are negative and statistically significant at the 5% level. This indicates that these crops behave as substitutes for each other in pairwise combinations for allocating land already in production. On the other hand, cross-price elasticities of wheat/maize, sunflower/rapeseed and other crops/soybean combinations are positive and statistically significant at the 5% level, suggesting that these crops behave as complements.

Finally, the model allows the comparison of the price change effect in one crop on the land quantities of another to that of the vice versa combination. The model identifies which crop price has more influence on land allocation of the other in pairwise comparisons. For two crops that are substitutes, it is possible to identify which crop's prices have the most influence on acreage, particularly in combinations of substitutes. For example, for a 1% price change (upward/downward) in soybean price, land allocated to rapeseed decreases/increases by 1.14%, but for a 1% price change in rapeseed, the land share of soybean changes 0.61%. This means that soybean price has more of an influence on hectares of rapeseed than rapeseed price has on hectares of soybean. In turn, a change in the price of rye has more influence on hectares of maize than it will the other way around. If the price of rye increases/decreases by 1%, hectares of maize will go up/down by 0.74% (compared to 0.21% in the reverse case). If the price of wheat goes up/down by 1%, then the hectares allocated to sunflower will decrease/increase by 0.42%. In comparison, hectares allocated to wheat will change only slightly in response to a 1% change in sunflower price; that is, it decreases/increases by only 0.08%. For a 1% upward/downward price change in maize, hectares of soybean will go up/down by 0.34%, whereas for a 1% upward/downward price change in soybean, hectares of maize will go up/down by 0.24%.

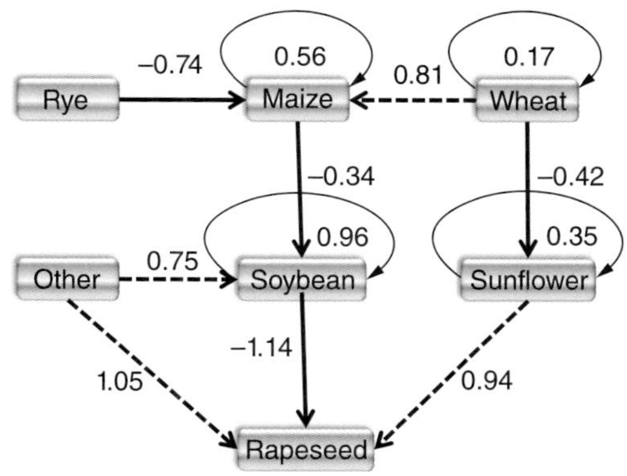

Fig. 6.8. Elasticity chart for selected crops in Russia. See text for explanation. (Figure is constructed from elasticities calculated by the authors.)

Discussion

A schematic diagram in Fig. 6.8 demonstrates graphically the conclusions of the analysis. The dynamic in the relationship between crop price and acreage is best represented by cross-price elasticities. In the diagram, the dotted lines represent the cross-price elasticity for crops that behave as complements, and the solid lines represent the cross-price elasticity for crops that behave as substitutes. The direction of the lines represents the crop price that is more influential on the acreage of the other crop in a two-crop combination (only the higher magnitude of the two possible elasticities is displayed). For example, rye price changes are more influential on maize acreage than vice versa; therefore, the direction of the solid line is from rye to maize, so the rye to maize (rather than maize to rye) price elasticity is displayed. Rye price changes have a significant influence on maize acreage, with maize competing with soybean for land, and in turn soybean competing with rapeseed. Soybean and other crops behave as complements in land allocation. Wheat competes for land with sunflower, while sunflower, rapeseed, maize and wheat behave as complements in respective combinations when it comes to land allocation.

The responsiveness of crop acreages to own prices is important. Price responsiveness is captured best by the own-price elasticity measure, which is represented by a thin circular line. Thus, most price-responsive acreages are those of soybean (0.96), maize (0.56), sunflower (0.35) and wheat (0.17).

Based on the full assessment of all results, we can confirm that wheat is still considered the major crop for Russia, just as maize is the major crop for the USA. The interesting dynamic is that in the USA, maize competes with wheat rather intensely; in Russia, however, maize and wheat behave as complements. This can be explained by a boom phase in the boom–bust cycle, as characterized by Schmitz (1995). Indeed, the dynamic of maize production expansion in Russia fits the description of the dynamics of the boom phase. Maize competes with rye and soybean in Russia due to comparable production and acreage. However, if maize production increased significantly in Russia, it could compete with wheat for land (Figs 6.5 and 6.6 signify a boom phase for maize in Russia).

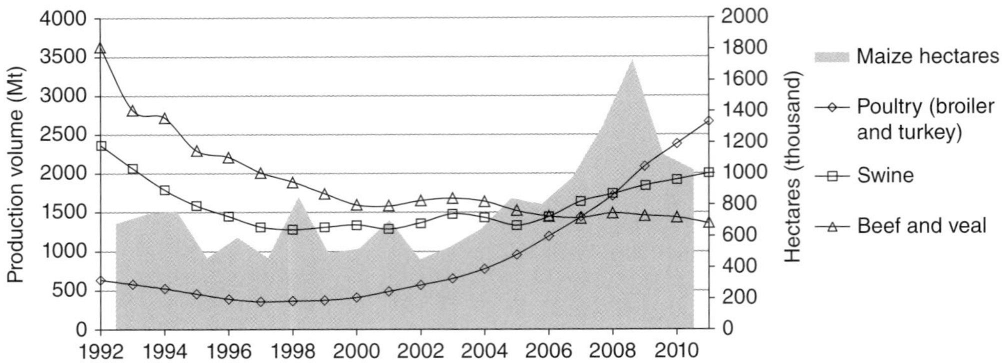

Fig. 6.9. Poultry, swine and beef production, and maize hectares, 1992–2011. (From data obtained from FAS/PSD, 2013.)

In addition, given that maize is a feed grain in animal production, maize production expansion coincides with animal production expansion (Fig. 6.9). Specifically, swine and poultry production have increased in Russia since 2000, while beef production has decreased. Figure 6.9 displays the dynamics of poultry, swine and beef production (plotted on primary axis) as compared to maize acreage expansion (plotted on secondary axis) for 1992–2011.

There is also a dynamic between maize and wheat. As discussed by Vorotnikova *et al.* (2013), maize production in the USA is expanding at the expense of soybean and wheat. This could possibly provide other countries, including Russia, the unique opportunity to step in to fill the gap when contracting wheat production in the USA.

With rapeseed prices increasing significantly, farmers in Russia have increased the acreage allocated to this crop; rapeseed acreage now competes with soybean acreage in land allocation. Both are oil crops. The acreage for sunflower seed (another oil crop) has increased the most in our eight-crop analysis. Land allocated to sunflower has been rising steadily since 1992, and sunflower seed now competes with wheat for acreage in Russia.

Conclusions

In general, land for maize, soybean and rapeseed production has increased the most significantly since 2005. Both the Energy Policy Act of 2005 (EPA, 2005) in the USA (Vorotnikova and Seale, 2013) and the National Priority Project in Agriculture of 2005 (MARF, 2005) in the Russian Federation (Liefert *et al.*, 2013) have influenced this situation.

If the USDA projection of a 5% increase in total grain land is accurate, then according to the results of our model, maize, barley, other crops, rye, wheat and soybean acreage will expand by 7.7%, 7.65%, 6.35%, 5.20%, 4.25% and 3.5%, respectively. Overall, when it comes to land allocation, we can conclude that Russian agriculture has become price responsive to the growing of various grain and oilseed crops.

References

Barten, A.P. (1964) Consumer demand functions under conditions of almost additive preferences. *Econometrica* 32, 1–38.

Barten, A.P. (1993) Consumer allocation models: choice of functional form. *Empirical Economics* 18, 129–158.

Clements, K.W. (1980) An aggregative multiproduct supply model. *European Economic Review* 13, 239–245.

EPA (2005) *The Energy Policy Act of 2005.* US Environmental Protection Agency (EPA), Washington, DC.

FAOSTAT (2013) FAOSTAT Website. Available at: http://www.fao.org/statistics/en/ or http://faostat3.fao.org/faostat-gateway/go/to/home/E (accessed 4 April 2014).

FAS/PSD (2013) FAS/PSD website. Available at: http://www.fas.usda.gov/psdonline/ (accessed 4 April 2014).

Hall, B.H. and Cummins, C. (2005) *TSP 5.0 Reference Manual*. TSP International, Palo Alto, California.

Keller, W.J. and van Driel, J. (1985) Differential consumer demand systems. *European Economic Review* 27, 375–390.

Laitenen, K. (1978) Why is demand homogeneity so often rejected? *Economics Letters* 1, 187–191.

Laitenen, K. and Theil, H. (1978) Supply and demand of the multiproduct firm. *European Economic Review* 11, 107–154.

Liefert, O., Liefert, W. and Luebehusen, E. (2013) *Rising Grain Exports by the Former Soviet Union Region: Causes and Outlook*. WHS-13A-01. Economic Research Service, US Department of Agriculture (USDA), Washington, DC.

Livanis, G. and Moss, C.B. (2006) Quasi-fixity and multiproduct firms. *Economics Letters* 93, 228–234.

MARF (2005) *National Priority Project for Agriculture of 2005*. Ministry of Agriculture of the Russian Federation, Moscow.

Meisner, J.F. (1979) The sad fate of the asymptotic Slutsky symmetry testing for large systems. *Economics Letters* 2, 231–233.

Schmitz, A. (1995) Boom/bust cycles and Ricardian rent. *American Journal of Agricultural Economics* 77(5), 1110–1125.

Theil, H. (1965) The information approach to demand analysis. *Econometrica* 33, 67–87.

Theil, H. (1977) The independent inputs of production. *Econometrica* 45, 1303–1328.

Theil, H. (1979) A differential approach to input–output analysis. *Economic Letters* 3, 381–385.

Vorotnikova, E. and Seale, J.L. (2013) Effects of relative price change on the land allocation dynamics among the top staple crops in the United States before and after the Energy Policy Act of 2005. *Proceedings of the American Agricultural Economics Association Conference*, Washington, DC. Availableat:http://ageconsearch.umn.edu/bitstream/143099/2/LandAllocationPaper-%20Jan19-SAEA.pdf (accessed 16 December 2014).

Vorotnikova, E., Asci, S. and Seale, J.L. (2013) Effect of relative price changes of top principal crops on US farm land allocation. *Proceedings of the Southern Agricultural Economics Association Conference*, Orlando, Florida. Available at: http://ageconsearch.umn.edu/bitstream/150455/2/PresentationPosterfinal2005-2.pdf (accessed 16 December 2014).

WTO (2012) WTO membership rises to 157 with the entry of Russia and Vanuatu. Available at: http://www.wto.org/english/news_e/pres12_e/pr671_e.htm (accessed 4 April 2014).

7 The Role of Production Cooperatives in Russian Agriculture

Svetlana Golovina,[1] Jerker Nilsson[2] and Axel Wolz[3*]
[1]*Kurgan State Agricultural Academy, Kurgan, Russia;* [2]*Swedish University of Agricultural Sciences, Uppsala, Sweden;* [3]*Leibniz Institute of Agricultural Development in Transition Economies, Halle (Saale), Germany*

Abstract
Agricultural production cooperatives in Russia have replaced some of the Soviet collective farms (*kolkhozy*) and state farms (*sovkhozy*). The cooperatives have been successful due to their strong market position and high market shares in several regions of Russia. This chapter examines why members (workers) choose cooperatives and why cooperatives are relatively strong. The workers of Soviet farms had assets that were best adaptable to large-scale agricultural units. There would have been high transaction costs if they had chosen other forms of organization. The performance of the cooperatives is a result of the members' large transaction-specific assets and labour-intensive production.

Introduction

The dissolution of the Soviet Union in 1991 gave rise to the need for transforming Russian agriculture. In a market economy, agricultural production can be performed by various types of organizations. One such organizational type is the production cooperative (Schiller, 1969). Other types are private/smallholder farms and commercial companies (investor-owned firms), where ownership is in the form of limited liability or joint-stock companies. The various organizational types compete with each other, with the more competitive types expanding at the expense of the less efficient ones. The agricultural production cooperative model conforms to the general definition of cooperatives; that is, an organization operating in the interest of users who control and own the organization (Dunn, 1988).

Some economists claim that agricultural production cooperatives cannot compete successfully with private/smallholder farms because they have difficulties when mobilizing sufficient financial resources – the members do not want to invest their own money into collective funds (Schmitt, 1991; Binswanger *et al.*, 1995; Allen and Lueck, 2005). Large-scale farms also have limited production cost advantages because tasks cannot be properly delegated and monitored (Pollak, 1985; Johnson and Ruttan, 1994).

The purpose of this chapter is to explain the role of agricultural production cooperatives in Russia. It examines why workers of the large Soviet farms chose production cooperatives, rather than private/smallholder farms, and analyses the prospects of this particular type of organization.

By focusing on the institutional side of agricultural production cooperatives through partial

*Corresponding author; e-mail: wolz@iamo.de

vertical integration between specific economic groups, the relationships between workers and their employers may be interpreted in terms of the transaction cost theory. The transaction-specific assets of the members may be so high that the members might have difficulty finding alternatives to this organizational type. Also, the fact that cooperatives are collectively owned and operated implies that property right problems affect these relationships. Unclear ownership rights contribute to members' risk aversion (Podgorbunskih and Golovina, 2005).

Hence, institutional theory may explain why production cooperatives within the Russian agricultural sector are relatively strong. Institutional theories are used widely in research dealing with marketing and supply cooperatives (Staatz, 1984; Ollila, 1989; Golovina and Nilsson, 2009). Based on these theories, researchers examine the existence of the cooperatives, their successes and their failures.

This chapter examines the institutional framework under which the production cooperatives are operating, analyses why members of the Soviet agricultural enterprises opt for production cooperatives and assesses the performance of production cooperatives.

The Market Position of Agricultural Production Cooperatives

In the Russian national statistics, agricultural production cooperatives are grouped together with joint-stock and limited liability companies as agricultural enterprises. The total number of agricultural enterprises has decreased steadily since 2000 (Table 7.1). Statistics indicate a concentration trend in Russian agricultural production similar to that in most market economies (Rosstat, 2010). Although the number of agricultural production cooperatives has decreased over time, they play a significant role in Russian agriculture.

The market share of production cooperatives fluctuated between 1995 and 2009 (Table 7.2). One reason may be the decline in the number of cooperatives because investments in agricultural

Table 7.1. Number of different types of agricultural enterprises in Russia (1995–2009).[a] (From Rosstat, 2010.)

Type of agricultural enterprise	1995	2000	2006	2007	2008	2009
Production cooperative	10,838	11,620	11,348	8,892	8,134	7,262
Open joint-stock company	1,963	1,987	1,769	1,582	1,494	1,461
Closed joint-stock company	2,716	2,870	2,679	2,133	1,861	1,776
Limited liability company	3,953	5,133	5,332	7,598	8,686	9,420
Other enterprises	7,423	5,990	4,145	3,764	2,805	2,563
Total	26,893	27,600	25,273	23,969	22,980	22,482

Note: [a]Data from 1 January 1995–2009.

Table 7.2. Percentage of market share by type of agricultural enterprise in Russia (1995–2009).[a] (From Rosstat, 2010.)

Type of agricultural enterprise	1995	2000	2006	2007	2008	2009
Production cooperative	32	33	35	36	33	33
Open joint-stock company	4	4	6	7	7	6
Closed joint-stock company	22	18	11	10	11	11
Limited liability company	28	31	39	39	41	42
Other enterprises	14	14	9	8	8	8
Total percentage	100	100	100	100	100	100

Note: [a]Data from 1 January each year.

production have been pursued primarily through limited liability companies (Wandel, 2011). However, limited liability companies do not seem to have become stronger compared to production cooperatives. For example, in 2006, production cooperatives constituted 45% of all agricultural enterprises and had a 35% market share compared to 2009, when they constituted around 32% of all agricultural enterprises and had a 33% market share.

One of the reasons production cooperatives constitute a large sector in Russian agriculture may be governmental action (Amelina, 2000; Lerman and Schreinemachers, 2005). In regions where agricultural production is important for the local economy, the government has been reluctant to register private/smallholder farms (Amelina, 2000; Lerman and Schreinemachers, 2005; Visser and Spoor, 2011). This may explain to some extent why there are so few private/smallholder farms in Russia.

Another reason why production cooperatives are a strong agricultural industry is because they are fairly labour-intensive (Minakov, 2007). In spite of fewer workers and decreased value of fixed assets, cooperatives have improved their overall production value (Rosstat, 2010). For example, 41.3% of the cooperatives made profits in 2002 (Minakov, 2007).

Institutional Background for Production Cooperatives in Russian Agriculture

Russian legislation demands that production cooperatives should be voluntary associations of citizens who become members to conduct joint production or other economic activities based on their personal labour, as well as their investments in shares. A production cooperative is formed at the decision of the founding members. When a former collective farm was transformed into a production cooperative, the workers themselves made this decision without formal interference from the government. However, it has been claimed that some collective farm directors used various strategies to shield members from the idea of becoming private/smallholder farmers. Such directors had an interest in maintaining the pre-reform status quo that assured them access to income, power and prestige (Allina-Pisano, 2002).

The legislation admits members of varying legal status. Basically, there are three types of members: (i) people who work in the cooperative; (ii) people who do not work in the cooperative but may be involved in other types of related activities; and (iii) legal entities that operate in the cooperative through representatives. The second and third groups of membership are not required for cooperatives. When included, their combined numbers must not exceed one-quarter of the total membership.

The legal status of non-working and legal entity members differs from that of working members. First, they are obliged to invest in two types of shares, basic and additional shares, to compensate for their lack of labour inputs. Second, they receive only a profit remuneration that depends on the amount of shares they possess. Working members have stronger rights because, in addition to their wages, they have the right to receive part of the cooperative's annual surplus (if any). The surplus is proportional to the amount of shares they own and is dependent on the quantity and quality of their labour.

The control system for cooperatives is similar to that of commercial organizations. There must be a general assembly as the supreme body and a board as the executive office, a chief executive officer (the chairman of a cooperative society) and a control body (an auditing committee). All members have one vote, regardless of the shares they own and their official position.

Cooperative assets comprise the share of capital paid in by members, accumulated profits from cooperative activities, credits and other sources. A cooperative society's equity capital consists of members' shares, but there may also be unallocated capital (indivisible funds). The existence of the indivisible funds differentiates cooperatives from other business organizations. They provide the cooperatives with a more stable financial status. Part of the indivisible funds may be transferred to the reserve fund intended to cover losses. The financial surplus that a cooperative earns, if any, is distributed (after taxes and other obligatory payments) between working and non-working members.

Decisions regarding new membership must be made by the general assembly. Any new member has to buy a share of the cooperative. The size of a share (both basic and additional, if any) paid for by the members can differ according to the time of joining; that is, when the cooperative is first being established or at a later stage. The shares may be paid in different ways: (i) cash, securities or other assets (e.g. buildings, premises, equipment and vehicles); (ii) rights of use of buildings; or (iii) results from intellectual activities. By the time a cooperative is registered or when a working member joins, at least 10% of the share of capital must be paid. The other two member categories have to pay their entire share as soon as they become members. Land and other natural resources may also constitute a share. The fact that the share capital can be paid with assets and not in cash is a consequence of how the assets of the former collective farms were distributed among the working members at the time of its transformation.

The members own their shares, but their ownership rights are demarcated. The decision-making power over the cooperative's entire assets lies in the hands of the cooperative's leadership. The cooperative can cultivate land that can be brought in by members in the form of share funds, transferred to the cooperative for rent or redeemed by the cooperative society. In the first and third cases, the land becomes the property of the cooperative.

The workers at the Soviet farms had two options when the cooperatives were created. They could transfer the whole collective farm into a production cooperative or they could take out just their share of the total equity and form a new cooperative from scratch. The cooperatives based on complete transformation initially had advantages over those starting from scratch in that they already had production means; did not change their production orientation; and had no problems with staffing, supplies of resources or the marketing of products.

Specific steps had to be followed in connection with the complete transformation of collective farms. When the owners of the land and the property shares (i.e. members and workers of former collective farms) wanted to establish a cooperative, they informed the enterprise's privatization commission that they wanted to use their share as an entrance fee to a new cooperative. The privatization commission was expected to allocate a corresponding part of land and assets to the new cooperative. However, there was no clear pattern to follow. If workers of collective farms wanted to establish a new (usually small-scale) cooperative from scratch, in general they received support from their former collective entities. After approval by the general assembly, the land could be transferred to those members as either jointly used and owned or jointly used and individually owned property. The area of land allocated to each member in the former enterprise was calculated by dividing the total land area by the number of workers, including other employees and pensioners.

Transaction Costs of the Cooperative Members

An industry can be organized by making changes in the size of firms, ownership structure, interrelationships between firms and production technologies. The field of science that explains how industries are organized is called industrial organization. One powerful theory within this field is transaction cost economics (Williamson, 1985). This may explain why production cooperatives exist in Russian agriculture.

The basis of transaction cost economics is that firms try to minimize their transaction costs, given that their production costs are unaffected. The transaction costs are the result of the firms' exchanges with each other and consist of all the resources that the trading parties use to make sure that safe transactions can be conducted. They result from efforts made before, during and after the transaction. Resource-consuming activities occur before the transaction to collect information about the market offers and price levels and to investigate prospective trading partners. During the transaction, the parties engage in negotiations, some of which take time and money. After the transaction occurs, the parties evaluate the contract to determine the next step of whether or not to continue to trade, to sue in court, to mediate the contract or to discontinue trade at the end of the contract (Niehans, 1987). Hence, the theoretical basis for this theory is that humans seek self-interests and are limitedly rational. Due to information asymmetry, there is always the possibility of partners using deception to reap benefits.

The focus of the transaction cost economics theory is on three determinants of the trading partners' transaction costs. These include the asset specificities, uncertainties and frequency of transactions (Staatz, 1984; Williamson, 1985).

Trading partners with transaction-specific assets have a limited degree of freedom that makes them vulnerable to dishonest trading partners. Hence, the partner with asset specificity wants to avoid a badly functioning market by using vertical integration, which is ownership in a partnering firm or achieving a long-term agreement to further the success of the entity. When transacting with each other, trade partners face uncertainties (e.g. the price of the product, the price of capital, future regulation that may affect the markets and whether or not the intended partner can be trusted). Uncertainties imply that the partner may get an unfair deal. The consequences of trade uncertainties may cause firms to avoid the market, to search for integrative solutions (e.g. alliances with reliable partners) or to seek ownership in a partnering firm to obtain less risky transactions. Transaction frequency relates to the existence of alternative trading partners. This involves a higher risk of being deceived, thereby explaining the interest in establishing an integrative solution.

After the transition of Soviet agriculture in 1991, rural workers received a share of land and other assets from the former collective enterprises. They could use these assets to invest in production cooperatives; to establish private/smallholder farms; or to make other investments and work in large-scale, investor-owned firms. The option of production cooperatives implies vertical integration (i.e. integration between workers and employers). Private/smallholder farming is a type of market solution because workers represent themselves in the market. Working in an investor-owned firm is another type of market solution in the sense that the workers bring only their labour, not their capital or other resources, to the market.

The agricultural cooperative choice is a way for workers to reduce their transaction costs within the labour market. All three determinants of transaction costs are applicable, especially the asset specificity of the workers, since their assets are linked to their former workplace (the Soviet collective farms). At the time of the transition, there were many forms of relationship asset specificities.

Agricultural fields and buildings are site-specific assets because they have a fixed location that cannot be moved to somewhere else. Farm workers have few choices because the situation for site-specific agricultural cooperatives is different from that of other types of production, where machinery and workers can be moved to other locations and raw products can be sourced elsewhere. If the workers move elsewhere, others will take their place within the cooperative. In reality, production cooperatives are just a continuation of the large-scale Soviet collective farming system.

At the time the cooperatives were established, the workers received shares in the new cooperative. The financial-specific assets received were mainly non-liquid because there was no well-functioning market if the cooperative members wanted to sell their assets. In 1991, establishing private/smallholder farms required large capital investments that most farm workers did not possess, such as registering a new firm, establishing procurement and marketing channels, and purchasing machinery and livestock. In addition, private/smallholder farmers had insufficient collateral to obtain credit loans due to cheap land prices.

Many physical-specific assets within the collective farms were indivisible because Soviet collective agriculture was highly mechanized and was capital-intensive. There was no way to divide the farm equipment equitably among potential private/smallholder farms (Herrold-Menzies, 2009). Because the workers received little money when leaving the village, it was difficult for them to invest in materials, technical resources and infrastructures. They also lacked institutions and organizations to support them to become private/smallholder farms (Wegren and O'Brien, 2002).

The worker's choice of organization had to be made within a few months. Even if the villagers tried to act rationally, time specificity meant that their decisions were not well deliberated, due to incomplete information and to limited resources (Golovina *et al.*, 2013).

Had the workers from the collective farms established private/smallholder farms, they would have lost the well-established social networks available with the Soviet collective farms. This future was not attractive to people who had lived all their lives in the same community. While the workers may not have had much trust in the collective farm itself, there was a good deal of social-specific capital assets within the familiar group of neighbours, friends and family. The established social networks among fellow workers were important when the workers decided to set up a production cooperative and when the workers wanted to influence the collective farm leadership (Golovina *et al.*, 2013). The production cooperatives provided the Soviet farm workers with substantial relationship-specific assets that dissuaded them from choosing private/smallholder farms or other organizational cooperative forms other than production cooperatives.

Because collective farms were characterized by specialization, most workers lacked both the diversified work skills and the entrepreneurial skills required to become private/smallholder farmers (Golovina *et al.*, 2013). Because of their lack of the appropriate skills, the workers were averse to risk-taking and preferred not to be employed independently. In addition, most workers lacked the necessary capital to start a private/smallholder farm.

When choosing between an agricultural production cooperative and a private/smallholder farm, most workers were averse to the risks and uncertainty of private/smallholder farm production. It is understandable that the workers were risk averse, as they had never experienced anything like private/smallholder farming and they were unaccustomed to making independent decisions. The unfavourable macroeconomic environment reinforced the risks surrounding agricultural producers. There were difficulties when enforcing ownership and contracts, as well as difficulties when collecting and processing information. When choosing between private/smallholder farms and large-scale agricultural enterprises, rural workers were worried about limited infrastructures (e.g. roads, houses and livestock), high production costs and uncertain revenues. Other risks concerned high transaction costs and the possibility of deceitful trading partners.

If the workers of the Soviet farms had decided not to continue in a production cooperative, they would have had few options for alternative employment opportunities in the villages where they lived. It would have been difficult for them to commute to other villages due to long distances and limited infrastructures. In short, the workers did not have much of a choice other than to choose the agricultural production cooperatives. Production cooperatives had the lowest transaction costs for the workers.

Assessments of the Agricultural Production Cooperatives

The Russian production cooperatives account for almost one-third of the total number of agricultural enterprises in Russia (Table 7.1). They produce one-third of all the agricultural products from agricultural enterprises and their market share is fairly stable (Table 7.2).

Because labour and the allocation of profits in the Russian production cooperatives are the basis for membership, the cooperatives have concentrated on labour-intensive production. This means that the cooperatives have fewer investments in fixed assets. They depend more on human labour than on machinery. This causes difficulties when mobilizing equity because the members prefer the profits being paid to them, rather than the profits being invested in the cooperatives.

Investment problems enabled the Russian production cooperatives to allow different forms of membership. Equity capital originates also from members who do not work in the cooperative but are involved in related activities and legal entities that operate in the cooperative through representatives. The cooperatives also obtain a stronger position in the local community because external parties have an interest in the financial success of the cooperatives (Bonin *et al.*, 1993).

The achievements of the Russian production cooperatives, like all other cooperative firms, are contingent on leadership. While production cooperatives have democratic governance formally, there is a chairman of the board, who has a decisive role almost like the chief executive officer in other enterprises. This can lead to abuse of power for self-enrichment.

However, the chairman lives and works among the members and knows everybody, and all members know the chairman. This social relationship may exert control on the chairman, in which case there may be opportunities for functional democratic control and good business operations. In spite of this, the chairman often makes decisions autonomously. The chairman is held responsible by the membership only when the yearly general assembly is held. This power asymmetry creates agency problems. The contracts between the cooperative and the members are, of necessity, incomplete and give rise to skewed relationships (Pencavel, 2001). Historical influences (e.g. from Tsarist and Soviet times) may underpin such a situation.

There are striking differences between Russian agricultural production cooperatives and Western-style agricultural cooperatives. In Russia, most assets are collectively owned, the work is conducted jointly and equality seems to be the guiding principle. In their survey, Golovina *et al.* (2013) found that members had no objections to the cooperatives' organizational structure.

After the agricultural production cooperatives were established, the members learned little about running private/smallholder farms (Golovina *et al.*, 2013). They continued in the same specialist jobs, so the same lock-in mechanism prevailed. The members incurred high transaction costs if they left the production cooperatives. Thus, it has become even more difficult to establish private/smallholder farms than it was during the first years after transition, when private/smallholder farms received significant government support.

Few members have become rich enough to leave the cooperatives, where they still enjoy the social relationships within the cooperative. If they were to move elsewhere, finding other employment and housing would be difficult (Golovina *et al.*, 2013). Also, insufficient capital of these members makes it difficult for them to redeem their investments from the cooperative.

Conclusions

Because the Soviet agricultural production was highly mechanized and capital-intensive, the standard of living among farm workers was relatively high. The workers had much to lose if they left the large agricultural enterprises after the break-up of the Soviet Union. Hence, many of them preferred to join cooperatives, rather than start private/smallholder farms. Many Soviet farms were fully or partly transformed into agricultural production cooperatives, while others were turned into corporations. Most production cooperatives are still in operation and play a significant role in agriculture. While they often have large market shares, their economic performance in terms of wage levels, investments and capital returns has not been impressive.

One explanation as to why the cooperatives were established is that the Soviet workers had a variety of relation-specific assets. They had low self-confidence, risk aversion, little entrepreneurial spirit, lack of financial means, strong social ties to one another and highly specialized skills. Because their assets were more adaptable to large agricultural enterprises than they were to private/smallholder farms, they really had no alternative.

During the years since the production cooperatives were established, the situation has remained unchanged. The members still have the same relation-specific assets. It has been financially difficult for them to become private/smallholder farmers without government support. The market conditions for private/smallholder farmers have become more difficult due to concentrated inputs and outputs in the markets and less profitability.

The market share of agricultural production cooperatives has been relatively stable and high. While the number of cooperatives and their aggregate memberships has declined, per member production has increased. These trends may indicate a striving for higher efficiency. However, their financial strength is still low, their investment ability is poor and they are characterized by a relatively low level of efficiency and profitability. Cooperatives have limited financial resources for newer farm technologies and agricultural inputs such as fertilizers, seeds, fuel and feeds.

The agricultural production cooperatives face structural obstacles, so it might be difficult for them to become more efficient. One such obstacle is the high degree of collectivism in terms of ownership, governance and operations. This hampers good performance, limits the formation of capital and does not provide strong incentives for development. Linked to this collectivism is the fact that these cooperatives have a strong focus on one type of production factor, namely labour. Because members receive their wages based on the amount of labour they accomplish per day, they do not have incentives to raise their efficiency.

The long-term survival of the Russian production cooperatives may be explained in terms of power. For example, fear of disturbing the existing power balance within the membership by increasing its heterogeneity may have far-reaching consequences. For this reason, cooperatives may be characterized by resistance to change of various kinds (i.e. conservatism). New production technologies and new financial solutions may also threaten the power balance.

Contributing to this problem is the fact that the equity capital of cooperatives is not subject to the forces of the open market; instead, it is decided by a criterion of the membership, where decisions tend to have political, rather than economic, considerations. Furthermore, cooperatives tend to be undercapitalized due to unclear property rights. Equity capital is owned jointly, whereby members are reluctant to invest.

A strong cooperative board and a strong chief executive officer may conduct measures that are not in perfect accordance with the wishes of the members. Due to the weak tradition of democratic decision making and co-determination for the working members, management has much power that might be even stronger than during the Soviet period, as they are no longer dependent on higher state authorities.

Hence, the same factors that affected why the workers of the Soviet agricultural enterprises decided to establish production cooperatives are still in operation. The major threats to the cooperatives include property-right problems, a skewed power balance and the unwillingness of the members to govern themselves or to put up individual capital investment in the cooperative.

In conclusion, agricultural production cooperatives play an important role in agricultural production in Russia. Their development path is linked closely to historical specificities and to the current economic, social and political situation. They face more problems ahead, but to date they have proven that they are able to survive hard times.

References

Allen, D. and Lueck, D. (2005) Agricultural contracts. In: Ménard, C. and Shirley, M. (eds) *Handbook of New Institutional Economics*. Springer, Dordrecht, the Netherlands, pp. 465–490.

Allina-Pisano, J. (2002) *The Post-Soviet Potemkin Village. Politics and Property Rights in the Black Earth*. Cambridge University Press, Cambridge, UK.

Amelina, M. (2000) Why Russian peasants remain in collective farms: a household perspective on agricultural restructuring. *Post-Soviet Geography and Economics* 41(7), 483–511.

Binswanger, H., Deininger, K. and Feder, G. (1995) Power, distortions, revolt, and reform in agricultural land relations. In: Behrman, J. and Srinivasan, T.N. (eds) *Handbook of Development Economics, Vol III*. Elsevier, New York, pp. 2659–2772.

Bonin, J.P., Jones, D.C. and Putterman, L. (1993) Theoretical and empirical studies of producer cooperatives: Will ever the twain meet? *Journal of Economic Literature* 31, 1290–1320.

Dunn, J.R. (1988) Basic cooperative principles and their relationship to selected practices. *Journal of Agricultural Cooperation* 3, 83–93.

Golovina, S. and Nilsson, J. (2009) Russian agricultural producers' views of top-down organized cooperatives. *Journal of Rural Cooperation* 37(2), 225–241.

Golovina, S., Nilsson, J. and Wolz, A. (2013) Members choice of production cooperatives in the Russian agriculture. *Post-Communist Economies* 25(4), 539–551.

Herrold-Menzies, M. (2009) The post-collective village: a tale of two transitions. *World Development* 37(1), 232–241.

Johnson, N.L. and Ruttan, V.W. (1994) Why are farms so small? *World Development* 22, 691–706.

Lerman, Z. and Schreinemachers, P. (2005) Individual farming as a labour sink: evidence from Poland and Russia. *Comparative Economic Studies* 47, 675–695.

Minakov, I.A. (2007) *Cooperative and Agroindustrial Integration in APK*. Kolose, Moscow (in Russian).

Niehans, J. (1987) Transaction costs. *The New Pelgrave. A Dictionary in Economics* 4, 676–680.

Ollila, P. (1989) Coordination of supply and demand in the dairy marketing system, with special emphasis on the potential role of farmer cooperatives as coordinating institutions. *Journal of Agricultural Science in Finland* 61, 137–321.

Pencavel, J. (2001) Introduction: Citizens, consumers, and workers. In: Pencavel, J. (ed.) *Worker Participation. Lessons from the Worker Coops of the Pacific Northwest*. The Russell Sage Foundation, New York, pp. 1–19.

Podgorbunskih, P. and Golovina, S. (2005) *Theory and Practice of Agricultural Enterprises Development: From the Classics to Institutionalism*. Zauralie Publishing, Kurgan, Russia (in Russian).

Pollak, R.A. (1985) A transaction cost approach to families and households. *Journal of Economic Literature* 23(2), 581–608.

Rosstat (2010) *Russian Statistical Year-book: 2009. Statistical Collection*. Rosstat, Moscow (in Russian).

Schiller, O.M. (1969) *Cooperation and Integration in Agricultural Production*. Asia Publishing House, London.

Schmitt, G. (1991) Why is the agriculture of advanced western economies still organized by family farms? Will this continue to be so in the future? *European Review of Agricultural Economics* 18(3–4), 443–458.

Staatz, J. (1984) A theoretical perspective on the behavior of farmers' cooperatives. Michigan State University, East Lansing, Michigan (dissertation).

Visser, O. and Spoor, M. (2011) Land grabbing in post-Soviet Eurasia: the world's largest agricultural land reserves at stake. *Journal of Peasant Studies* 38(2), 299–323.

Wandel, J. (2011) Business groups and competition in post-Soviet transition economies: the case of Russian 'agroholdings'. *Review of Austrian Economics* 24, 403–450.

Wegren, S.K. and O'Brien, D.J. (2002) Introduction: Adaptation and change: old problems, new approaches. In: O'Brien, D.J. and Wegren, S.K. (eds) *Rural Reform in Post-Soviet Russia*. John Hopkins University Press, Baltimore, Maryland, pp. 1–20.

Williamson, O.E. (1985) *The Economic Institutions of Capitalism: Firms, Markets, Relational Contracting*. Free Press, New York.

8 Agricultural Cooperative Development in Kazakhstan and Ukraine

David Sedik[1]* and Zvi Lerman[2]
[1]*Food and Agricultural Organization (FAO), Budapest, Hungary;*
[2]*Hebrew University of Jerusalem, Rehovot, Israel*

Abstract

The agricultural service cooperative is a means for incorporating small farmers into value chains that include larger downstream middlemen, processors, markets and grocery chains. This chapter describes the development of agricultural service cooperatives in the large wheat producing countries of Kazakhstan and Ukraine, and analyses the reasons for the lagging development there, with a review of policy alternatives for development of the cooperative movement. While there appears to be widespread recognition by both governments of the usefulness of agricultural service cooperatives, support policies focus on subsidies for input purchases and subsidized loans, thus breeding a class of false cooperatives that are established only for the purpose of gaining access to cheap money with no regard for cooperative principles. Neither government has succeeded in creating a supportive business environment for service cooperatives, with tax disincentives as an important barrier to their development. Neither of the governments of Kazakhstan or Ukraine has informed the rural population adequately on the cooperative idea and its benefits.

Introduction

The purpose of this chapter is to review the evidence on agricultural service cooperatives in Kazakhstan and Ukraine and to consider the reasons for their lack of development compared to North America and Western Europe. For example, although many farms in North America and Western Europe are members of at least one cooperative, only 0.4% of farms in Ukraine are members of a cooperative (Table 8.1).

A number of reasons are often cited for the dearth of self-help organizations in ex-socialist countries (Akimbekova, 2010; Korinets, 2013; Tomich, 2013). Joining an agricultural service cooperative presents problems, such as limited start-up capital, few capable managers and limited understanding of the principles and benefits of membership.

In this chapter, we show that there are potentially good economic reasons for farmers to join agricultural service cooperatives in Kazakhstan and Ukraine. Despite potential economic benefits, support for the agricultural service cooperative idea is low in Kazakhstan and Ukraine, based on past experience with the Soviet cooperative model and on the lack of experience with alternatives. There are several research studies on the distrust of cooperation in post-socialist economies. Gijselinckx and Bussels (2012) investigated the lack of cooperation in the ex-socialist countries of Europe and found a high correlation

*Corresponding author; e-mail: david.sedik@fao.org

 Transition to Agricultural Market Economies (eds A. Schmitz and W.H. Meyers)

Table 8.1. The development of agricultural service cooperatives in selected countries, 2007–2010. (From USA: USDA, 2012, and USDA/NASS, 2013; Ukraine: Korinets, 2013, pp. 36 and 38, and State Statistics Service of Ukraine, 2011, p. 10; France: Eurostat, 2013, and Filippi, 2012, p. 14; Hungary: Szabó, 2012, p. 23; Italy: Bono, 2012, p. 19; Kazakhstan: Conception, 2012.)

Country	Year	Number of service cooperatives	Number of cooperative members	Number of farms	Portion of farms with cooperative members (%)
Hungary	2007	58	20,177	626,300	3.2
Italy	2008	5,800	900,000	1,679,400	54
Kazakhstan	2010	300	N/A	1,850,000[a]	N/A
Ukraine	2010	801	21,521	5,300,000	0.4
USA	2010	2,310	2,200,000	2,200,000	100
France	2010	2,900	500,000	516,100	100

Notes: [a]The sum of peasant farms (170,000) plus an estimate of the number of household plots (1,680,000, assuming that each rural household has a household plot). Data for peasant farms from *Agriculture in Kazakhstan 2007–2011* (Statistical Agency of Kazakhstan, 2012a); data for rural households from *2009 Population Census of Kazakhstan: Households* (Statistical Agency of Kazakhstan, 2012b).

between social capital and cooperative member intensity (percentage of farmers who were cooperative members), but found no discernible correlation between general deep-rooted cultural values and member intensity. Social capital is defined as the trinity of networks, norms of reciprocity and trust (Gijselinckx and Bussels, 2012). Indicators of general deep-rooted cultural values were based on research by Hofstede *et al.* (2010). Lissowska (2013) came to similar conclusions in her analysis of social attitudes toward cooperation in European countries based on the European Social Survey, where Lissowska found that transition countries in Central and Eastern Europe had less faith in the cooperative model due to past experience with the socialist version of cooperation.

Overcoming the legacy of the past requires a reimagining of the cooperative model on the part of farmers and policy makers. Farmers need to experience the liberal democratic model of cooperation, while policy makers need to recognize that the role of the state when organizing service cooperatives should be minimal. Public policy should provide a supportive enabling environment for cooperatives based on the model of self-help style, grass-roots cooperation. This has been a difficult task for the governments of Kazakhstan and Ukraine.

To understand how the governments in Kazakhstan and Ukraine have dealt with this task, we consider two aspects of the enabling environment. First, we consider government policies toward cooperatives in Kazakhstan and Ukraine. Second, we analyse legislation to gauge the extent of state support for agricultural service cooperatives and how legislation could be improved.

The Need for Agricultural Service Cooperatives in Kazakhstan and Ukraine

Agriculture in Kazakhstan and Ukraine is based primarily on smallholders (peasant farms and household plots). In 2011, the smallholder farm sector produced 70% of gross agricultural output (GAO) in Kazakhstan and 55% in Ukraine (Statistical Agency of Kazakhstan, 2012a; State Statistics Service of Ukraine, 2012). Small farmers everywhere in the world face constraints when accessing market services, such as difficulties with access to: (i) sales channels for market products; (ii) supply channels for farm inputs, farm machinery and transportation equipment; (iii) information and advisory services that are essential for raising productivity and efficiency; and (iv) the credit resources required to finance short-term working capital and long-term investment needs. These constraints combined create what is sometimes referred to as *the curse of smallness* – a trap that prevents smallholders from exploiting their inherent productivity advantages fully due to barriers of access to markets (Abele and Frohberg, 2003).

In developed market economies, private intermediaries often compete to buy and sell agricultural commodities from farmers, or farmers themselves market their own production. In Kazakhstan and Ukraine, rural private intermediaries often act as monopsonistic buyers in the region where they operate. In such cases in market economies, farmers organize service cooperatives whose function it is to correct market failure by allowing alternative access to markets without relying on private service intermediaries. Such cooperatives can cover the whole field-to-market value chain, including joint farm input purchases, water distribution and irrigation (through water user associations), machinery pools for field work, processing at packing houses and transportation of agricultural products to market. They can also provide agricultural extension and market information services, as well as veterinary and artificial insemination services, all of which are essential for the improvement of both crop and livestock production. What service cooperatives do not do is engage in primary agricultural production as a collective of farmers.

When dealing with a service cooperative, the market effectively deals with a relatively large entity that combines many smallholders into a single negotiating position. Access difficulties imposed by smallness are negated by service cooperatives that conduct market transactions for a large number of small farmers simultaneously.

Kazakhstani and Ukrainian Policies on Agricultural Service Cooperatives

Kazakhstan

The latest official policy on cooperatives in Kazakhstan is summarized concisely in the *Kazakhstan Strategy 2050* presented by the President in 2012 (Kazakhstan 2050, 2014). The policy was to create a system of support and incentives for small and medium-sized businesses to stimulate entrepreneurship through associations and cooperatives. A more detailed exposition of Kazakhstani government policies for agricultural service cooperatives can be found in Conception (2012). This basic document outlines three main policy directions: (i) improved legislative base; (ii) cooperative information and training; and (iii) financial support through low interest rate credit loans, tax breaks and government subsidies for inputs and investments.

Attempts to improve the legislative base continue, mainly through the efforts of the agricultural lobby, but very little seems to be happening with regard to providing information about cooperatives and training for producers and potential managers. The main policy focus is on financial support, including credit and subsidies. Two models of government support observed in Kazakhstan that rely on state-controlled companies are the Agricultural Credit Corporation and rural social entrepreneurial corporations. As an arm of the Ministry of Agriculture, the Agricultural Credit Corporation was established in 2001 to grant 5- to 7-year loans at 5% per annum as seed money for the creation of rural service cooperatives (not including the water user associations). Social entrepreneurial corporations (*Natsional'naya kompaniya Sotsial'no-predprinimatel'skaya korporatsiya*) are part of a regional network of state–private partnerships (service procurement centres; *servisno-zagotovitel'nye tsentry – SZTs*) that provide agricultural services to rural service cooperatives and to agricultural producers. The participation of service cooperatives in equity capital is a necessary condition for the creation of service procurement centres.

By 2009, the Agricultural Credit Corporation of Kazahstan had distributed 8 billion Kazakhstani tenge (KZT) (US$50–65 million, depending on exchange rates) in subsidized agricultural loans (Akimbekova, 2010; Conception, 2012). It is argued that the availability of low interest rate credits and other subsidies intended specifically for service cooperatives encourages the creation of false cooperatives that are established only for the purpose of gaining access to subsidized credits and inputs with no regard for true cooperative principles. This trend may be responsible for the large discrepancy in cooperative statistics from different sources and for the large gaps among the number of registered and active cooperatives. Service procurement centres duplicate the basic functions of service cooperatives, with no attempt to adhere to cooperative principles. They appear to be a carry-over from the Soviet model of enlargement that characterized the agroindustrial complex, and their goal is apparently profit maximization, as is evident from the high prices they charge for services to cooperatives and farmers (Conception, 2012).

Ukraine

While the government of Ukraine has consistently declared support for agricultural service cooperatives, implementation of this support has been far less consistent. More importantly, when government programmes do provide support, the focus is on subsidies rather than on establishing cooperatives. Instead of education programmes and legal assistance for forming new cooperatives, government policy has concentrated on special agricultural value-added taxes (VATs), which strongly skew the beneficiaries by distributing the largest share of subsidies to a relatively small number of large producers rather than to tens of thousands of smallholders.

While the Presidential Decree *On Means for Development of the Cooperative Movement and Strengthening of its Role in the Market Reform of the Ukrainian Economy* (President of Ukraine, 2000) promised government support for the organization of credit lines and guaranteed bank loans for service cooperatives, no funding for service cooperatives has been made available from the central government (Korinets, 2013, p. 56). In contrast, *The Programme for the Development of Agricultural Service Cooperatives for 2003–2004* (Ministry of Agrarian Policy and Food of Ukraine, 2002) promised support for cooperatives only through support from domestic and foreign investors, share contributions of cooperative members and funds from local budgets. Ukrainian Law *On the Main Pillars of State Agrarian Policy to 2015* (Parliament of Ukraine, 2005) noted that one of the main priorities of the state agrarian policy was state support for developing competitive agricultural production based on cooperation and integration. The *State Special Programme for the Development of the Ukrainian Village to 2015* (Cabinet of Ministers of Ukraine, 2007) also envisioned state support for agricultural service cooperatives with no apparent budget resource availability (Korinets, 2013, p. 57).

The first substantial funds made available from the central government for the development of agricultural service cooperatives were in accordance with the Cabinet of Ministers Decree No 557, *The State Special Economic Programme for the Support of the Development of Agricultural Service Cooperatives to 2015* (Cabinet of Ministers of Ukraine, 2011). While this decree planned for the creation of cooperatives through support for enabling legislation, funding infrastructure and removing obstacles for cooperative development, the programme only made funds available for the purchase of agricultural machinery for cooperatives (Korinets, 2013, p. 56).

In 2011, the Ukrainian government again declared its support for agricultural cooperatives. President Yanukovich noted that the cooperatives *should become an important foundation of the structural reform of agriculture* (Korinets, 2013, p. 57). Support was continued into 2012 with the new version of the law *On Agricultural Cooperation* (Parliament of Ukraine, 1997) signed into law by the President on 19 January 2013. According to this legislation, 5 million hryvnia (UAH) were to be made available for the purchase of agricultural machinery by cooperatives. In addition, in 2013, the Minister of Agrarian Policy (Nicolai Prisiazhniuk) noted that agricultural cooperation would help small producers become more competitive in the world market (RBK-Ukraine, 2013).

Enabling Environment: Classification and Characteristics of Cooperatives

The International Cooperative Alliance (ICA) defines a cooperative as an autonomous association of persons united voluntarily to meet their common economic, social and cultural needs and aspirations through a jointly owned and democratically controlled enterprise (ICA, 2013). The persons who voluntarily unite to form a cooperative are usually referred to as members, or member owners, and one of the key attributes of membership is active participation in the activity of the cooperative – whether it be economic, business or social. A cooperative is a legal entity and, in a certain sense, it is an analogue of a shareholder corporation. There are, however, some fundamental differences between a cooperative and a corporation (Table 8.2).

Table 8.2. Comparative attributes of a cooperative and a shareholder corporation. (From authors' estimations.)

Attribute	Cooperative	Corporation
Owners	Members	Shareholders/investors
Owners' objective	Use of services provided by cooperative	Earning income
Organization's objective	Maximize members' benefits from working with cooperative	Maximize corporate profits
Voting rights	One member, one vote, regardless of share contribution	Number of votes proportional to number of shares (i.e. share contribution)
Income distribution rules	Income distributed to members in proportion to participation in cooperative activities	Income distributed to shareholders in proportion to number of shares held

Perhaps the main difference concerns the organization's objective – while business corporations aim to maximize their profit, cooperatives aim to maximize the benefits that members derive from their participation in cooperative activities, including lower prices paid for inputs and services and higher prices received for products (Cobia, 1989). The Western cooperative paradigm distinguishes between production cooperatives, service cooperatives and consumer cooperatives based on their functional characteristics.

Production cooperatives are cooperatives in which members are jointly engaged in the production process, regardless of whether or not it is the production of manufactured goods, agricultural commodities or services. Production cooperatives are based on member labour and they sell their output to outsiders. The main function of production cooperatives is to improve the well-being of their members by providing jobs and working conditions that are superior to what would otherwise be available in the market. In agriculture, it is often argued that by allowing members to pool their fragmented smallholdings into large farms, production cooperatives may exploit economies of scale and achieve higher efficiency. Yet many researchers have shown that agricultural production cooperatives are substantially less efficient than are private/smallholder farms. As a result, production cooperatives are a small minority among organizations that produce manufactured goods, agricultural products or services. According to ICA data (2013), production cooperatives account for less than 5% of all cooperatives in the world.

Service cooperatives are the largest category of cooperatives. They are cooperatives that provide services to their members or producers who act independently. In contrast to the minor role of production cooperatives in market economies, service cooperatives in many countries account for a large share of transactions in the relevant economic sector. For instance, agricultural marketing, processing and supply cooperatives are major players in markets for farm products and farm inputs in North America, Western Europe, Japan and South-east Asia. In the USA, for example, agricultural cooperatives handle about 30% of the total farm marketing volume and 28% of the total supply purchases of the farmers (Mather *et al.*, 2004). In the European Union, the share of agricultural cooperatives is even larger in countries such as the Netherlands, Denmark, Ireland and Sweden, where 70%–80% of farm products are marketed through cooperatives and cooperatives account for 50%–70% of all farm input purchases (Van Bekkum and Van Dijk, 1997). Service cooperatives may actually employ some of their members as workers, but most employees (and even most managers) are hired outsiders.

Consumer cooperatives are trading firms that sell consumer goods primarily to their members at advantageous prices. They are basically a variety of supply cooperatives outside of agriculture, but they are treated as a separate class because of their numerical importance in the world. The largest groups of consumer cooperatives are the cooperative grocery stores and supermarkets, but cooperative stores that sell clothes, housewares, appliances and even cars to their members at advantageous prices also fall in the category of consumer cooperatives. These consumer cooperatives, unlike other service cooperatives, are most likely to deal with non-members (third parties) on a regular basis, which is their strategy of achieving growth comparable with other firms in the highly competitive consumer services sector. Consumer cooperatives charge a higher price on their transactions with non-members.

farm services such as product marketing and input supply to household plots and to provide consumer services through trading outlets to the rural population in general.

Kazakhstani and Ukrainian Legislation on Agricultural Service Cooperatives

The USSR law on cooperation

The first mention of cooperatives in Kazakhstani and Ukrainian legislation actually pre-dates independent Kazakhstan and Ukraine, occurring in the 1988 former Soviet law *On Cooperation in the USSR* (Decree of the Council of Ministers of the USSR, 1988). This law set the precedent for subsequent legislation in that it considered two main types of cooperatives: production (*proizvodstvennyi*) and consumer (*potrebitel'skii*). However, the conceptual distinction between production and consumer cooperatives (Article 3.2, *On Cooperation in the Union of Soviet Socialist Republics*) is not mutually exclusive. Production cooperatives produce goods and products, and render paid services. They operate for the production, procurement, processing and marketing of agricultural production. Consumer cooperatives serve the demands of their members and other citizens in trade and service industries as well as they serve the demands of their members for housing, parking, garden plots, social-cultural activities and other services. Consumer cooperatives may develop various production activities that are not necessarily limited to processing and may include subsidiary agricultural enterprises.

The Decree of the Council of Ministers of the USSR (1988) is usually credited with being the first step toward private enterprise and market economy in the former USSR under the guise of cooperatives that are defined as new organizational forms free from strict state control but without adherence to Western cooperative principles. Yet Articles 45 and 46, *On Cooperation in the USSR*, of the 1988 law (Decree of the Council of Ministers of the USSR, 1988) also perpetuated the traditional Soviet system of state-controlled consumer cooperation (*sistema potrebitel'skii kooperatsii*) that was designed primarily to provide

Kazakhstan cooperative law

In Kazakhstan, the basic definition of cooperatives as a corporate form of organization appears in the Civil Code, which was originally adopted in 1994 and then amended repeatedly. Unlike Ukraine, Kazakhstan has never adopted the term 'service cooperative' in its legislation and continues to use the traditional Soviet terms for production cooperatives (*proizvodstvennyi kooperativ*) and consumer cooperatives (*potrebitel'skii kooperativ*). Despite the retention of the traditional Soviet term 'consumer cooperative', this concept in Kazakhstan has been detached from the sweeping concept of the rural consumer cooperation system (the Soviet model *potrebsoyuz*), which is intended on paper to serve the entire rural population at the level of household plots, providing collection or marketing of produce and extending consumer goods sales through a network of stores in villages.

A detailed definition of the for-profit production cooperative was included in the 1994 version of the Civil Code (Articles 96–101), which was basically consistent with the Western concept of production cooperatives, whereby labour was performed by members (Kazakhstan Civil Code, 1998). A new definition of the not-for-profit consumer cooperative was added in 1998 (Kazakhstani Civil Code, Article 108, 1998), which was identical to the Western concept of service cooperative, whereby labour was performed by non-members (Article 2, *Law of Non-commercial Organizations*, Ministry of Agriculture of the Republic of Kazakhstan, 2001); this principle is explicitly specialized to service cooperatives in its Article 14(4).

Five cooperative-related laws were passed between 1995 and 2003, elaborating on the basic definitions in the Kazakhstan Civil Code (1998). These are the *Law of Production Cooperatives* (1995); the *Law of Rural Consumer Cooperation* (1999); the *Law of Agricultural Partnerships* (2000); the *Law of Consumer Cooperatives* (2001); and the *Law of Rural Consumer Cooperatives of Water User*

Associations (2003). These laws define the main types of production and service/consumer cooperatives; thus, there is no ambiguity as to cooperative type in Kazakhstani legislation. The only quirk is that the Russian term 'consumer cooperative' (*potrebitel'skii kooperativ*) should be consistently rendered as 'service cooperative' in English. A comparative analysis of the five laws reveals that they mostly repeat the same cooperative principles and attributes. All cooperatives are voluntary associations of users who have a common goal and who engage in common activities; the cooperative is created by a general assembly of its founding members; the assets of a cooperative represent the sum total of members' contributions (shares) and cumulative retained profits; and all cooperatives primarily serve their members but may also engage in transactions with non-members. The *Law of Agricultural Partnerships* holds a somewhat special place, as it explicitly stipulates that an agricultural partnership sells its services to members at cost, so that no profit is created (Ministry of Agriculture of the Republic of Kazakhstan, 2000). The services provided by an agricultural partnership to its members include marketing, storage, processing of farm products, supply of farm inputs, etc. An agricultural partnership is thus essentially a classical service cooperative.

Ukraine cooperative law

Based on the principles of the International Cooperative Alliance, the basic law on agricultural cooperatives in Ukraine represents an immense improvement over the 1988 USSR law. The concept of consumer cooperative was replaced with the Western concept of service cooperative (*obsluzhivaiushchii kooperativ* in Russian, *obslugovuiuchii kooperativ* in Ukrainian), and the conceptual distinction between production and service cooperatives is clear. An agricultural production cooperative is a for-profit organization made up of agricultural producers who are physical persons. An agricultural service cooperative is a not-for-profit organization made up of agricultural producers who are either physical or legal persons. The addition of *On Cooperation* in 2004 defined the basic legal framework for agricultural service cooperatives in Ukraine (Parliament of Ukraine, 2004). The laws for agricultural cooperatives have been amended many times to improve the enabling environment for cooperatives in Ukraine.

Taxation of Cooperatives

A cooperative is a legal body that deals with other legal bodies (agricultural enterprises and private/smallholder farms) and also with smallholders who are not registered for tax purposes (e.g. household plots). The cooperative as a legal body is subject to taxation that includes both profit tax and VAT. Smallholders, on the other hand, are exempt from these taxes or avoid them in various semi-legal ways. As a result of the taxes that cooperatives pay, a smallholder dealing through a cooperative is likely to net less from his marketing transactions (or pay more for purchasing transactions) than the amounts he would have received (or paid) by dealing directly with buyers and suppliers. These tax distortions place cooperative members at a disadvantage compared to those who operate independently.

Conceptual issues of taxation on cooperatives

Although in principle cooperatives are non-profit organizations, their financial reports may show an accounting profit at the level of revenues and expenses (surplus). This accounting profit is created because the cooperative initially may have underpaid its members for products delivered (expenses too low) or overcharged them for inputs supplied (revenues too high). In effect, the accounting profit is the result of internal pricing decisions within the cooperative and is not profit in the usual economic sense of the term.

Cooperatives do not know in advance, at the time of the actual transaction, how much to pay to members for product deliveries and how much to charge for input supplies and other services. Financial settlements with members are handled in two phases. In the first phase, payments and charges are the initial advances paid. In the second phase, final settlement is made at the end of the period when the cooperative financial statements have been prepared. As a result, a service cooperative usually presents a financial statement that shows a positive surplus (accounting profit) or a deficit (accounting loss) according to the scheme in Table 8.3.

Gross income (*dokhod*) represents the surplus or deficit that is further allocated, via three channels: (i) the cooperative reserve fund or other capital funds (retained earnings); (ii) dividends distributed to members in proportion to their share of contribution to the cooperative capital; and (iii) patronage refunds in proportion to members' use of cooperative services. The sum of the three amounts distributed to members is equal to gross income. Patronage refunds are determined as the difference between gross income and the first two allocations.

Conventional interpretation of the tax code will require the cooperative to pay tax on the full gross income at applicable rates. Recognition of the special nature of patronage refunds in cooperatives (as second-wave adjustment of initial overpayment or underpayment to members) suggests that this component of gross income should not be taxable. If this occurs, cooperatives will be required to pay tax only on the share of gross income retained in reserve funds and other capital funds. The amount retained in the reserve funds and other capital funds will be shown as net after deduction of the appropriate taxes.

When buy-and-sell transactions occur between legal bodies, the seller (e.g. farmer) charges a VAT on his or her sales and the buyer deducts from his or her ultimate VAT liability the amount of VAT charged by the seller. As an example, the buyer pays the seller US$100 for his or her tomatoes plus US$20 VAT. Reselling the tomatoes for US$110, the buyer will add US$22 VAT, but his or her net VAT obligation will be only US$2 (US$22 less US$20 paid to the original seller). If, however, the seller is a physical person (a physical body) who is a smallholder not registered for tax purposes, he or she will not pay VAT, but will charge the buyer US$100 for the tomatoes without adding VAT. In this way, when the buyer (a legal body) sells the tomatoes for US$110 plus US$22 VAT, there is nothing to offset against this VAT and the seller's actual VAT obligation is US$22. The buyer will naturally attempt to charge this extra cost back to the original seller, paying less than US$100 for the tomatoes. The smallholder is thus at a clear disadvantage when selling his or her products to a legal body (such as a cooperative).

A similar chain can be traced for purchase transactions. The seller of inputs (a cooperative) charges VAT on the sales transaction. A private/smallholder farmer (legal body registered for tax purposes) can offset the amount of VAT against his or her cumulative VAT obligations. For a smallholder, the actual cost is the cost of input plus the amount of VAT that he or she must absorb. A smallholder would be better off by dealing with a small private supplier who does not charge VAT rather than dealing with the cooperative.

It would be fair to exempt all transactions between cooperative members and the cooperative from VAT altogether. Cooperative laws and the tax code should recognize that cooperative members do not sell their production to cooperatives, but only transfer it to the cooperative for marketing. In this respect, the cooperative is an extension of the farm itself.

Table 8.3. Schematic profit and loss statement of a cooperative. (From authors' compilation.)

Marketing service cooperative	Supply service cooperative
Revenues:	Revenues:
Sales of members' products to third parties	First-wave payments from members
Costs:	Costs:
First-wave payments to members	Paid to suppliers
Operating costs of the cooperative	Operating costs of the cooperative
Revenues – costs = Gross income (surplus/deficit)	
Less allocation to reserve fund and other funds	
Less distribution in proportion to share contribution	
Difference available for distribution as patronage refunds	

Kazakhstan: taxation of cooperatives

The current Kazakhstani tax code contains a special tax regime for agricultural producers (agricultural enterprises and private/smallholder farms) and agricultural (or rural) service cooperatives. Legal entities opting for the special tax regime are understood to be small agricultural producers. They are entitled to follow a highly simplified tax return system with minimum accounting requirements; plus they pay only 30% of the standard tax rate on all basic taxes (VAT and corporate income tax, land tax, land-lease payments to the state, property tax, social tax, vehicle tax, etc.). Since the tax code explicitly puts agricultural service cooperatives in the same category with other agricultural producers, there are no blatant tax distortions for cooperatives. Possible distortions may arise for operators of household plots. They pay only personal income tax, and their agricultural activities are exempt from taxes. If they join a service cooperative, the cooperative (a legal entity) will be obligated to calculate VAT on their transactions, albeit at the concessionary rate.

Another possible distortion stems from Article 448 (paragraph 3) of the Kazakhstani tax code (Conception, 2012). According to this curious article, a rural service cooperative is not entitled to follow the special tax regime if its members are also members of other service cooperatives or if it has subsidiary enterprises (this provision was originally intended to preclude large agricultural enterprises from enjoying the special regime tax concessions).

Ukraine: taxation of cooperatives

Beginning in 2013, Ukrainian law *On Changes to Ukrainian Law on Agricultural Cooperation*, No 5495-VI, explicitly recognized the non-profit status of agricultural service cooperatives (Parliament of Ukraine, 2012b). Because the Ukrainian tax code does not recognize the non-profit status of these cooperatives, agricultural service cooperatives are required to pay profit tax on their gross income (Table 8.3), like other legal entities such as joint-stock companies. A partial solution to this issue (introduced into the tax code in 2013) was reached whereby the profit tax for service cooperatives was assessed on its net income after subtracting all mandatory and dividend payments, meaning that the profit tax was assessed on the net income available for distribution as patronage funds. Mandatory and dividend payments are defined as allocations made to reserve funds and other funds and payments made as share contributions, or dividends (Table 8.3; *On Changes to the Tax Code of Ukraine for Regulation of Certain Taxation Issues*, No 5412-VI, Parliament of Ukraine, 2012a). This compromise still subjects farmers who join cooperatives to double taxation, but the tax base has been reduced considerably. A bill to amend the tax code to ensure that agricultural service cooperatives are considered non-profit entities was introduced into the Ukrainian Rada in March 2012 by the Communist parliamentary faction, but has not been passed.

Service cooperatives, like all legal bodies, are subject to VAT in Ukraine, including VAT on transactions with their members. Thus, in Ukraine, smallholders are at a disadvantage when buying and selling from cooperatives.

A further disincentive to join a cooperative is that by joining a service cooperative, any agricultural producer who is a legal person risks losing the tax benefits that accrue by virtue of his or her agricultural producer status. This risk is a powerful disincentive for agricultural producers to buy and sell through agricultural service cooperatives. The status of service cooperatives as legal persons subject to both profit taxes and VAT sets them apart from agricultural producers subject to a special tax regime that leaves them nearly untaxed. The two elements of the special regime for agriculture are the fixed agricultural tax (FAT) and the agricultural VAT (AgVAT). The FAT is a flat-rate tax introduced in lieu of profit, land and a number of other taxes. Its rate varies from 0.09% to 1% of the normative value of farmland, depending on the farmland's type and location. In 2010, FAT resulted in an average tax payment of roughly 6 UAH/ha (US$0.75/ha). This meant that farm profits in Ukraine were virtually untaxed (Nivievskyi, 2012). Under the AgVAT, agricultural producers accumulate their own VAT payments, but do not transfer them to the government. Instead, they are required to deposit VAT payments in a special account to use for their own input purchases, such as for farming equipment and seeds. Thus, the AgVAT virtually allows agricultural businesses to avoid VAT payments (Nivievskyi, 2012).

Conclusions

This short review of the status of agricultural service cooperatives in Kazakhstan and Ukraine has shown the difficulties encountered by both governments when encouraging the formation of agricultural service cooperatives. While the governments publicly recognize the usefulness of cooperatives, policies designed to support them focus on input purchase subsidies and subsidized loans. While these policies have helped some, they also seem to have bred a class of false cooperatives, established only for the purpose of gaining access to subsidized credits and inputs, with no regard for true cooperative principles.

Furthermore, neither government has succeeded when creating a supportive business environment for service cooperatives because of tax disincentives. The VAT system in both Kazakhstan and Ukraine discriminates against smallholders, making service cooperatives unattractive for precisely the set of farms for which these organizations can be most helpful. In Ukraine, moreover, even farms that are registered legal entities risk their special tax status by buying or selling through service cooperatives.

References

Abele, S. and Frohberg, K. (eds) (2003) Subsistence agriculture in Central and Eastern Europe: how to break the vicious circle. In: *Studies on the Agricultural and Food Sector in Central and Eastern Europe*, Vol 22. IAMO, Halle, Germany.

Akimbekova, G.U. (2010) Agricultural cooperation in Kazakhstan: problems and possible solutions. *Ekonomika APK (Kiev)* 9, 161–166.

Bono, P. (2012) Support for farmers' cooperatives: Italy. Available at: http://edepot.wur.nl/244801 (accessed 15 January 2014).

Cabinet of Ministers of Ukraine (2007) *State Special Programme for the Development of the Ukrainian Village to 2015*. Cabinet of Ministers of Ukraine Decree from 19.09.2007, No. 1158. Cabinet of Minister of Ukraine, Kiev, Ukraine. Available at: http://zakon1.rada.gov.ua/laws/show/1158-2007-%D0%BF (accessed 15 January 2014).

Cabinet of Ministers of Ukraine (2011) *State Special Economic Programme for the Support of the Development of Agricultural Service Cooperatives to 2015*. Resolution 7.07.2011 No 557, Cabinet of Ministers of Ukraine, Kiev. Available at: http://zakon1.rada.gov.ua/laws/show/557-2009-%D0%BF (accessed 15 January 2014).

Cobia, D. (ed.) (1989) *Cooperatives in Agriculture*. Prentice Hall Publishers, Englewood Cliffs, New Jersey.

Conception (2012) *Conception of the Development of Agricultural Cooperation in Kazakhstan*. Kazakhstani Research Institute of Agricultural Economics, Almaty, Kazakhstan.

Decree of the Council of Ministers of the USSR (1988) *On Cooperatives in the USSR*. Council of Ministers, Moscow.

Eurostat (2013) Agricultural holdings, 2000–2010. Available at: http://epp.eurostat.ec.europa.eu/statistics_explained/index.php?title=File:Agricultural_holdings,_2000-2010.png&filetimestamp=20121105110459 (accessed 15 January 2014).

Filippi, M. (2012) *Support for Farmers' Cooperatives: Country Report France*. Wageningen University, Wageningen, the Netherlands.

Gijselinckx, C. and Bussels, M. (2012) *Support for Farmers' Cooperatives; EU Synthesis and Comparative Analysis Report: Social and Historical Aspects*. Wageningen University, Wageningen, the Netherlands.

Hofstede, G., Hofstede, G.J.and Minkov, M. (2010) *Cultures and Organizations: Software of the Mind*, 3rd edn. McGraw Hill, New York.

ICA (2013) Co-operative identity, values, and principles. International Cooperative Alliance (ICA), Brussels. Available at: http://ica.coop/en/whats-co-op/co-operative-identity-values-principles (accessed 15 January 2014).

Kazakhstan 2050 (2014) *Kazakhstan Strategy 2050*. Available at: http://kazakhstan2050.com/ (accessed 15 January 2014).

Kazakhstan Civil Code (1998) Kazakhstan Civil Code website (English). Available at: http://adilet.zan.kz/eng/docs/K940001000_ (accessed 15 January 2014).

Korinets, R. (2013) *Agricultural Service Cooperatives in Ukraine*. FAO REU Policy Studies on Rural Transition No 2013-6. FAO Regional Office for Europe and Central Asia. Available at: http://www.fao.org/docrep/018/ar593r/ar593r.pdf (accessed 15 January 2014).

Lissowska, M. (2013) The deficit of cooperative attitudes and trust in post-transition economies. *EAEPE Papers in Evolutionary Political Economy*, No 10. Available at: http://eaepe.econ.tuwien.ac.at/pepe/papers/PEPE_10.pdf (accessed 15 January 2014).

Mather, J.W., DeVille, K.C., Gessner, A.L. and Adams, C.C. (2004) *Farm Marketing, Supply and Service Cooperative Historical Statistics*. USDA Cooperative Information Report 1. US Department of Agriculture (USDA), Washington, DC.

Ministry of Agrarian Policy and Food of Ukraine (2002) *Programme for the Development of Agricultural Service Cooperatives for 2003/2004*. Available at: http://zakon2.rada.gov.ua/laws/show/1858-2002-%D0%BF (accessed 15 January 2014).

Ministry of Agriculture of the Republic of Kazakhstan (1995) *Law of Production Cooperatives*. Ministry of Agriculture, Astana.

Ministry of Agriculture of the Republic of Kazakhstan (1999) *Law of Rural Consumer Cooperation*. Ministry of Agriculture, Astana.

Ministry of Agriculture of the Republic of Kazakhstan (2000) *Law of Agricultural Partnerships*. Ministry of Agriculture, Astana.

Ministry of Agriculture of the Republic of Kazakhstan (2001) *Law of Consumer Cooperatives*. Ministry of Agriculture, Astana.

Ministry of Agriculture of the Republic of Kazakhstan (2001) *Law of Non-commercial Organizations*. Ministry of Agriculture, Astana.

Ministry of Agriculture of the Republic of Kazakhstan (2003) *Law of Rural Consumer Cooperatives of Water User Associations*. Ministry of Agriculture, Astana.

Nivievskyi, O. (2012) Agri-food policy review: Tackling the dilemma of agricultural taxation: Moving towards a comprehensive and transparent approach (Issue 9). Available at: http://www.ier.com.ua/files/Regular_products/Agro_review/2012/APD_AFPR_2012_9_en.pdf (accessed 15 January 2014).

Parliament of Ukraine (1997) *On Agricultural Cooperation*. Law of Ukraine No 469/97. Available at: http://zakon2.rada.gov.ua/laws/show/469/97-%D0%B2%D1%80 (accessed 15 January 2014).

Parliament of Ukraine (2004) *On Cooperation*. Law of Ukraine No 1087-15, as amended to 2014. Available at: http://zakon2.rada.gov.ua/laws/show/1087-15 (accessed 15 January 2014).

Parliament of Ukraine (2005) *On the Main Pillars of State Agrarian Policy to 2015*. Law of Ukraine 2005 No 2982-IV. Available at: http://zakon1.rada.gov.ua/laws/show/2982-15 (accessed 15 January 2014).

Parliament of Ukraine (2012a) *On Changes to the Tax Code of Ukraine for Regulation of Certain Taxation Issues*. Law of Ukraine No 5412-VI. Available at: http://zakon1.rada.gov.ua/laws/show/5412-17 (accessed 15 January 2014).

Parliament of Ukraine (2012b) *On Changes to Ukrainian Law on Agricultural Cooperation*. Law of Ukraine No 5495-VI. Available at: http://zakon4.rada.gov.ua/laws/show/5495-17 (accessed 15 January 2014).

President of Ukraine (2000) *On the Means for Development of the Cooperative Movement and Strengthening of its Role in the Market Reform of the Ukrainian Economy*. Presidential Decree 06.06.2000 No 767/2000. Available at: http://zakon2.rada.gov.ua/laws/show/1348/2000 (accessed 15 January 2014).

RBK-Ukraine (2013) More than 83% of local governments support the formation of agricultural cooperatives. RBK-Ukraine, Kiev, Ukraine. Available at: http://www.rbc.ua/rus/news/economic/obedinenie-v-selhozkooperativy-uzhe-podderzhali-bolee-28022013130900 (accessed 15 January 2014).

State Statistics Service of Ukraine (2011) *Main Agricultural Characteristics of Rural Households in Rural Areas*. State Statistics Service of Ukraine, Kiev.

State Statistics Service of Ukraine (2012) *Agriculture of Ukraine, 2011: Statistical Yearbook*. State Statistics Service of Ukraine, Kiev.

Statistical Agency of Kazakhstan (2012a) *Agriculture in Kazakhstan 2009–2011 Statistical Yearbook*. Astana, Kazakhstan. Available at: http://www.stat.gov.kz (accessed 15 January 2014).

Statistical Agency of Kazakhstan (2012b) Population Census of Kazakhstan 2009: Households. Astana, Kazakhstan. Available at: http://www.stat.gov.kz (accessed 15 January 2014).

Szabó, G.G. (2012) *Support for Farmers' Cooperatives: Country Report Hungary*. Wageningen University, Wageningen, the Netherlands.

Tomich, I.F. (2013) Initsiativa 'Rodnoe selo'—posledniaia nadezhda selian, *Gazeta.ua*, 22.03.2013. Available at:http://gazeta.ua/ru/articles/life/_tomich-iniciativa-rodnoe-selo-poslednyaya-nadezhda-selyan/488810 (accessed 15 January 2014).

USDA (2012) *Cooperative Statistics 2011*. USDA Report 72. USDA, Washington, DC.

USDA/NASS (2013) *Farms, Land in Farms, and Livestock Operations 2012 Summary*. USDA, National Agricultural Statistics Service, Washington, DC.

Van Bekkum, O. and Van Dijk, G. (1997) *Agricultural Cooperatives in the European Union*. Van Gorcum, Assen, the Netherlands.

9 The Strategy of Innovative Development of Russian Agriculture

David Epstein*
North-West Research Institute of Agricultural Economics, St Petersburg, Russia

Abstract

The chapter describes and characterizes the adverse situation in the area of innovative development of Russian agriculture. A system of the necessary institutes and conditions for successful functioning of innovation processes in agriculture is revealed, and ways to create it are offered. The purposes, goals and specific measures of a new strategy and the directions of significant changes to the agrarian policy related to accession to the WTO are offered that ensure the creation of an effective system of innovative development of agriculture at both the federal and regional levels.

Introduction: Problem Statement

If one takes a closer look at the existing Russian policies describing the innovation-based economic strategies for the country, its agriculture and regions, one will notice a general lack of macroeconomic consideration for innovation in agriculture and agribusiness as a whole. For instance, the programme *Innovative Russia 2020* fails to even mention agriculture (The Strategy of Innovative Development, 2012). The research and development (R&D) initiatives in Russian agriculture are of a general nature only; for example, the national project on developing agribusiness in Russia addresses the challenge of increasing investments through subsidizing loans, but does nothing to facilitate innovative development. *The National Programme for the Development of Agriculture by 2020* outlines a plan for subsidizing projects promoting innovation without specifying key concrete directions of innovative development. The programme discusses biotechnologies in general terms, but does not analyse the ongoing problem of insufficient funding of innovations or the replacement of existing equipment fleets with imported or Russian-made machines. *The Strategy of Innovative Development of the Agro-industrial Complex of the Russian Federation until 2020* (Ministry of Agriculture, 2011) considers the advancement of institutions that deliver innovative development, but completely overlooks the current problem of raising the demand for innovations and strengthening the weak government support for their transfer.

With Russia's accession to the World Trade Organization (WTO) in 2012, the overall strategy for innovative development of the country and its agriculture must be improved and supplemented by strategies on the regional level. This lends relevancy to such issues as to how this

*E-mail: epsteindb@gmail.com

strategy should be designed, what indicators were used to assess innovative development, what the goals and objectives of the strategy are and what the core measures and features on the regional level are.

The Current State of Innovation Processes

Russia's existing framework of indicators of innovative development is incomplete. Available statistics do not reflect fully the real situation in the area of innovation and, in particular, do not include information on innovations in the agricultural sector. In fact, the Ministry for Industry and Trade did not classify goods, works or services as innovation products or as high-technology products until 2013 (RG.RU, 2013).

For this reason, we have analysed the trend data of general and particular indicators of agricultural output and productivity that are indirectly linked to innovation processes. Such indicators include labour productivity, crop yield, productivity of livestock farming and consumption of various types of means of production per unit of output (i.e. mineral fertilizers, gasoline, diesel fuel, feed and electric energy).

Findings indicate rapid increases in production efficiency and labour productivity since 1999 in Russia as a whole and in its northwestern regions in particular. Agricultural enterprises in the Northwestern Federal District (NFD) were selected for this research based on the geographical specialization of the Research Institute.

This growth of production efficiency and labour productivity could not have been possible without a transfer of innovations. The innovation processes in Russian agriculture and in the NFD have allowed labour productivity to grow by 3.6% (in Russia) and 4.8% (in the NFD) annually. This rate was virtually unheard of in Soviet times; at least, it was unheard of in the agricultural sector.

Figure 9.1 illustrates the trends in labour productivity in Russia, the Central Federal District (CFD) and the various regions of the NFD. Kaliningrad and Novgorod almost doubled their labour productivity in the Russian agricultural sector between 2006 and 2010.

Likewise, resource costs have been decreasing since 1998 (Fig. 9.2). From 1999 to 2010, the per rouble (RUB) cost of agricultural output dropped by 42%, in comparable prices that corresponded to an annual decrease of 3.9%.

Consider two specific examples. The per rouble consumption of gasoline of gross output in 2010 decreased by 65% from 1998 and by 79.6% from 1990. In 2010, agricultural enterprises consumed less per unit energy of gross output: 77.8% less than in 1998 and 52.4% less than in 1990.

Given that most agricultural enterprises are small and rarely comprise departments capable

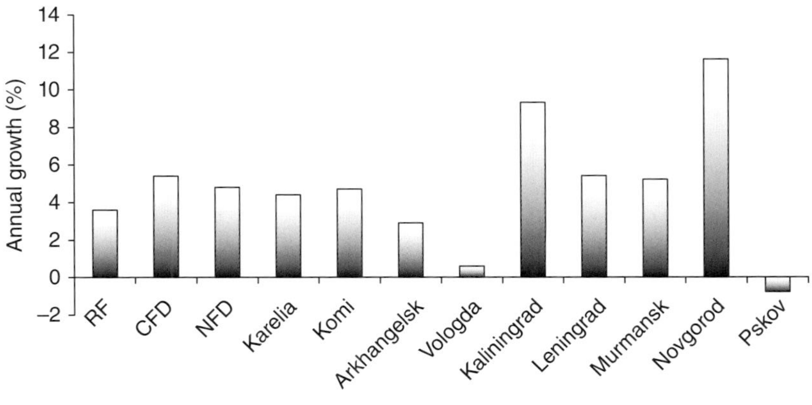

Fig. 9.1. Annual growth in labour productivity in the Russian agricultural sector (per cent), 2006–2010. (From author's calculations.)

of developing innovations independently, as well as the fact that Russian applied sciences all but ceased functioning for lack of funding between 1990 and 2005, we conclude that the innovation process in agriculture has been driven primarily by the acquisition of new equipment and technologies that has been linked to investment volumes. Accordingly, the growth of investment and demand for innovative equipment and technologies will continue to be the key factor in the innovative development of Russian agriculture. Let us look at data for investments in comparable prices (Table 9.1).

Table 9.1 shows uneven trends for capital investment in Russian Federation agriculture in 2006 prices. Investments grew by 18% in 2007, but then declined because of the economic downturn, first dropping by 1% in 2008 and then slumping to 24% in 2009. Numbers continued to shrink in 2010 and rebounded only in 2011, growing by 14% that year but still remaining 13% below the 2006 levels.

The CFD investment trends tell a different story. After 3 years of decline (38.5% from 2008 to 2010), investments rose sharply by 88% in 2011, exceeding the 2006 levels by 35%. In 2011, profitability in grain and livestock almost doubled, from 5% to 9.5%.

In the NFD, investment growth began in 2010, registering in 2011 at a 31% surplus compared to the 2006 levels. However, the trends varied by region. For example, the Arkhangelsk and Kaliningrad Oblasts saw increases, but investment in the Pskov Oblast decreased. Investment grew the most in the Novgorod Oblast in both 2008 and 2009 (four and five times greater than

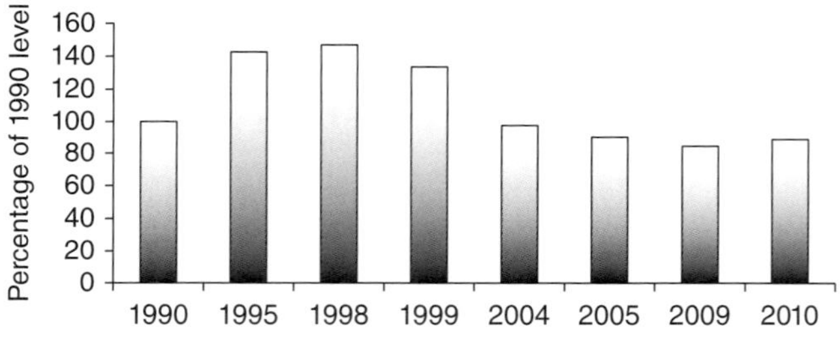

Fig. 9.2. Russian resource expenses in comparable per rouble prices of gross output as a percentage of the 1990 level. (From author's calculations.)

Table 9.1. Trend data for Russia in billions of roubles for capital investments in the agriculture, fish and wildlife complex in 2006–2011, 2006 prices. (From author's compilations.)

Region	2006	2007	2008	2009	2010	2011	2011 to 2006
	Billion roubles						Per cent
Russian Federation	223.4	263.4	251.4	191.7	170.4	194.5	87.1
CFD	51.5	60.2	54.5	38.4	37.0	69.7	135.3
NFD	13.6	15.4	17.1	14.7	16.5	17.8	131.0
Republic Karelia	0.70	0.81	0.96	0.33	0.43	0.71	100.7
Republic Komi	1.94	2.05	1.91	1.37	1.38	1.56	80.5
Arkhangelsk Oblast	1.63	1.84	1.46	2.03	4.89	2.75	168.5
Vologda Oblast	2.06	2.38	1.99	1.14	1.36	1.36	65.9
Kaliningrad Oblast	0.64	0.70	1.57	0.84	0.61	0.98	153.8
Leningrad Oblast and St Petersburg	5.69	6.71	6.19	5.47	6.34	6.07	106.7
Murmansk Oblast	0.22	0.25	0.36	0.49	0.16	0.16	74.9
Novgorod Oblast	0.44	0.46	1.62	2.05	1.49	1.08	245.4
Pskov Oblast	0.32	0.36	0.58	0.46	0.19	0.22	68.0

2007, respectively). However, despite a period of decline, by the end of 2011, investments were still 150% above the 2006 levels. This upswing was brought about by a change in agricultural policy made by a newly appointed governor, who built his economic policy around agricultural growth, especially boosting poultry and pork production. Yet, in 2011, investment in Russian agriculture as a whole remained 40% below its 1990 levels.

Investment growth in agriculture from 2005 to 2011 was restrained by low profitability levels of agricultural enterprises (9%–10%) and high interest rates (14%–20%). New strategies need to be implemented to bolster the demand for innovations. The main focus must be on facilitating investment activity in the agricultural sector through more domestic production, affordable credit loans and better technology.

The Necessary Elements for Successful Functioning of Innovation Processes

It is advisable to follow a systematic approach when identifying the necessary elements for the successful functioning of innovation processes. Through our analysis we were able to determine the following four categories of necessary elements: (i) framing criteria of economic conditions; (ii) building a supply of innovations; (iii) fostering a demand for innovations; and (iv) factoring in readiness of enterprises/entrepreneurs to adopt innovations (Fig. 9.3).

For framing the criteria of economic conditions, the Russian government needs to play an active, constructive regulatory role in the economy. Ways to do this include the following: stabilizing a economic and financial policy that facilitates economic growth; providing the agribusiness sector with favourable macroeconomic conditions; supporting the science and innovation processes; and guaranteeing the protection of property, accessibility and fair justice for all through transparency in government.

Several elements are needed for building a supply of innovations. An industrial system needs to be implemented for creating, financing and distributing innovative projects and solutions. It is important for the industrial system to employ the fundamental and applied sciences when creating innovative projects and solutions. The government must be involved by financing projects and by offering low customs duties on imported technologies and equipment.

Fostering a demand for innovations involves creating a steady, high-profitability level for the production and marketing of agricultural products through affordable lending for agricultural producers and for the implementation of

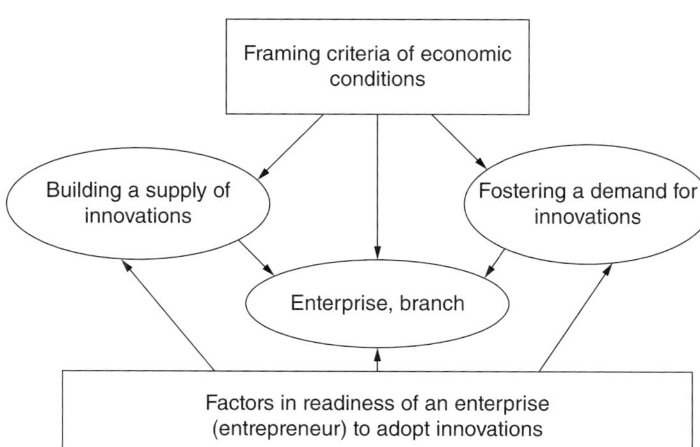

Fig. 9.3. The necessary elements for successful functioning of innovation processes in Russia. (From author's calculations.)

new technologies. Sectoral and regional training systems are needed to educate agricultural producers about implementing innovations.

Factoring the readiness of an enterprise to adopt innovations consists of properly training agricultural producers and providing sufficient funds for implementing innovations. Enterprise management should play an active role in innovation-related activities, and it should be qualified appropriately for analysing information related to innovations and to new technologies.

In addition to the four categories of elements outlined above, certain institutions should be involved and should interact regularly to share their needs and capabilities. For example, technology platforms organized by universities or R&D institutes could serve in this capacity. Creating and funding a network system of institutions for the innovative development of Russian agriculture (as well as most other economic sectors) would help solve the problems of insufficient scientific research funding and non-collaboration among the scientific organizations, educational institutions and enterprises (Surin and Molchanova, 2008).

In our opinion, there are several major gaps in the Russian innovation framework. First, no developed link for creating innovations exists (i.e. an applied science with a system for creating innovations as solutions ready to be deployed by end users and complemented by an implementation link). Second, government funding of fundamental science and applied sciences is insufficient (expenditures for science and engineering in Russia, as a share of the gross domestic product (GDP), are one-half of those in the European Union (EU) countries on average), and private capital is barely involved in science funding. Third, farms lack the financial means that lead the demand for real innovations. And fourth, industries that supply the means of production for the agricultural sector are only marginally competitive as a result of a prolonged absence of government support (e.g. the Russian mechanical engineering, animal breeding and seed farming industries). These gaps are characteristic of the Russian economy as a whole and of the Russian agricultural sector in particular. The strategy of innovative development must be aimed at resolving these funding problems.

Important Choices when Designing the Strategy of Innovative Development

There are some options from which to choose when shaping the federal and regional strategies of innovative development. One important choice to make when designing the strategy of innovative development is what to develop first, such as supplying own innovations (created in-country/in-region or in one's own enterprise) versus allowing external innovations (from other countries/regions) access to the market. Because Russia is a large country, a certain priority should be given on the national economic level to funding the development of a potential homemade supply of innovations via subsidies, customs and tax policy (this priority does not exist in practice yet), although this policy should not, of course, shut out imported innovations from competing. At the same time, on the sectoral level (in the agricultural sector in particular), the governing bodies that are limited in their financial means are forced to channel the better part of their budgets toward raising the demand for innovations – regardless of their source of origin – as opposed to building the supply of innovations. However, the existing scientific capacity for the supply of innovations aimed at developing agriculture, including that of the Russian Academy of Agricultural Sciences, must receive more funding, which should be increased on a regular basis.

On the macro level, it is necessary to identify the prospective agricultural industries that are capable of competing on the world market if properly supported. Strategically, prospective Russian agricultural industries include grain production, flax fibre production, linen fabric manufacturing, poultry production, pork production, milk production and egg production. For the north-western regions, these branches are flax fibre production and milk production. Government programmes should be used to speed up innovative development and efficiency improvements first and foremost in the strategically prospective industries.

Another decision to make is whether or not to support large investments in existing enterprises or actively to facilitate the creation of new, substantially larger enterprises based on new technologies. In our opinion, on the federal and regional levels, special focus should be placed on supporting larger enterprises based on cutting-edge technologies in industries where Russia considerably lags behind the rest of the world and is forced to import large volumes of food products. This concerns the production of pork, beef, turkey, agricultural animal feed and greenhouse vegetables.

Goals of the Strategy of Innovative Development

The goals that the strategy of innovative development should attain are to: (i) make changes in the macroeconomic policy that facilitate rapid innovative developments in the economy; (ii) create a comprehensive network system of institutions to deliver innovative developments; (iii) expand the role of and increase the funding of the fundamental and applied sciences as the basis for growth in innovative development; (iv) create an environment that allows for steady profitability by agricultural producers as the basis for the growth of potential demand for innovations; and (v) formulate measures for supporting industries that supply the means of production for agriculture, including Russian-made agricultural engineering, animal breeding and seed farming.

A list of measures designed specifically to enhance the strategy of innovative development of agriculture on the federal level includes: (i) strengthen and develop macroeconomics, the fundamental sciences and the applied sciences; (ii) improve funding to ensure the support for innovation in the economy and the agricultural industries; (iii) improve funding for agricultural producers and agricultural sciences; and (iv) enhance the planning and regulation of the innovation process at the level of the Russian Ministry of Agriculture.

Key Measures of the Strategy of Innovative Development on the Federal Level

There are several key measures of the overall strategy used to strengthen and develop macroeconomics on the federal level. First, transition from a passive macroeconomic policy that is reliant mainly on financial and fiscal tools to an active industrial and agrarian policy entailing systematic government support of major industries and applying the tools of government planning and investment. Second, increase the competitiveness of Russian production by amending the monetary and exchange rate policy; this can be accomplished by funding the Russian economy to meet the needs of businesses and by eliminating artificial support of the rouble to allow its exchange rate to drop to a level that would make Russian industry and agriculture more competitive. Third, invest a substantial part of the existing surplus federal reserves into real production sectors and socially significant industries within the country. Fourth, support profitability at a level sufficient to allow sustainable production growth. And fifth, increase competitiveness up to the level of developed countries in strategically important industries, including agriculture.

To strengthen and develop the fundamental science, Russia needs to transition from a policy of selective funding of certain fields of science to systematically funding all the sciences. The government needs to increase substantially science and R&D funding to approach the level of the developed countries (expressed as a share of GDP) by including a *science and innovation* tax of at least 2% (exclusive of VAT) within the sales prices of significant science-based products (e.g. sophisticated technical equipment, electronics, power, extractable resources and their resulting products, and military industrial products). The government also needs to provide significant fiscal relief to encourage businesses to fund R&D research. In cooperation with the sectoral science academies, Russia needs to develop a system for objectively rating the activities of government science organizations and their personnel, taking into account their profiles (fundamental or applied) and their branch specifics. There should be a regular evaluation of the performance of science organizations based on committees composed of both international scientists and recognized representatives of the corresponding branches of science. Expanding the role of scientists and academic councils should be considered when selecting the topics of research.

To improve funding to ensure support for innovation and the agricultural industries requires creating government subsidies for: (i) ready-to-use innovative solutions in key sectors of the economy; (ii) accessible information on effective innovative technology recommendations, technology development programmes and innovative solutions; and (iii) industries that supply Russian-made means of production directly to agricultural producers (tractors, combine harvesters, seeds, breeding animals, feed additives and others), provided that the quality of these products is increased.

To improve funding for agricultural producers and agricultural science, Russia should: (i) ensure more accessible loan financing for agricultural and food producers; (ii) increase government support for agriculture to a level that would guarantee the profitability necessary for sustainable production growth (in particular, raise the corresponding amber box support to US$9 billion, which is the level approved by the WTO; (iii) create a fund for financing agrarian science and its innovation within agribusinesses based on proceeds from a newly created 1% surcharge on the retail prices of food products; and (iv) amend the federal laws governing the status of the cities of Moscow and St Petersburg to permit them to provide budgetary support to the agricultural producers of their respective oblasts in such amounts as necessary for the sustainable provision of food products to these cities.

To enhance the planning and regulation of the innovation process on the level of the Russian Ministry of Agriculture, the government should cooperate with agrarian science representatives to create accessible information on cutting-edge technologies that are recommended for region-wise implementation; subsidize 50% of the cost of the distribution of cutting-edge technologies; and task science organizations with the development of recommended lists of agricultural technologies and systems of Russian-made and imported machinery for the federal districts.

The Goals and Key Measures of the Strategy of Innovative Development on the Regional Level

Strategies of innovative development should penetrate to the regional (oblast, krai or republic) level so as to ensure a consistent innovation policy across the country. The strategy of innovative development for a given region should attain the following goals: (i) receive government support on a regional level for streamlining the innovative development of the region's overall economy and agriculture in particular; (ii) develop a regional system of institutions that deliver innovative development, including innovative development of agriculture; (iii) significantly improve the funding of fundamental science and applied sciences through purchases financed from regional budgets; (iv) raise the efficiency of innovation processes in the region's agricultural sector and ensure that they run according to plan; (v) subsidize agricultural producers to encourage potential demand for innovations in agriculture; and (vi) improve the skills of agricultural producers and provide them with accessible information on innovative developments and on cutting-edge technologies.

To attain the goals and objectives of the strategy of innovative development of agriculture on the regional level, we examine measures aimed at strengthening the systematic execution of innovative processes, improving funding, promoting the distribution of new innovative technologies, implementing personnel development and integrating science into agricultural production. In those constituent territories of the federation where higher education institutions specializing in agriculture or agricultural R&D institutes operate, regional authorities should facilitate the set-up, functionality and funding of so-called technology platforms (e.g. conferences, regular meetings, exhibitions, etc.) to discuss the issues of innovative development of the industry and ways to cooperate when resolving such issues. Regional authorities should cooperate with representatives from both science and business to develop a list of effective innovative technologies recommended for distribution in each of the key agricultural industries of the given region, including large-scale solutions supported by the government. In addition, they should cooperate with scientists from higher education institutions and agricultural R&D institutes to develop and subsidize programmes directed at implementing effective technologies and large-scale innovation solutions that may be funded from regional and federal budgets involving private and government partnerships. A programme should be designed to develop an agricultural advice service to align it with the goals of the programme for implementing effective technologies. Last, but not least, regional authorities should cooperate with organizations engaged in effective management personnel training for the agricultural sector and for higher education institutions.

Consideration of Regional Features when Designing Programmes of Innovative Development for Regions

The strategies commonly formulated for all regions should be supplemented taking into account the specific features of each region. The features that may affect concrete measures of the agricultural innovation strategy of a region include natural and climatic differences affecting the potential of a region's agriculture and its realistic goals; the level of technological and innovative development of the region's agriculture compared with other regions of Russia; the volume of investments in agriculture in relation to production volumes; the degree of advancement of main institutes of innovative development; the volume of possible government support in relation to production volumes compared to other regions; the views of the region's administration on government support for agriculture; differences in the patterns of agricultural production (i.e. the relative contributions of agricultural enterprises, individual household producers within agricultural enterprises and private/smallholder farming operations to production volumes); and differences in the sectoral structure of agricultural production.

To factor in the specific features of each region in concrete terms, whether it be oblast or republic, it is advisable to start by selecting a set of quantitative indicators that characterize each feature. For example, the natural and climatic differences may be characterized by the following indicators: average year-round temperature, the number of days with air temperature over 10°C, amount of precipitation and volume of per capita agricultural products in the region. Indicators for characterizing the differences in the level of technical and innovative development of agriculture in the regions include fixed assets value per worker, labour productivity by agribusiness sector, productivity by sector and depreciation of assets. Indicators for characterizing the differences in volumes of investment in agriculture include per rouble investments of gross or commodity output per worker, per hectare of farmland, and so on.

For quantifying the degree of advancement of key institutions that deliver innovative development in agriculture, Russian statistics provide no suitable indicators. Indirect assessment is possible based on data from regional agricultural advice services and government agencies that play a notable role in innovative development. As such, we used the following indicators in the north-western parts of Russia: R&D organizations and researchers with academic degrees (per million population); costs per researcher (per million roubles); patents for inventions and cutting-edge technologies developed and used (per billion roubles in R&D costs); innovative activity of organizations (percentage of all activities); expenditures on technology innovations (percentage of internal R&D costs); and volume of innovative goods, works and services (percentage of total).

Based on these indicators, ranks may be assigned per oblast and added together to obtain a cumulative ranking of regions. The same method may be applied to other sets of indicators. For instance, indicators for quantifying the volume of possible government support as a ratio of production inputs and output volumes include volume of government support per rouble of output, support per worker or support per hectare of farmland.

A suitable indicator for characterizing the view of a region's administration on government support for agriculture may be obtained by contrasting the volume of government support in an oblast as a ratio of the oblast's budget. Thus, the index of the view of a region's administration on government support I_{vags} and the indicator of the possibility of that government support increasing are related as follows:

$$I_{vags} = \frac{\text{Subsidies from regional budget as a percentage of all expenditures from regional budget}}{\text{Gross value added of agriculture as a percentage of the region's GDP}}$$

Indicators for characterizing the differences in the structure of agricultural production are the contributions of the three key agricultural production sectors (agricultural enterprises, individual farming operations and farm households) to the total volume of agricultural production. Another significant indicator is the share of production of agroholding companies in the total production of all Russian agricultural enterprises. In addition, indicators for quantifying the differences in the sectoral structure of agricultural production include the share of crop farming and animal farming in the total commodity output and the share of each main product sector in the structure of commodity production.

With the important features described in the above fashion, the formulated criteria may be used to divide the oblasts under consideration into groups. For each such group, suggested changes may be recommended to form a model of regional strategy on the basis of typical strategy. In practice, we have derived three groups. For example, to factor in the natural and climatic differences, three groups of oblasts were defined. Group 1 oblasts are able to meet most of their population's needs in the main types of agricultural products due to their favourable natural and climatic conditions. Group 2 oblasts are able to meet only some specific, urgent needs of their population in select types of agricultural products. Group 3 oblasts occur somewhere between Groups 1 and 2 based on their natural and climatic features.

For the group with the least favourable conditions, we estimate which goals of the strategy of innovative development are likely to be impacted by the group's specific features and how they may be influenced. Based on this estimated impact, we determine how the goals and measures of the strategy should be adjusted to ensure that investments are worthwhile. For example, comparing the NFD region's oblasts in terms of agricultural per capita output, we derived three groups of oblasts according to the formulated hypothesis of the impact of natural and climatic conditions on agriculture (Fig. 9.4).

Group 1 (regions that are able to meet most of their population's needs in the main types of agricultural products) includes Vologda/Kaliningrad/Leningrad Oblast, Novgorod Oblast and Pskov Oblast (Table 9.1).

Group 2 (regions with unfavourable climate where local production is able to meet only some specific, urgent needs of the population) includes the northernmost oblasts: Karelia, Komi, Arkhangelsk and Murmansk. In this group, agricultural per capita output is 50% or less of the country average. Local production in these regions meets some of the specific needs of schools and special institutions, but most of the food products need to be imported. The general provisions of the policy of innovative development must be adjusted for these regions to develop specific innovations suitable for their climates. Special stress should be put on developing local breeding and seed farming technologies, as well

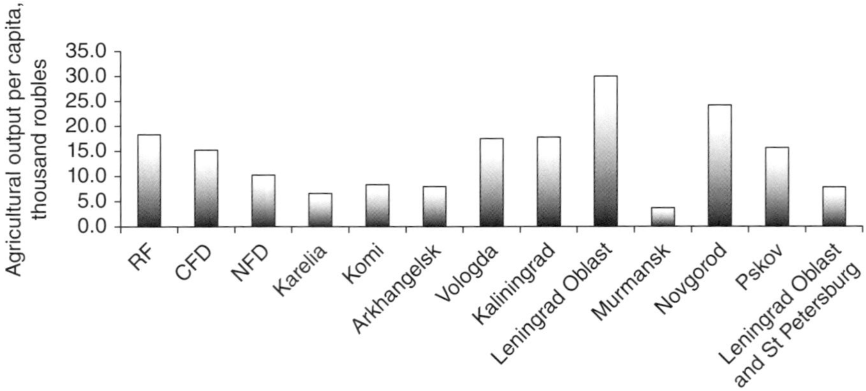

Fig. 9.4. Russian agricultural per capita of population output in 2010, in thousands of roubles. (From author's calculations.)

as transferring technologies that take into account the natural and climatic differences.

Group 3 includes the remaining oblasts (i.e. the city of St Petersburg and the part of the Leningrad Oblast that surrounds the city) and has the status of a separate constituent territory of the NFD. Some of its local production meets the food requirement of the area and some does not. It would be advisable for St Petersburg, as a significant investor and consumer of vast volumes of food products, to help the innovative development of the agriculture of the Leningrad Oblast.

Changes in Agrarian Policy Required to Accelerate Innovative Development of Russian Agriculture

Above, we have formulated a number of conclusions concerning the agrarian policy, namely a strategic approach to innovating agriculture, expanding support for agrarian science, raising the profitability of the agricultural sector and shifting innovation subsidy efforts from subsidizing loan interest charges to subsidizing direct equipment costs. The approach is limited by the subsidy restraints of Russia's WTO agreement. According to these restraints, government support will decrease from US$9 billion in 2013 to US$4.5 billion by 2018. Of approximately RUB 370 billion (US$11.8 billion in 2013) of federal and regional support for agriculture and the fishing industry, roughly RUB 190 billion (US$6 billion) will go to agricultural enterprises. Roughly RUB 140 billion (US$4.5 billion) are for WTO amber box subsidies. While this would exceed the scheduled limit for 2018, agricultural producers will still look forward to the expansion of government support. Furthermore, private/smallholder farms will also receive subsidies. Up to 70% of the subsidies to agricultural enterprises apply to loan interest charges (Russian bank interest rate loans for agricultural producers range from 16% to 25%, which is five to ten times greater than those for farmers from the EU and the USA). The best way out of this dilemma is to shift support for innovations into other areas permitted by the WTO.

In addition, Russia will need to increase its use of subsidies. For example, annual adjustments will need to be made for subsidizing investments in crop and livestock production. Types of subsidies include: (i) infrastructure development subsidies to improve roads, water and gas and electric power for helping agricultural enterprises substantially lower operating costs and raise profitability; (ii) insurance subsidies (includes income insurance); and (iii) rural development subsidies for building schools and hospitals to keep skilled workers and young people from leaving the rural areas for urban areas and to increase productivity and improve the finances of agricultural production.

References

Ministry of Agriculture (2011) *The Strategy of Innovative Development of the Agro-Industrial Complex of the Russian Federation until 2020*. Russian Academy of Agricultural Sciences, Ministry of Agriculture, Moscow.

RG.RU (2013) *RG–Federal Issue No 6035*. Ministry of Industry and Trade of the Russian Federation. Available at: http://www.rg.ru/2013/03/20/kriterii-dok.html (accessed 21 April 2014).

Strategy of Innovative Development, The (2012) The strategy of innovative development of the Russian Federation until 2020 (*Innovative Russia 2020*). Available at: http://government.ru/gov/results/17449/ (accessed 21 April 2014).

Surin, A.V. and Molchanova, O.P. (2008) *Innovative Management*. Infra-M, Moscow.

10 The Emergence of Agroholdings and Patterns of Land Use in Ukraine

Volodymyr Lapa, Taras Gagalyuk and Igor Ostapchuk*
Ukrainian Agribusiness Club, Kiev, Ukraine

Abstract

Between 2007 and 2009, Ukraine's large agricultural holdings (agroholdings) trended upwardly (some years by 1 million hectares). Since then, agroholdings over 10,000 ha have decreased significantly. What are the prospects for future developments of agroholdings in Ukraine? Is further land consolidation expected? What will be the impact of changes on the agricultural sector? This chapter answers these and other questions relevant to the development of Ukraine's agriculture in general and of agroholdings in particular.

Agroholding Issues

An agroholding is defined as a type of land bank – a consolidated set of parent and controlled subsidiary agricultural companies – containing 10,000 ha or more of agricultural land. Between 2007 and 2009, there was a steady upward trend of agroholdings in Ukraine, with some agroholdings acquiring over 1 million ha (Mha) (Fig. 10.1). Since 2009, however, there has been a decrease in the consolidation of land resources by large agroholdings, due to key land, financial, administrative/regulatory and local issues.

Land issues include land reform and land-use limitations. Increases in the normative and pecuniary value of agricultural land make land more expensive (Parliament of Ukraine, 2011a). For example, increasing Ukraine's Consumer Price Index (CPI) to 1.75 will cause additional losses for producers to the amount of more than 4.5 billion hryvna (UAH) annually. Limited areas of land and the transfer of tenancy do not require large investments, so as a rule most farms do not sell the land lease rights.

Financial issues in Ukraine include difficulties with financing and credit resources on both the domestic and foreign markets, and weak markets for international investments. For example, in 2011, only four agricultural companies (or six, if including secondary public offerings such as Kernel and AgroGeneration) conducted initial public offerings.

Issues with administrative and regulatory governmental policies in the agrarian sector (e.g. export limitations) have created problems for agroholdings. The penalties for violations can be very costly.

Local issues include climate instability and changing corporate strategies. In some cases, priority has been given to optimizing production processes, creating systemic business strategies and developing the infrastructure, rather than

*Corresponding author; e-mail: ostapchuk@ucab.ua

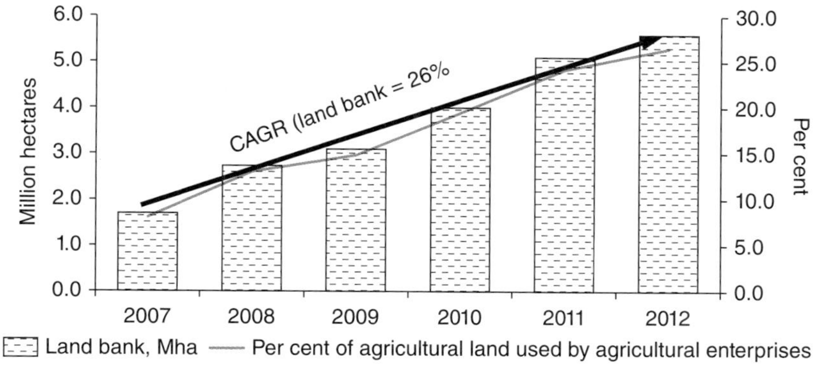

Fig. 10.1. The dynamics of changes in Ukrainian agroholdings (areas more than 10,000 ha), 2012. (From UCAB, 2011.)

purchasing assets. In other cases, the expansion policy was abandoned in favour of the need to maintain existing assets in a changing environment.

Despite the relative inertness of consolidating land resources in Ukraine, a number of large mergers should be noted. For example, in 2011, the agricultural company Ukrlandfarming bought several agroholdings: Dakor, Rise, Agro Alpha and Olimpex-Agro (totalling 370,000 ha). The result of this rapid expansion was that Ukrlandfarming became the largest farmland tenant in Ukraine, with a total agroholding of about 530,000 ha.

Another major player in agroholdings in 2011 was Kernel, which acquired a controlling stake in UkrRos, Enselko, Inter-Agro-Capital and Stiomi. These acquisitions allowed Kernel to increase its agroholdings by 330,000 ha. In addition, Kernel purchased the Ukrainian Black Sea Industry in 2012. This acquisition strengthened the position of the company in the oilseed processing market and significantly increased Kernel's storage capacities.

In January of 2012, Landkom (one of the first agroholding companies in Ukraine) was purchased by Alpcot Agro, which had assets in Russia and Ukraine. The acquisition of Landkom by Alpcot Agro was regarded as an opportunity for the additional diversification of land assets. The merger of the two companies made Alpcot one of the top 20 largest Ukrainian land users.

Traditionally, the growth of the majority of an agroholding was carried out through the acquisition of private/smallholder farms. But, as we have noted, the number of such operations in the reporting period have decreased significantly since 2009. This is due to a gradual digression from the established trend of land consolidation (investments), when agroholdings were considered liquid assets. Most likely in the short term, the focus will be on corporate strategies to increase a company's presence in a particular product segment, the development of infrastructure, or regional concentration.

The evolution of agroholdings confirms the thesis of the shift toward a more balanced approach. Apart from a few large corporate mergers in 2011, there have been a number of structural reorganizations. The most significant was a change of ownership at Illich-Agro GC, with the new owners creating the HarvEast Group (the shareholders of the holding were represented by the SKM Group and the Smart-Holding Group). One of the results of this reorganization was the optimization of assets (including land) that resulted when reducing the agroholding by almost 10% that year.

From a 2012 study in which we examined the activities of 78 agroholdings of more than 10,000 ha each, we rated the top ten agroholdings in Ukraine (Table 10.1). The size of the top ten agroholdings varied from 119,400 ha to 481,800 ha. In our overall analysis, the 78 agroholdings totalled 5.6 Mha, which was nearly 26% of all Ukrainian agricultural land under the control of agricultural enterprises and farms. Comparing our analysis to the 2012 public data on the land bank, we noted the controlled share of agroholdings was 29%.

Table 10.1. Land bank rating of the top ten of 78 agroholdings examined in Ukraine, 2012. (From UCAB, 2013.)

	UCAB data		Public data		Planned changes	
	1000 ha	Per cent	1000 ha	Per cent share of land	1000 ha	Year
NCH Capital	481.8	7.3	–	–	–/–	–
UkrlandFarming	470.3	32.4	532.0	23.7	–/500	–
Kernel	331.9	39.2	330.0	78.4	330/330	2013
Mriya	298.2	23.0	295.0	23.0	650/650	2015
Ukr. Agrarian Investment	259.6	2.5	260.0	0.0	300/300	2013
MHP	254.4	9.2	280.0	12.0	–/500	2014
Astarta	237.9	9.6	245.0	14.0	–/400	2015
HarvEast (Illich-Agro)	203.9	–9.8	220.0	0.0	–/–	–
Agroton	128.8	–3.6	150.0	–0.7	–/200	2013
Valars Group (Valinor)	119.4	–0.7	–	–	–/150	2013

Of the top ten agroholdings in Ukraine in 2012 (Table 10.1), only HarvEast reduced its land assets because of the need for optimizing production standards. In general, the problem of formalization of land relations between companies and landowners was typical for many agroholdings. As a consequence, the expansion in many agroholdings was due to the legalization of land tenure rather than to the acquisition of land – renting rather than buying land.

In the short term, the further expansion of agroholdings will depend on the final decision of land reform laws. Despite potential legal difficulties, the increased global demand for agricultural products forces companies to seek opportunities actively to expand their land assets now, rather than at a later date, when the market may be more expensive.

It should be noted that potential buyers would be more selective in terms of acquired assets that promote further development of mergers and acquisitions in the market. The support for this trend will be provided by the high cost of credit resources. Therefore, it is possible that some of the companies, especially medium-sized companies, will choose improved production efficiency over land expansion.

In this regard, we note that the difficulty of raising additional funding will be compounded, based on adverse market conditions, European Union (EU) economic instability and the emergence of more attractive investment countries such as Ukraine. According to the corporate plans of the companies, despite these factors, agroholdings could increase by almost 6.7–7 Mha.

Market development also depends largely on the development of the state-operated land bank, which in the first stage will control the retail market (buying and leasing agroholdings). It is possible that in the future, the land bank will boost the development of legislation on land consolidation to form land arrays (large area/grid of land).

Specialization

In 2011, the general specialization of agroholdings did not change significantly. Crop production was dominated by the highly profitable cultivation of grain and oilseed crops, and by consolidation and asset developments in the pork and dairy production sectors.

However, in some markets between 2010 and 2012, there were significant changes with the direct participation of agroholdings (Fig. 10.2). For example, in the sugar industry, consolidation decreased significantly the number of non-core market players and independent plants. Some long-term support for the sugar industry has been provided by the Verkhovna Rada Company of Ukraine through the 2012 Law of Ukraine No 10572-1 (*On Amendments to Some Laws of Ukraine on the Production and Use of Motor Fuels Containing Bio-components*) concerning the production of ethanol for gasoline (Parliament of

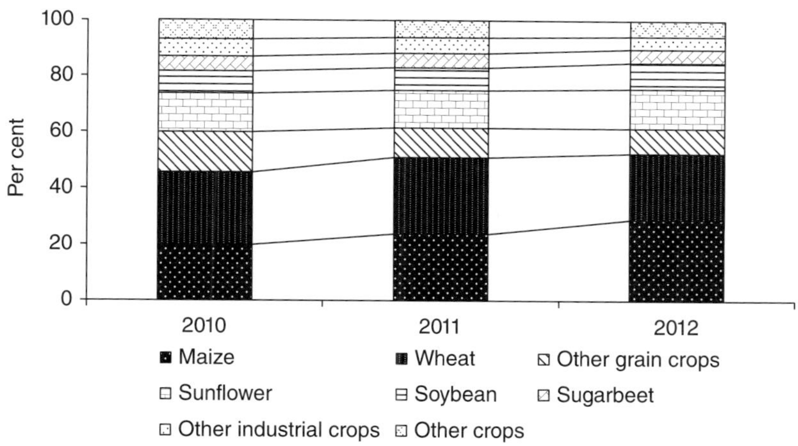

Fig. 10.2. Percentage of structural changes of crop areas of agroholdings. (From UCAB, 2011; authors' calculations.)

Ukraine, 2012d). In particular, the document stated that, by 2016, gasoline produced/sold in Ukraine should contain at least 7% ethanol.

The major players in the Ukrainian sugar market are Astarta, Ukrlandfarming and Mriya (T-Tsukor). Astarta, the leader in the sugar market in Ukraine, expanded its facilities in 2011 by acquring Novoivanivsky, a sugarbeet factory in the Kharkiv region. The result of the massive consolidation of sugar assets is the fact that the major sugarbeet producers in Ukraine are among the largest land users (Astarta, Agroprominvest, Ukrlandfarming, Kernel and Mriya). In 2012, these companies provided 46% of the total production of sugar in Ukraine.

In the grains market, some transformations were also observed. The majority of agroholdings increased their market share in cereals by shifting their production priorities from wheat to maize. This shift can be explained by the high economic efficiency of maize in 2010–2012, with a steady demand present in domestic and foreign markets. Also, such tendencies were promoted by active governmental regulation of the wheat market, especially regarding the export component. One of the examples of the change in priorities is demonstrated by the MHP Company, which increased its share in the grains market from 71% in 2010 to 73% in 2012 (during this period, the MHP Company reduced its wheat production acreage from 28% to 23%, while increasing its maize production acreage from 37% to 42%). In addition, the Ukrainian Agrarian Investments Company increased its overall grains production acreage from 57% to 62% (wheat acreage decreased from 33% to 20%, while maize acreage increased from 11% to 35%).

In oilseeds production, the cultivation of sunflower seeds still remains a priority for the majority of agroholding companies. There is an increasing trend for the expansion of soybean acreage. In particular, this trend is for agroholding companies with strong livestock production. The following companies serve as an example: Kernel, with a ratio of sunflower to soybean acreage of 14:14% in 2011 compared with 15:11% in 2010; the MHP Company, with a ratio of 12:5% in 2011 compared with 15:3% in 2010; Astarta, with a ratio of 7:16% in 2011 compared with 10:14% in 2010.

In general, a peculiar feature of the previous years, in which the majority of agroholdings were multi-branched structures without a key specialization, remains unchanged. The main trend has been diversification and finding activities with a synergy effect. For example, in 2011, the Industrial Milk Company (IMC), which specialized in the production of milk and dairy products, significantly increased its agroholdings from 39,000 ha to 60,000 ha. Moreover, according to the semi-annual financial report for 2012, the share of milk sales in the income structure decreased to 10% compared to 27% in 2011. The greater share of the proceeds was formed by maize sales (55%). The company plans to continue diversification of its operational activities.

Degree of vertical integration

The characteristic feature of the vertical integration of agroholdings is increasing the infrastructure for storage, which is quite natural, given the plans of agroholding companies to increase the production of grains and oilseeds. Most companies either increased storage capacity or announced plans to expand them. For example, Kernel, with warehouse storage facilities totalling 2.7 million tonnes (Mt) in 2012, plans to build another six warehouses with a storage capacity of 400,000 t by 2014. Ukrlandfarming, with about 1 Mt of storage capacity, announced plans for the construction of two new warehouses with total storage capacity of about 800,000 t. Other companies, such as Nibulon, Svarog and Loture, also plan to increase their storage capacities.

Besides the production of grains and oilseeds, another sector in which changes can be clearly observed in vertical integration is the sugar industry in Ukraine. Without exception, all the agroholdings in the sugar processing industry formed a vertical closed-loop production under a single holding company. This integration reduced the costs due to links between the primary producers and processors, thereby increasing operational efficiency.

In general, for the vast majority of agroholding companies, the situation of full vertical integration is typical, starting from crop production, storage and refinement to market standards, with subsequent sales on both the domestic and export markets. In this case, we do not take into consideration the major livestock agroholding companies for which crop production is the basis for their feed supply. Among the exceptions, there are a number of companies that are also the largest processors, such as Kernel, which is the largest producer of sunflower oil in Ukraine.

At the same time, it was common for a company to optimize its operations through vertical integration. For example, in 2011, Agroton sold its assets for the production of flour, pasta, bread, bakery products and dairy products to focus exclusively on the production of primary products.

Despite broad regional expansion, some territorial priorities can be observed. Agroholdings have actively developed the Forest-Step region, especially the Zhitomir and Chernihiv regions. This is due partly to the historically low rental rates for land in these regions. Thus, the Chernihiv region was the leader, both in relative and absolute figures, in the growth of agroholdings in 2012, at an increase of 26.5%, or 108,000 ha. In the Zhitomir region, the increase was 6.2%, or 12,000 ha. An additional incentive for increasing agroholdings in the region includes natural and climatic conditions. For example, climatic changes make it possible to grow commercially lucrative crops such as maize while lowering the risk level of agricultural production.

At the same time, a new trend should be noted – increasing the presence of agroholdings in the regions that, for some reason (quality of land, climatic conditions), was not the object of active development. In particular, one should mention the development of the central and southern areas of Ukraine. Strengthening the positions of agroholdings in these regions can be explained. First of all, there has been a reduction of free land resources in the traditional regions of Ukraine. It should be noted that the regional presence of agroholdings depends directly on the corporate plans for business development.

The role of agroholdings in Ukraine's large commodity agricultural production becomes more influential. Thus, according to the results of 2011, the share of agricultural holdings (without households) was 30.9% for crop production and 58.4% for livestock production (Fig. 10.3).

The domination of agricultural holdings in Ukraine's livestock production sector in 2011 is explained by the high concentration in the production of poultry meat and eggs. Active development of large commercial pork production in recent years has also contributed to the strengthening of the positions of agroholdings. Current trends allow us to assume that the dairy sector would move gradually into the sphere of the interests of large-scale commercial producers.

Despite the relatively small share of agroholdings in crop production in general (31%), they dominate in some agricultural industries. Thus, the consolidation of the sugar industry has occurred as a result of agroholdings owning the sugarbeet refineries. Thus, they have loaded the processing capacities at the expense of own-produced sugarbeets. As a result, about 60% of the sugarbeet produced by agricultural enterprises are grown by agroholdings.

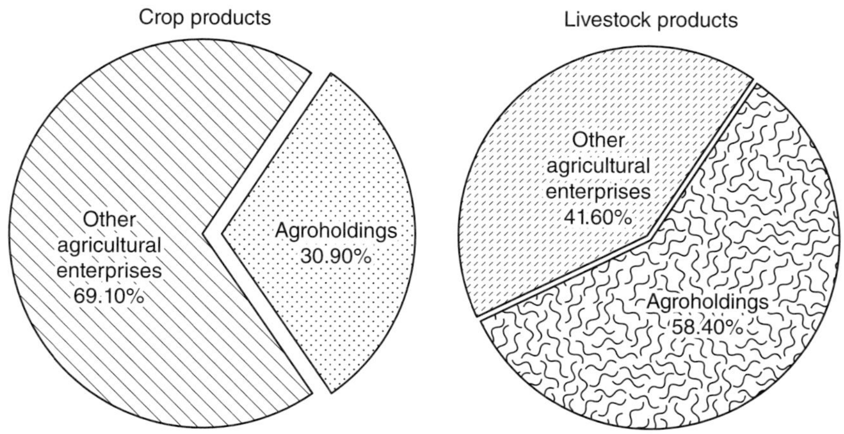

Fig. 10.3. Share of Ukrainian agroholdings in gross agricultural output in 2011 (without households). (From UCAB, 2011; authors' calculations.)

Among agricultural enterprises in 2014, 45% of maize production is provided by agroholdings that have the resources to acquire quality maize seeds and production technologies, which ultimately has positive effect on increasing performance indicators. In addition, agroholdings account for over 33% of oilseed production volumes of rapeseed and soybean.

Since 2012, a series of laws and regulations have been adopted that have had a major effect on agribusinesses in Ukraine. This chapter briefly describes the main changes that have occurred.

Land market

The transfer of ownership to land parcels designated for commercial farming is not possible in Ukraine, due to the prohibition (moratorium) of land sales effective since 2000. Cancellation of the moratorium has been postponed several times, due in part to adoption of the 2011 Laws *On the Land Market* and *On the State Land Cadastre* (Parliament of Ukraine, 2011b,c). Note that a cadastre is defined as a public record, survey or map of the value, extent and ownership of land as a basis of taxation. The Ukrainian Law *On the State Land Cadastre* provides an open cadastral map to identify the location of a land parcel. It should be noted that the Ukrainian cadastral map contains a large number of inconsistencies and technical errors.

Draft law on farmland turnover

After extended discussions, the original draft law *On Farmland Turnover* (Parliament of Ukraine, 2012f) was revised to provide more restrictions relating to ownership title to farmland. Pursuant to the draft law, foreign individuals, as well as foreign and Ukrainian legal entities, are prohibited from acquiring ownership of farmland in the territory of Ukraine. Only the state and citizens of Ukraine can buy farmland.

The revised draft law also stipulates that Ukrainian citizens can own a maximum of 100 ha of farmland. This restriction is quite controversial because 100 ha are insufficient for commercial farming under the conditions of the Ukrainian market. Also, the revised draft law restricts the maximum area of leased land to no more than 100,000 ha generally, or 20%, for a minimum lease term of 8 years and a maximum lease term of 50 years. The revised draft law is expected to undergo extended discussions and amendments before being submitted to Parliament.

The revised draft law determines the potential buyers of farmland in Ukraine as the executives of rural, town and city councils; determines the government-owned state land bank; and determines the citizens of Ukraine who are registered on the official website of the central executive body that implements the state policy on land relations and who meet the requirements. To be eligible, a citizen must be: (i) recorded as a resident for at least 3 years in villages located within 20 km of the land to be sold; or (ii) a private/smallholder farmer; or (iii) someone experienced in agriculture or engaged in agricultural production and registered as an individual entrepreneur in farming.

Foreigners, stateless persons, foreign governments, international governmental organizations, non-governmental organizations and legal persons CANNOT acquire ownership of agricultural land in excess of their inheritance rights under Ukrainian law.

General changes in land laws

New procedures for the registration of rights to immovable property, including land parcels, have come into force. While there is no need to double register rights to a land parcel (such registration is now carried out by a unified state registration service that is a subdivision of the Ministry of Justice of Ukraine), very often problems will arise during the registration of lease rights to land parcels if ownership is not already registered. Most state and municipal lands do not have registered ownership title, which results in delays during the registration of lease rights.

Turnover of land and land reform

The 2012 law of Ukraine *On Amendments to the Land Code of Ukraine* (Parliament of Ukraine, 2012b) continues the restraint on the alienation of farmland (the so-called moratorium). At this time, the moratorium has been prolonged until the law on the turnover of farmland goes into force in 2016.

Beginning in 2013, lands of state and communal property are to be considered demarcated. Demarcation criteria are defined in the 2012 law *On Amendments to Some Legislative Acts of Ukraine Regarding the Delimitation of State and Municipal Property* (Parliament of Ukraine, 2012e). These amendments, among other things, redistribute powers between the bodies of state power (instead of state administrations), meaning that the State Agency of Land Resources of Ukraine and its territorial bodies will manage the agricultural land as state property.

Amendments to Articles 135–139 of the Land Code of Ukraine No 5077-VI (Parliament of Ukraine, 2012b) legislatively regulate the procedure for conducting land auctions. The amendments allow the designated authorities to sell land plots of state and communal property and also allow them the legal rights to sue.

The 2012 law *On Amendments to the Law of Ukraine on Banks and Banking Activity* (Parliament of Ukraine, 2012c) created the State Land Bank, which was registered as a public joint-stock company, with registered capital of UAH 120 million. It can be expected that the State Land Bank will extend credit, at an interest rate below the market price, to small and medium-sized domestic agricultural producers that use land as collateral.

In 2013, the law *On the State Land Cadastre* (Parliament of Ukraine, 2012c) came into force. According to this law, land plots are subject to state registration in the State Land Cadastre, which includes quantitative and qualitative characteristics of the land, including the cadastral number, location, description of the boundaries, area, intended use and normative monetary valuation. The official website of the State Agency of Land Resources of Ukraine now features the Public Cadastre Map, which shows the boundaries of land plots and displays some data of the State Land Cadastre relative to it by cadastral number.

New procedure for the registration of rights to immovable property

Land is immovable property as defined in the law *On State Registration of Proprietary Rights to Immovable Property and their Encumbrances* (Parliament of Ukraine, 2004). Under the new order, the right to lease land can be registered only after the registration of ownership right to the relevant land, and that could prove challenging in practice.

Framework legislation

Decree No 1185 allowed the Cabinet of Ministers of Ukraine to increase the normative land evaluation to a ratio of 1.756 in 2012, which consequently increased the normative monetary value of land (Parliament of Ukraine, 2011a). This, in turn, has a proportional impact on land rent for the land, the size of which is set as a percentage of the relevant land plot normative value.

In 2013, the law *On Agricultural Receipts* (Parliament of Ukraine, 2013a) was enacted. This law introduces agricultural receipts that are title documents recording the debtor's absolute obligation to supply agricultural products (agricultural commodity receipts) or to pay money (agricultural financial receipts). The debtor's obligation is secured by a pledge of future crops from a particular tract of land. Agricultural receipts are certified by a notary, and then recorded in a special register. This law is designed to replicate a successful Brazilian practice in Ukraine and to facilitate the access of agricultural enterprises to material resources.

The 2012 law *On Amendments to Certain Laws of Ukraine on the Guarantee Fund of Performance under Warehouse Receipts for Grain* (Parliament of Ukraine, 2012a) regulates the establishment and operation of the Guarantee Fund of Performance under Warehouse Receipts for Grain, which is jointly liable for the obligations of the grain warehouse and has the duty to ensure the protection of the rights and interests of the owners of warehouse grain documents. This law provides for the mandatory participation of grain warehouses in the Guarantee Fund.

The regime of the taxation of agricultural producers that are legal entities may soon change. Thus, the Decree of the President of Ukraine *On the National Action Plan for Prosperous Society, Competitive Economy, Effective State* No 128 (Parliament of Ukraine, 2013b) provides for the submission to the Verkhovna Rada of Ukraine of a draft law on introducing amendments to the Tax Code of Ukraine concerning the reform of the fixed agricultural tax and the revision of exemptions from the payment of value-added taxes (Parliament of Ukraine, 2013b).

The 2012 law of Ukraine *On Specifics of Insuring Agricultural Goods with State Support* (Parliament of Ukraine, 2012i) provides for government support in the form of subsidies to agricultural producers to compensate for part of the insurance premium on agricultural products.

In 2012, the law *On List of Permits in the Area of Economic Activity* (Parliament of Ukraine, 2012g) was extended to include the following documents: (i) the consent for the import into Ukraine of plant protection products; (ii) the certificate of conformity of agricultural products and raw materials of plant origin with respect to the content of residual amounts of pesticides, agrochemicals and heavy metals; and (iii) the certificate of compliance with allowed use of pesticides and agrochemicals (only plant growth regulators) in agricultural products and raw materials of plant origin. The procedure for issuing these permits is established by the Law of Ukraine *On Protection of Plants* (Parliament of Ukraine, 2012h).

Conclusions

After a steady upward trend in Ukraine between 2007 and 2009, land acquisitions by agroholdings began declining in 2009, due to financial concerns and regulatory problems. Thus began the process of small company land acquisitions. Despite the moratorium on the selling of agricultural land, some positive changes have taken place. In particular, many companies have abandoned the practice of active, sometimes chaotic, consolidation of land that was influenced by the investment component, when the land was considered a liquid asset. Instead, the majority of companies focus on the quality of their production resources to strengthen their presence in particular segments and to strengthen their production resources on the development of their infrastructures. Companies with large debt loads that are selling with a good discount as donors are excluded from this list.

Formalizing land resources between companies and landowners is a problem for many agroholdings due to land tenure registration legislation. The active development of agroholdings by the land bank can be observed in the Zhytomyr (108,000 ha) and Chernihiv (12,000 ha) regions due to historically low rent prices for agricultural land compared to other regions of Ukraine.

In the agricultural sector, crop production is dominated by highly profitable grain and oilseed crops. For example, agricultural enterprises have increased maize production because maize is highly profitable and is in high demand on both the domestic and foreign markets. Livestock production focuses mainly on pork and dairy production.

Some changes have been observed in the sugar industry based on the Ukrainian Law *On Amendments to Some Laws of Ukraine on the Production and Use of Motor Fuels Containing Bio-components* (Parliament of Ukraine, 2012d) to increase bioenergy production to meet demand. The impact of this law on Ukraine's economics is very high, with the likelihood of higher agricultural production costs and higher fuel prices. Agroholdings are increasing storage capacities and are increasing integrated production, especially in the sugar industry.

In 2013, the largest agroholding in Ukraine was Ukrlandfarming, with a total land bank of about 530,000 ha. It is a large diversified agribusiness. Based on estimates by the Ukrainian Agribusiness Club, the total land bank of agroholdings in Ukraine was more than 6 Mha in 2013, representing more than 27% of total Ukrainian agricultural land.

References

Parliament of Ukraine (2004) Law *On the State Registration of Proprietary Rights to Immovable Property and Their Encumbrances*. Parliament of Ukraine, Kiev.

Parliament of Ukraine (2011a) *On Amendments to the Methodology of the Normative Pecuniary Valuation of Agricultural Land and Populated Areas*. Land Use Law 1185. Parliament of Ukraine (Verkhovna Rada), Kiev.

Parliament of Ukraine (2011b) *On the Land Market*. Parliament of Ukraine, Kiev.

Parliament of Ukraine (2011c) *On the State Land Cadastre*. Parliament of Ukraine, Kiev.

Parliament of Ukraine (2012a) *On Amendments to Certain Laws of Ukraine on the Guarantee Fund of Performance under Warehouse Receipts for Grain*. Parliament of Ukraine, Kiev.

Parliament of Ukraine (2012b) *On Amendments to the Land Code of Ukraine*. Parliament of Ukraine, Kiev.

Parliament of Ukraine (2012c) *On Amendments to the Law of Ukraine on Banks and Banking Activity*. Parliament of Ukraine, Kiev.

Parliament of Ukraine (2012d) *On Amendments to Some Laws of Ukraine on the Production and Use of Motor Fuels Containing Bio-components*. Law of Ukraine No 10572-1. Parliament of Ukraine, Kiev.

Parliament of Ukraine (2012e) *On Amendments to Some Legislative Acts of Ukraine Regarding the Delimitation of State and Municipal Property*. Parliament of Ukraine, Kiev.

Parliament of Ukraine (2012f) *On Farmland Turnover*. Parliament of Ukraine, Kiev.

Parliament of Ukraine (2012g) *On List of Permits in the Area of Economic Activity*. Parliament of Ukraine, Kiev.

Parliament of Ukraine (2012h) *On Protection of Plants*. Parliament of Ukraine, Kiev.

Parliament of Ukraine (2012i) *On Specifics of Insuring Agricultural Goods with State Support*. Parliament of Ukraine, Kiev.

Parliament of Ukraine (2013a) *On Agricultural Receipts*. Parliament of Ukraine, Kiev.

Parliament of Ukraine (2013b) *On the National Action Plan for Prosperous Society, Competitive Economy, Effective State*. Law of Ukraine No 128. Parliament of Ukraine, Kiev.

UCAB (2011) *AgriSurvey: Largest Agricultural Companies in Ukraine: On the Verge of Restructuring*. Ukrainian Agribusiness Club (UCAB), Kiev.

UCAB (2013) *AgriSurvey: Largest Ukrainian Crop Holdings*. Ukrainian Agribusiness Club (UCAB), Kiev.

11 Large-Scale Dairy and Poultry Production in Russia: Level and Trends of Development

Vladimir N. Surovtsev, Yulia Nikulina and Uliana Saigusheva*
North-West Institute of Economics and Organization of Agriculture, St Petersburg, Russia

Abstract

This chapter considers the reasons and consequences of the changes taking place from the beginning of the economic reforms in 1990 until 2013 in Russia's dairy farming and poultry farming sectors. It examines production concentration and intensification processes, investments and the changing demographics of the dairy and poultry industries in Russia in general and in the Leningrad region in particular.

The Role of Dairy Production in the Agricultural Sector of Russia and the Leningrad Region

Dairy and poultry production in Russia

Dairy production is one of the most important sectors of Russian agriculture. The main trends taking place in this sector are described in this chapter based on a case study of the Leningrad region, a well-developed agricultural area. Traditionally, dairy production has been one of the main food sources of protein in Russia, as well as a source of organic fertilizers for crop production. The dairy industry has played an essential role in ensuring employment for the rural population in many regions of Russia.

Poultry production has become the most technologically advanced agricultural industry in Russia. It is adopting progressive technologies to ensure a sustainable growth in domestic poultry meat and egg consumption. Broiler meat and eggs are a cheap source of protein, accounting for over 27% of the total protein consumption in Russia (Ministry of Agriculture of the Russian Federation, 2001–2012).

Dairy and poultry production in the Leningrad region

By the late 1980s, an industrial type of agriculture was formed in the Leningrad region, focused on providing milk, poultry and eggs. The agroindustrial market reforms in Russia's Leningrad region affected the highly specialized enterprises negatively due to production concentration and intensification. Government financial intervention in the agricultural sector in the region was minimal, aimed mainly at developing and

*Corresponding author; e-mail: skarapejas@yandex.ru

modernizing production and expanding the nuclear herd stock. Government incentives were focused on creating production growth. Regional agriculture faced difficulties in remaining competitive because of high pressure from the global market.

Market pressure, accompanied by lack of government support for weak producers, led to a tough competitive environment that motivated agricultural companies to increase production efficiency. Unprofitable and low-margin agricultural producers were forced either to leave the market or to develop new strategies and search for investors or other financing sources that allowed them to intensify production and adopt innovations. Severe economic selection caused negative social and environmental consequences, such as greater unemployment, depressed rural income and increased environmental pollution.

In 2012, the Agroindustrial Complex of Leningrad Region included 232 large- and medium-scale agricultural enterprises with a 76% share of total agricultural output. In comparison, there were 796 private/smallholder farms and 90,400 individual household producers involved in agricultural production in the Leningrad region in 2012, with a 21.8% share of total agricultural output (1.7% by private/smallholder farms and 20.1% by household plots), valued at 14.3 billion roubles (RUB) (Committee for Agroindustrial Complex of Leningrad Region, 2009–2013). The wage earned in agricultural enterprises was the main source of income for people employed in agriculture in the Leningrad region.

While the share of population employed in agriculture decreased, the total output of agricultural enterprises increased from RUB 16.6 billion in 2006 to RUB 44.7 billion in 2012, signifying a high growth rate of labour efficiency. In 2012, the annual revenue per employee rose to RUB 1.8 million. The average monthly salary increased from RUB 8300 in 2006 to RUB 22,900 in 2012. This salary growth was in line with the growth rate of product sales per employee for the same period. In 2012, the average nominal salary per person employed in dairy production was RUB 20,700/person as compared to RUB 29,500/person employed in poultry production (Committee for Agroindustrial Complex of Leningrad Region, 2009–2013).

In 2012, dairy and poultry production were the leading agricultural industries in the Leningrad region. That year, farms of all types produced 570,000 t of milk (526,000 t from agricultural enterprises), with average milk yield of 7233 kg, making the Leningrad region one of the largest milk producing areas in Russia. In 2010, the Leningrad region ranked first in total egg production output (9%) and second in total poultry meat production output (6%) (Serova and Karlova, 2010).

An Overview of Dairy Farming and Poultry Production in Russia

Traditionally, dairy farming in Russia was the most important agricultural sector. In 1990, at the beginning of the Soviet transition reforms, there were 20.8 million dairy cows, but by 2012 there were only 8.9 million (Russian Union of Dairy Enterprises, 2013). At the same time, dairy milk production decreased from 55.7 million t (Mt) in 1990 to 31.9 Mt in 2012 (Table 11.1).

Table 11.1. Key indicators of milk production in Russia, selected years, 1990–2012. (From Federal Service of State Statistics, selected years, 1990–2013.)

	Years			
Indicators	1990	2000	2010	2012
Livestock on farms of all types (millions)	57	28	20	20
Dairy cows (millions)	21	13	9	9
Milk production on farms of all types (Mt)	56	32	32	32
Marketed milk, agricultural enterprises (Mt)	40	13	13	14
Cow milk productivity, agricultural enterprises (kg/cow/year)	2783	2341	4189	4985

Note: Mt = million tonnes.

Poultry production in Russia has intensified due to large industrial poultry farms focusing on egg and broiler production. Also, because of low production risks and the rapid adoption of best practices in selective breeding and feeding, Russia's poultry farms have increased their investment appeal, attracting investments worth around RUB 250 billion. Those funds were used to modernize existing facilities and to build new facilities (Ministry of Agriculture of the Russian Federation, 2013). In 2010–2012, the annual growth in egg production was about 3%, while poultry meat production was about 18%. As a result, poultry meat production in 2012 more than doubled compared to 1990 (Table 11.2).

Poultry production has shown the highest growth rate among all animal breeding sectors. Between 2001 and 2011, poultry numbers increased by 1.8 times and cattle numbers decreased by 40% (Fig. 11.1). This situation brought significant structural changes to meat production in Russia. As a result, the poultry meat market share in total meat output was 45% in 2012 and the cattle meat market share was 21% (Fig. 11.2).

The Role of Various Farming Enterprises in Dairy and Poultry Production

In 2013, there were three types of farming entities in the Russian agricultural sector: agricultural enterprises/large private companies (usually open joint- or close joint-stock companies), private/smallholder farms and household plots (after work hours, not part of employment). By definition, household plots are owned by rural populations, who are not required to form legal entities or are not required to pay taxes, except for land taxes. Until the mid-1960s, they were a main source of income for collective farm employees.

In 1990–1995, the share of small-scale farming in total agricultural output grew rapidly through increased production output. Small-scale farming can remain sustainable during a period of economic transformation due to the ability to respond promptly to market and environmental changes and to respond to the availability of mobile and flexible resources. Beginning in 2000, Russia introduced government support and an agrarian protectionism policy. For example, the national priority project, *The Development*

Table 11.2. Poultry numbers and poultry output in Russia, selected years, 1990–2012. (From Federal Service of State Statistics, selected years, 1990–2013.)

	Years					
Indicators	1990	2000	2005	2010	2011	2012
Total poultry, million birds	660	341	358	449	473	495
Eggs, billion eggs	37	24	27	31	33	33
Poultry, carcass weight (Mt)	1259	502	1094	2516	2858	3233

Note: Mt = million tonnes.

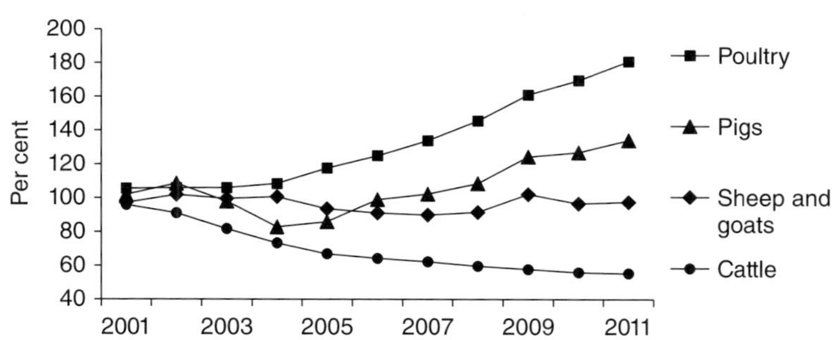

Fig. 11.1. Percentage share of livestock and poultry in agricultural enterprises in Russia, selected years, 2001–2011. (From Federal Service of State Statistics, selected years, 1990–2013.)

of the Agroindustrial Complex (Ministry of Agriculture of the Russian Federation, 2005) was adopted in 2006, and the *Government Programme for the Development of Agriculture and Regulation of Agricultural Commodities, Raw Materials, and Food Markets for 2008–2012* (Ministry of Agriculture of the Russian Federation, 2007) was adopted in 2008. With government support, large agricultural enterprises have been the fastest to adopt innovations and to modernize production facilities, thus making their products more competitive compared to small producers.

Economies of scale vary by industry (Surovtsev *et al.*, 2009). Institutional problems in Russia, including complicated land relations, have had different impacts on dairy and poultry production. Difficulty when transferring agricultural lands from inefficient owners to efficient owners has had a greater effect on the development of large-scale dairy production than it has had on poultry production.

Industry-specific and institutional peculiarities of dairy production in Russia resulted in a situation in which small-scale producers with low livestock yields accounted for almost 50% of total milk output. For example, in 2012, household plots accounted for 48.3% of the total number of cows and 48.4% of gross milk output in Russia. Since 2000, the share of household plots in the total number of cows and milk production has decreased, while the share of agricultural enterprises and private/smallholder farms has increased (Figs 11.3 and 11.4; Federal Service of State Statistics, various years).

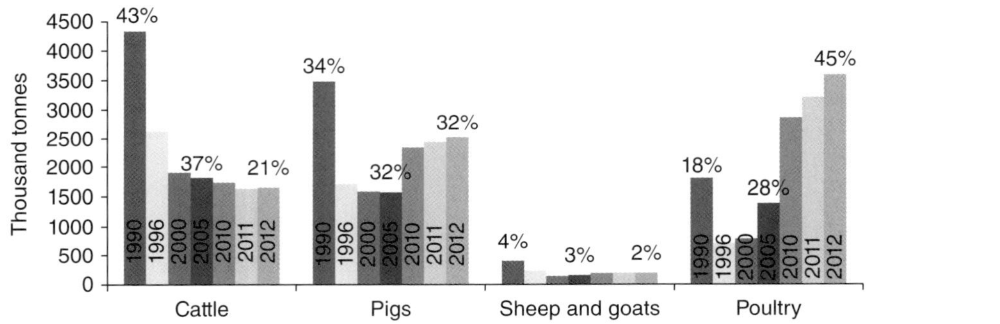

Fig. 11.2. Percentage share of livestock and poultry for slaughtering (in live weight, thousand tonnes) in total output of all types of farms in Russia, 1990–2012. (From Federal Service of State Statistics, selected years, 1990–2013.)

Fig. 11.3. Milk output of agricultural producers in Russia (million tonnes), 1990–2012. (From Federal Service of State Statistics, selected years, 1990–2013.)

Both poultry and egg production are concentrated on large agricultural enterprises with high numbers of animals (Figs 11.5 and 11.6). The main producers of poultry meat are big broiler poultry farms, with a share of around 90% of total output. The output share for farms producing other types of meat (geese, duck, turkey and guinea fowl) is quite low and accounts for only 0.8% of total poultry output.

The Leningrad region's agricultural enterprise share in total dairy and poultry output is higher than in other areas in Russia because the region began intensifying its milk and poultry production much earlier than did other regions. In 2012, there were 115 agricultural enterprises in the Leningrad region that had developed dairy production, accounting for 92% of total milk production. Household plots contributed 7% and private/smallholder farms contributed 1% to total milk output (Figs 11.7 and 11.8) (Committee for Agroindustrial Complex of Leningrad Region, 2012).

In 2011, agricultural enterprises accounted for the majority of total output for poultry and egg production in the Leningrad region (Table 11.3). For the production of non-traditional species of birds (e.g. quail, turkey, duck and goose), the regional government programme for the development of private/smallholder poultry farms in the Leningrad region in 2010–2020 was adopted.

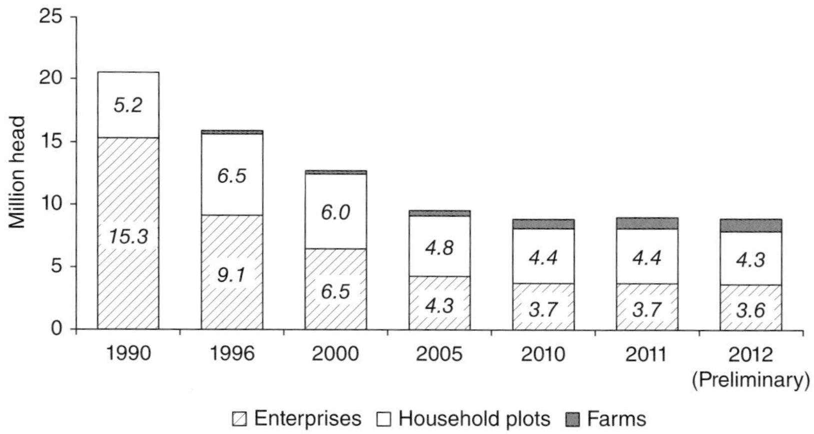

Fig. 11.4. Number of cows in agriculture in Russia (million head), 1990–2012. (From Federal Service of State Statistics, selected years, 1990–2013.)

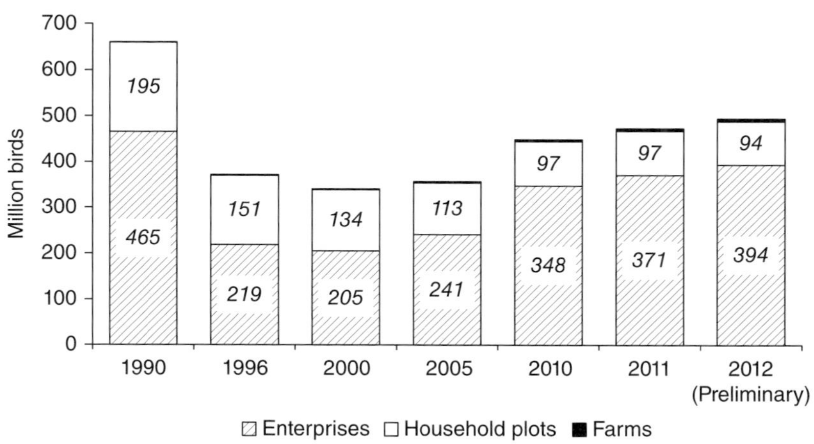

Fig. 11.5. Poultry numbers in farming enterprises in Russia (million birds). (From Federal Service of State Statistics, selected years, 1990–2013.)

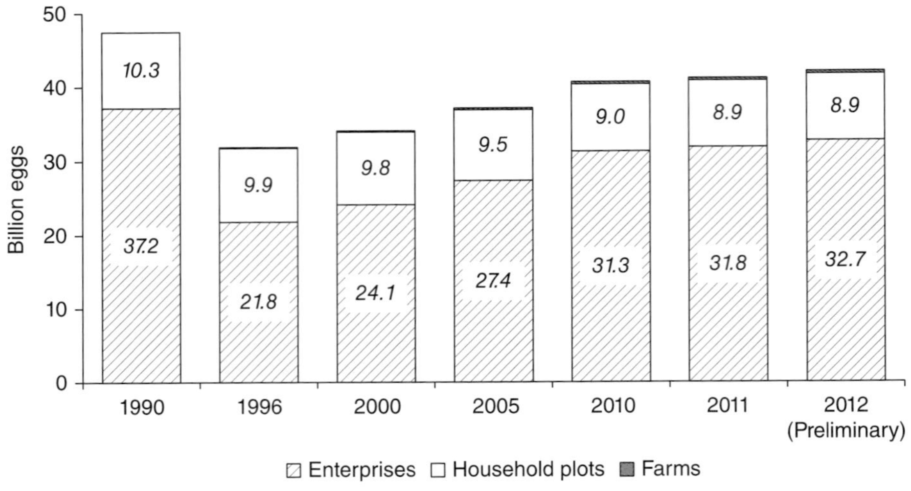

Fig. 11.6. Egg production in farming enterprises in Russia (billion eggs). (From Federal Service of State Statistics, various years, 1990–2013.)

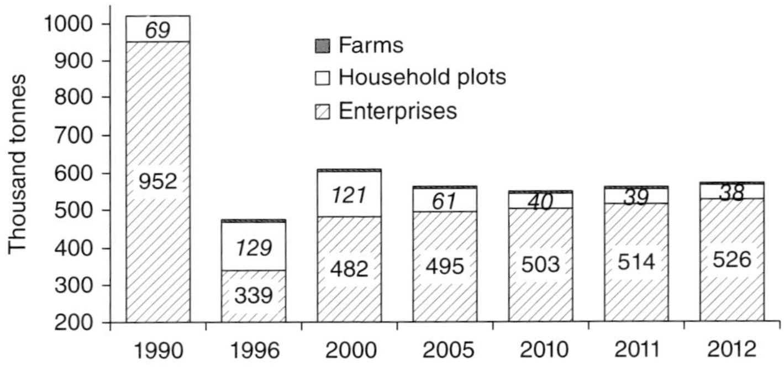

Fig. 11.7. Milk output of agricultural producers in Russia's Leningrad region (thousand tonnes). (From Committee for Agroindustrial Complex of Leningrad Region, 2012.)

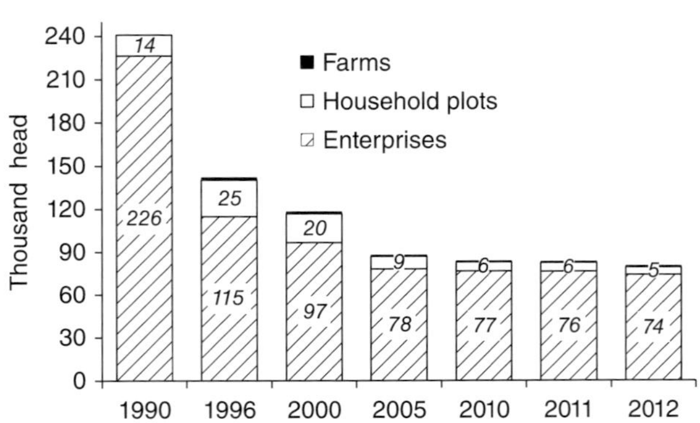

Fig. 11.8. Number of cows in agriculture in Russia's Leningrad region (thousand head), 1990–2012. (From Committee for Agroindustrial Complex of Leningrad Region, 2012.)

Table 11.3. Poultry production output and poultry stock in Russia's Leningrad region, selected years, 1990–2011. (From Committee for Agroindustrial Complex of Leningrad Region, 2012.)

Indicators	Years				
	1990	1996	2000	2007	2011
	Output				
	Eggs (million eggs)				
Total output	2120	1642	1973	2252	2901
Agricultural enterprises	1968	1476	1876	2218	2870
Individual household plots	152	164	95	33	30
Private/smallholder farms	–	2	2	1	1
	Poultry meat (thousand tonnes)				
Total output	78	19	49	131	214
Agricultural enterprises	74	18	47	130	213
Individual household plots	4	1	2	1	1
Private/smallholder farms	–	0	0	0	0
	Poultry stock (million birds)				
Total output	22	11	14	20	25
Agricultural enterprises	20	9	13	20	25
Individual household plots	2	2	1	0	0
Private/smallholder farms	–	0	0	0	0

Development of Investment Activities in Dairy and Poultry Production

Budget support and agrarian protectionism policies have had different impacts on the investment flow in dairy and poultry production in the three types of farming establishments in Russia. Subsidizing interest rates for investment and short-term credits are the main tools of the Russian government support for agricultural producers within the national priority project titled *The Development of Agroindustrial Complex* (Ministry of Agriculture of the Russian Federation, 2005) and *the Government Programme for the Development of Agriculture and Regulation of Agricultural Commodities, Raw Materials, and Food Markets for 2008–2012* (Ministry of Agriculture of the Russian Federation, 2007). For example, in poultry production, investments received through interest rate subsidies gave access to the latest technologies to make poultry producers more competitive within the global food market (Table 11.4).

Russian dairy production, characterized by long-term turnover, low production and management efficiency, requires huge amounts of land for pastures and forage conservation. Over the 2008–2011 period, 336 new dairy farms were created and 784 existing farms were reconstructed (Table 11.5).

From 2006 to 2012, investment projects in the agricultural sector of the Leningrad region were implemented using credits (73%) and each farm's own funds (27%). In 2012, investment credits subsidized in the Leningrad region totalled RUB 18.6 billion (2.6% of total credits in the country). During this time in Russia, the Leningrad region ranked eighth in subsidized investment credits and third in the number of agricultural producers receiving the credits.

From 2006 to 2012, improvements in dairy production in the Leningrad region included an improved infrastructure and new technology such as computerized milking equipment and robotic voluntary milking systems (VMSs) (Committee for Agroindustrial Complex of Leningrad Region, 2012). New equipment improved labour efficiency in dairy production in the Leningrad region. Dairies raising cows in free-stall barns with milking parlours or robotic VMSs reduced their direct man-hour labour costs to 0.6–0.8/100 kg of milk (the average for agricultural enterprises in the Leningrad region is 1.7 man-hours/100 kg of milk).

Poultry production in the Leningrad region has received the major share of agricultural investments (47.6%), while dairy production has received 18.1%. The total value of investment projects in poultry production beginning from 2006 was nearly RUB 16 billion. The share of

Table 11.4. Increase in poultry for slaughtering at newly built or reconstructed broiler facilities in Russia, 2008–2011. (From Committee for Agroindustrial Complex of Russia, 2001–2012.)

Indicators	Year				Total 2008–2011
	2008	2009	2010	2011	
Number of new facilities	22	17	18	9	66
New facility, slaughtered poultry (live weight, thousand tonnes)	92	76	81	49	298
Number of reconstructed facilities	63	44	43	31	181
Reconstructed facility, slaughtered poultry (live weight, thousand tonnes)	28	50	101	65	244

Table 11.5. Increase in milk output at newly constructed or reconstructed farms in Russia, 2008–2011. (From Agroindustrial Complex of Russia, various years, 2001–2012.)

Indicators	Year				Total 2008–2011
	2008	2009	2010	2011	
Number of new facilities	109	73	75	79	336
New facility, milk output (thousand t)	120	134	110	88	452
Number of reconstructed facilities	368	176	148	92	784
Additional milk output due to reconstructed facilities (thousand t)	92	56	30	19	197

poultry raised in reconstructed facilities in 2006 was 2.1% compared to 14.7% for cattle. By 2011, the share of poultry raised in reconstructed facilities was 34.8% compared to 26.8% for cattle (Table 11.6).

Table 11.6. Percentage share of animals in reconstructed housing in Russia's Leningrad region, 2006–2011. (From Committee for Agroindustrial Complex of Leningrad Region, 2012.)

Years	Livestock		
	Cattle	Pigs	Poultry
		Per cent	
2006–2008	15	21	2
2009	20	30	16
2010	24	27	26
2011	27	35	35

Concentration and Intensification of Dairy and Poultry Production

Investment support available to the agricultural sector in Russia has contributed to increasing the concentration of funds to agricultural enterprises that are more likely to get loans. These loans help agricultural enterprises adopt new capital-intensive, resource-saving technologies to the detriment of smaller producers.

Concentration

Poultry production has been concentrated mainly on large-scale enterprises (total output is more than 20,000 t of birds and 200 million eggs/year) with closed-loop cycle production, accounting for 75% of poultry meat and 55% of egg production in 2011 (Ministry of Agriculture of the Russian Federation, 2013).

Concentration has been even more obvious in poultry farming in the Leningrad region, owing to the economies of scale. In 2011, the two largest egg production farms accounted for 67% of total egg output, and three broiler poultry farms produced 93% of total poultry meat output (Committee for Agroindustrial Complex of Leningrad Region, 2012).

In contrast, Russia's largest dairy enterprises received only 10% of their proceeds from selling milk (Serova and Karlova, 2010). The milk processing market in Russia has been dominated by Danone-Unimilk and Wimm-Bill-Dann; together they accounted for 44% of total milk products in 2011. In the Leningrad region, agricultural enterprises with high livestock concentration accounted for the main share of the milk produced in 2011 (Fig. 11.9).

Large agricultural dairy enterprises (more than 800 cows) in the Leningrad region accounted for more than 50% of dairy cows and 60% of total milk production in 2011. The number of farms with an average annual livestock above 1000 cows has been growing rapidly. Modernized technologies to reduce labour and resource intensity in dairy production have made large enterprises more efficient and have eliminated much of the competition from smaller farms.

Moving from geographically dispersed farms to large farming complexes with free-stall barns and higher concentrations of cattle have contributed to rising management efficiency, reduced service costs and labour and improved milk quality and safety. With production and herd concentration estimated to increase in the future, less efficient small producers may be eliminated from the market and from government support programmes.

Intensification

Globalization has led to resource intensity and has increased intensification of production. In the Leningrad region, agricultural enterprises that intensify production contribute the major share of milk output (Fig. 11.10). The largest livestock breeding companies breed 77% of the cattle in the Leningrad region and account for 82% of total milk output (Committee for Agroindustrial Complex of Leningrad Region, 2009–2013).

Dairy production in Russia is converting its cattle feeding system to fodder rations based on cow milk productivity; lactation period; and enriched compound feeds, feed additives and biological feed products. Over the 1995–2010 period, feed consumption needed for milk production decreased by 33% and milk yield increased by 50%. The rise in milk productivity is due mainly to an increase in the share of feed concentrate (Federal Service of State Statistics, selected years, 1990–2013).

The key sector-specific issue restraining the development of dairy production in Russia is insufficient feed production. Intensified feed production is important to Russia to aid cows when digesting cheaper bulk feed, especially on breeding farms (Committee for Agroindustrial Complex of Leningrad Region, 2013). Underdeveloped technological infrastructure (Table 11.7) has impacted the development of feed production negatively. In addition, the percentage of machinery modernization at agricultural enterprises in Russia in 2011 was 3.4% for tractors, 5.3% for combine harvesters and 6.4% for forage harvesters (Ministry of Agriculture of the Russian Federation, 2001–2012).

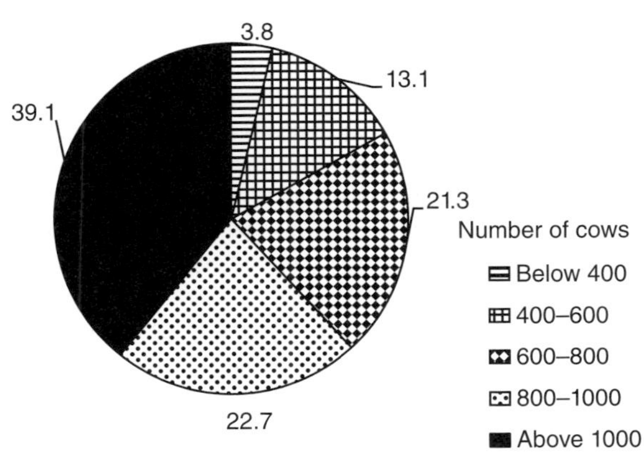

Fig. 11.9. Percentage share of milk output produced in agricultural enterprises with different herd sizes in Russia's Leningrad region, 2011. (From Committee for Agroindustrial Complex of Leningrad Region, 2009–2013.)

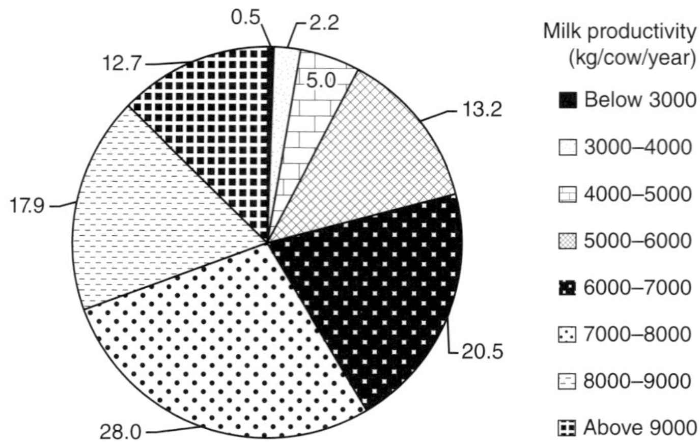

Fig. 11.10. Percentage share of agricultural enterprises (with different milk productivity) in total milk output in Russia's Leningrad region in 2011. (From Committee for Agroindustrial Complex of Leningrad Region, 2009–2013.)

Table 11.7. The number of forage harvesters (%), equipment amortization and purchases (units) at agricultural enterprises in Russia, selected years, 2000–2010. (From Agroindustrial Complex of Russia, 2001–2012.)

	Years		
Indicators	2000	2005	2010
Annual purchases of new harvesters (1000 units)	59.6	33.4	20.0
Annual purchases of new harvesters (%)	3.3	3.3	4.1
Annual amortization of old harvesters (%)	9.5	10.9	8.1

In the Leningrad region, the structure of harvested fodder has changed. In terms of nutrient value, the share of silage has grown from 33% to 66%. On Leningrad regional farms, consumption of fodder per 1 kg of milk was 0.85 fodder units in 2009 compared to 1.11 fodder units in 1990. In 2010, agricultural enterprise fodder accounted for the major share of the milk prime cost at 63%, including 36% of purchased fodder (Committee for the Agroindustrial Complex of Leningrad Region, 2009–2013). The high portion of fodder concentrate in cow rations increases the prime cost of milk, affects the cows' health, affects the cows' reproductive functions and increases costs of herd reproduction.

The period when cows are effectively used at the regional breeding farms has been reduced. The number of calves per 100 cows decreased from 76 in 2006 to 73 in 2011 (Tyurenkova and Proshina, 2013). Low reproduction indicators do not allow additional profit from selling pedigreed young stock to breeding farms and serve as obstacles to increasing livestock at regional farms.

Due to the globalization of the grain market and to adverse weather impacts (e.g. severe drought), the prices for grain and compound feed have increased. With support from agricultural companies when purchasing more modern machinery and equipment, all poultry production operations have been automated, contributing to lower labour costs.

In Russia, over the 1990–2010 period, fodder consumption per ten eggs produced fell by 30% and fodder consumption per 1 kg of broiler weight gain was reduced by 46%. Moving to advanced technologies has helped the poultry production industry to compete in the global market in terms of labour efficiency and resource intensity (Federal Service of State Statistics, various years, 1990–2013).

Laying hens producing over 300 eggs/year, accounting for 83% of total egg production in Russia, indicates a high-level intensity of poultry production. Broilers with weight gain of more than 50 g, achieving commercial weight in less than 40 days, account for 47% of total poultry meat.

Between 2000 and 2012, the average annual laying ability of each bird on a poultry farm in the Leningrad region increased from 315 eggs in 2000 to 330 eggs in 2012, and the average daily weight gain of broilers increased from 30 g to 60 g (Committee for Agroindustrial Complex of Leningrad Region, 2009–2013). The Russian poultry egg production industry is now extending its range of products and premium-quality commodities (e.g. eggs with vitamin enrichment). In 2012, the share of these products in total regional egg production was 16%.

Sufficient concentration and intensification in dairy and poultry production has become a requirement to reduce fixed costs. Increased marketed commodities will help to ensure returns on investment in labour and innovative technologies.

Conclusions

Dairy and poultry production in Russia are important sectors of agricultural production. The two sectors have adopted innovative technologies through investments and credit resources, and through the increase of production intensification and concentration. This has resulted in lower resource intensity and lower costs, improved quality of products, stronger logistics and marketing, higher production efficiency and increased output and profit. Since 2000, Russian poultry production has followed a decreasing cost development scenario, while dairy production continues to follow an increasing cost development scenario, with very different outcomes.

References

Committee for Agroindustrial Complex of Leningrad Region (2001–2012) *Statistics Digests of the Russian Federation*. Agroprogress, Moscow.

Committee for Agroindustrial Complex of Leningrad Region (2009–2013) *Characteristics of the Agroindustrial Complex of Leningrad Region in 2008–2012*. St Petersburg, Russia.

Committee for Agroindustrial Complex of Leningrad Region (2012) *Government Programme for Agricultural Development in the Leningrad Region for 2013–2020*. St Petersburg, Russia.

Federal Service of State Statistics (1990–2013) *Major Indicators of Russian Agriculture in 1990–2013*. Federal Service of State Statistics, Moscow.

Ministry of Agriculture of the Russian Federation (2001–2012) *Agroindustrial Complex of Russia in 2000; 2001; 2002; 2003; 2004; 2005; 2006; 2007; 2008; 2009; 2010; 2011*. Ministry of Agriculture of the Russian Federation, Moscow.

Ministry of Agriculture of the Russian Federation (2005) *The Development of the Agroindustrial Complex.* Ministry of Agriculture of the Russian Federation, Moscow.

Ministry of Agriculture of the Russian Federation (2007) *Government Programme for the Development of Agriculture and Regulation of Agricultural Commodities, Raw Materials, and Food Markets for 2008–2012*. Ministry of Agriculture of the Russian Federation, Moscow.

Ministry of Agriculture of the Russian Federation (2013) *Industry Programme Development of Poultry Production in the Russian Federation for 2013–2015*. Ministry of Agriculture of the Russian Federation, Moscow.

Russian Union of Dairy Enterprises (2013) *Yearbook: Dairy Industry in the World and in the Russian Federation*. Russia Union of Dairy Enterprises, Moscow.

Serova, E. and Karlova, N. (2010) *The Russian Federation: Review of the Dairy Sector*. FAO Investment Centre, Paris.

Surovtsev, V.N., Nikulina, Y.N., Ponomarev, M.A., Chastikova, E.N., Lukichev, P.M. and Belyakov, A.V. (2009) *Development of Agriculture in the Leningrad Region, Conditions of Transformation Economy*. Swedish University of Agricultural Sciences (SLU), Uppsala, Sweden.

Tyurenkova, E. and Proshina, O. (2013) Estimation of herd reproduction impact upon dairy farming efficiency using information technologies. In: Haritonov, V.D. (ed.) *Proceedings of Conference on Dairy Farming in the World and in the Russian Federation*. Milk Union of Russia, Moscow, pp. 59–62.

12 Does Agroholding Membership Increase Productivity and Efficiency in Russian Agriculture? Evidence from Agroholdings in the Belgorod Oblast

Konstantin Hahlbrock[1]* and Heinrich Hockmann[2]
[1]*VitalFields.com, Hamburg, Germany;* [2]*Leibniz Institute of Agricultural Development in Transition Economies (IAMO), Halle, Germany*

Abstract
There has been little research on the effects of group membership on the productivity and efficiency of agricultural enterprises in Russia. This chapter analyses whether farms in the Belgorod Oblast of Russia that belong to agroholdings (entities composed of groups of agricultural enterprises) perform better than do independent farms (individual agricultural enterprises that are independent of agroholdings). We calculate partial land and labour productivity, total factor productivity and technical efficiency scores for these two categories of farms: agroholdings and independent farms. The results are used to decompose total factor productivity (TFP) into a scale effect, a technological change effect and a technical efficiency effect. Total factor productivity growth and efficiency for agroholdings exceeds that for independent farms.

Acknowledgement

The authors would like to thank the DFG for financial support under project HO 1302/6-2.

Introduction

The Russian agricultural sector faces essential challenges due to the undercapitalization of its agricultural enterprises. The appearance of agroholdings may help to bridge the liquidity constraints and thus increase the productivity and efficiency of Russian agricultural enterprises. In this chapter, we consider an agroholding as a company that has control over the resources of several agricultural enterprises, and we consider an independent farm as a single corporate/commercial farm that does not belong to an agroholding.

The impact of membership in agroholdings on enterprise performance in terms of productivity and efficiency has been the source of controversy in the literature. Moreover, there has been little research on the effects of group membership in terms of the productivity and efficiency of agricultural enterprises in Russia. This gap can be explained by the fact that Russian agroholdings are a complex and new phenomenon, so empirical time-series data are often unavailable. We contribute to this discussion by analysing whether farms that belong to

*Corresponding author; e-mail: konstantin@vitalfields.com

agroholdings perform better than do independent farms. We concentrate on performance indicators (partial productivities), structural indicators (size, specialization) and capacity utilization. We investigate how these indicators vary with respect to organizational form. Moreover, we examine how group affiliation affects the production possibilities of agricultural enterprises.

This chapter provides an overview of existing studies of the performance of agroholdings in comparison to the performance of independent farms in Russia. The differences in production using firm-level data from the Belgorod Oblast are measured by structural indicators, as well as by partial land and labour productivities for the different farm types. A production function is estimated within the framework of stochastic frontier analysis. The results are used to decompose total factor productivity (TFP) into a scale effect, a technological change effect and a technical efficiency effect. These indicators are further examined to determine whether group affiliation has any effect on this.

Literature Review: Farm Performance with Respect to Group Affiliation in Russia

Previous studies on the efficiency of agricultural enterprises in Russia have focused on regional development since the dissolution of the former Soviet Union in 1991 (Liefert, 2007). These studies estimate production functions for crop and animal production fields and regions by using either stochastic frontier analysis or data envelopment analysis. The research findings usually show significant technical and allocative inefficiencies in agriculture.

Another group of empirical studies deals with differences in farm organization and the impact on production and productivity. Generally, vertical coordination in the agrofood supply chain is an important and growing phenomenon in transition countries in Europe (Swinnen *et al.*, 2005). Relatively little research compares the efficiency of agroholdings and independent farms. One major reason for this could be the lack of information regarding the membership of individual agricultural enterprises in their parent companies (agroholdings). Therefore, the consequences of the appearance of agroholdings in the agrofood sector are still not well understood.

There is evidence that agroholdings increase their investments in agricultural production technology (Hockmann *et al.*, 2005). Vertical integration in the agricultural supply chain generally brings notable investments, and this flow of capital facilitates the purchase of modern machinery and allows farms to introduce the most advanced technology (Dries *et al.*, 2009). Rylko and Jolly (2005) show that being affiliated with an agroholding increases investments in assets such as machinery, and increases access to input markets and to credit. Additionally, vertical integration brings new management abilities to the farming sector. However, that might not be the case for all oblasts. For example, by observing partial productivities and cost-effectiveness indicators, Gataulina *et al.* (2006) found that agricultural enterprises within agroholdings in the Orel Oblast worked less efficiently that did independent farms.

Empirical work on the differences between agroholdings and independent farms is rare. For the Belgorod Oblast, Hockmann and Kopsidis (2007) carried out an empirical analysis of agroholdings and independent farms, concluding that many agroholdings were less efficient than were independent farms. Efficiency scores of agroholdings calculated in a study by Hockmann *et al.* (2009) showed that, from 2001 to 2003, technological and managerial innovations introduced by agroholdings did not necessarily move agricultural production on an efficient trajectory.

The studies mentioned above did not find a holistic approach that could explain the difference in the performance indicators of agroholdings and independent farms. It has not been ascertained as to why agroholdings do not perform as well as independent farms, although agroholdings have better access to input markets and have higher input quality. It is uncertain whether this issue is related to transaction costs or whether it represents a needed challenge for Russian agriculture.

Differences between Organizational Structures in the Belgorod Oblast

In this section, we use agricultural firm data from the Rosstat Agricultural Registry provided by Golohvastov–Agroconsult. This unbalanced panel of 524 farms located in the Belgorod Oblast contains data for 2001, 2004 and 2007, and includes 210 indications covering gross production (acreage, yields, average annual livestock and gross volume of production); marketing (sales volumes and sales revenue); and cost structures of the farms. The data are supplemented with regional price indices and with the FIRA-Database (2001–2007) containing information about agroholding affiliation. In 2001, 82% of the arable land in the Belgorod Oblast was controlled by these farms, compared to 67% of the arable land in 2007. Due to the high share of arable land cultivated by the farms that belong to agroholdings, the data can be regarded as representative of the agricultural enterprises in the Oblast. The relevant indicators are the use of arable land, workers in agriculture and the share of revenue in constant prices. This allows the calculation of the following indicators: labour productivity (gross production value per worker); land productivity (gross production value in crop production per hectare of sown area); land-to-man ratio (worker and crop/100 ha of sown area); and quantity of milk per kilogram, per cow (kg/cow).

The first approach

The examination of the whole sample suggests that the agricultural enterprises increased the amount of arable land by 1000 ha from 2001 to 2007 (Fig. 12.1). The increasing confidence interval over time suggests that farms developed heterogeneously with regard to controlled land. This heterogeneity can be observed for agroholdings and for independent farms. The land controlled by agroholdings increased significantly (on average, from 3500 ha to more than 5000 ha).

On average, agroholdings employ more workers than do independent farms. This is not surprising because agroholdings cultivate more land than do independent farms. The number of workers employed by agricultural enterprises has decreased slightly, thus indicating a reduction of employment in the agricultural sector of the oblast. Surprisingly, private/smallholder farms greatly reduced their work (20%), whereas agroholdings decreased their workforce (12%).

Figure 12.2 illustrates production specialization in the oblast, private/smallholder farm and agroholding levels. At the oblast level, cattle production decreased and pork production increased. Agroholdings reduced their grain production 20% between 2001 and 2007. To a great extent, this was compensated for by increased revenue shares from cattle, pork, poultry and eggs. The extent of factory-style specialization of poultry and pork production within agroholdings increased and cattle production decreased. This result was in line with the theory of Allen and Lueck (2003), who argued that these specializations were less subject to severe transaction cost disadvantages and principal agent problems than they were to other production activities.

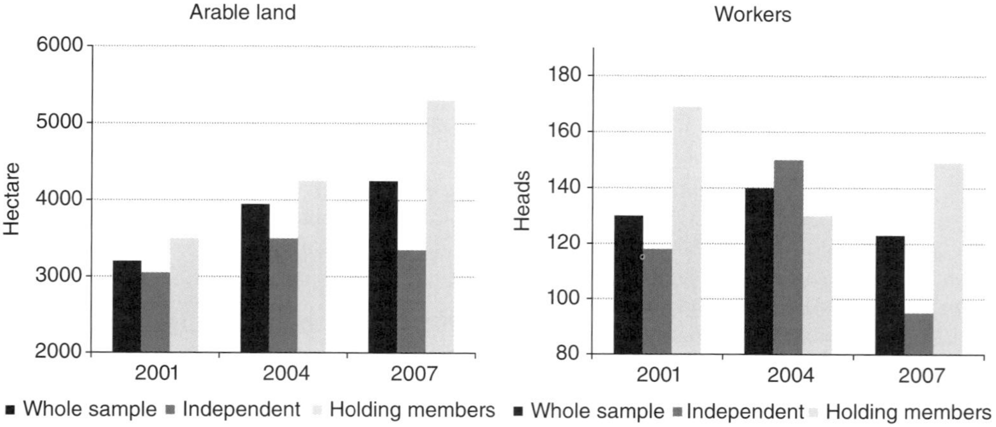

Fig. 12.1. Arable land and employed workers by organizational form in Russia, 2001, 2004 and 2007. (From authors' calculations.)

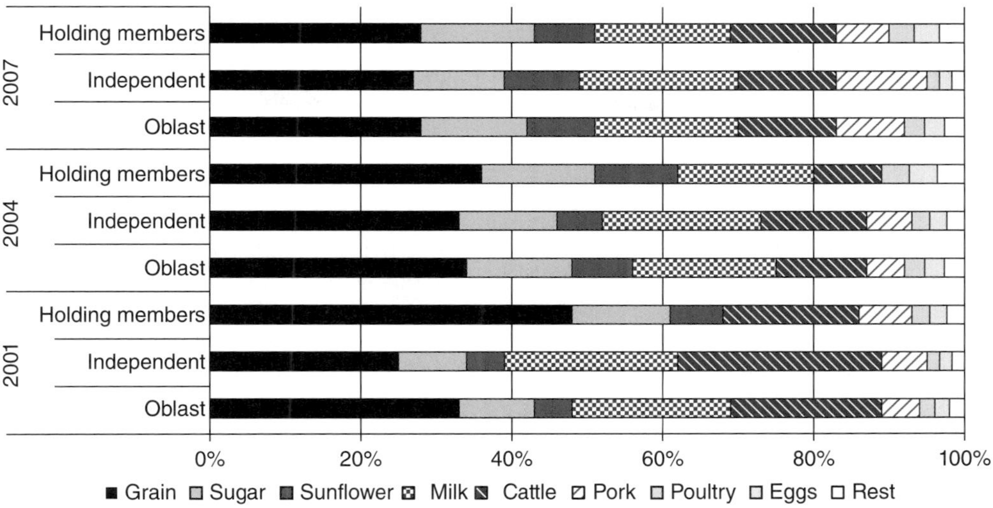

Fig. 12.2. Specialization in production by organizational form in Russia, 2001, 2004 and 2007 (measured in revenue shares in constant prices). (From authors' calculations.)

Partial productivity

Labour productivity in Russia increased greatly in 2001, 2004 and 2007. In the Belgorod Oblast, this productivity approximately doubled between 2001 and 2004, and tripled between 2004 and 2007 (Fig. 12.3). There were no significant production differences among the oblasts, private/smallholder farms and agroholdings over time. However, the wide range of productivities in the three groups implied that there were heterogeneous developments in the data, regardless of the type of organizational form.

Land productivity was rather constant over the years, with heterogeneous development within the groups examined. The partial productivity of the land-to-man ratio (Fig. 12.3) shows that farms increased their labour efficiency due to the fact that one worker was used to farm 40 ha in 2001 and that number increased to 80 ha in 2007. The falling ratio of land-to-man in agroholdings was largely responsible for this trend. Milk production per cow increased over the observed period. In all years, agroholdings were more productive than were private/smallholder farms. On average, milk production per cow for agroholdings was about 500 kg higher than it was for private/smallholder farms. Other indicators of partial productivity did not express significant differences between agroholdings and private/smallholder farms. For instance, the differences in grain yields were quite small, and changes could be explained largely by climatic conditions.

Total Factor Productivity

Partial productivity cannot present a coherent picture of the performance of agroholdings and private/smallholder farms because the impact of capital and material inputs (e.g. seed, fertilizer and feedstuff) is missing. In addition, partial productivity measures considered in isolation can provide a misleading indication of overall productivity. In this chapter, parameters that add up to TFP will illustrate the undistorted performance of the respective farms.

Estimation procedure

A stochastic frontier analysis (SFA) was conducted to estimate farm technical efficiency. We chose the double heteroscedastic model (Kumbhakar and Lovell, 2000):

$$\ln y_{it} = f(\mathbf{x}_{it}, q_{it}, h, t) + v_{it} - u_{it}$$

$$v_{it} \sim N(0, \sigma_{v,it}) \text{ with } \sigma_{v,it} = e^{\delta_o + \mathbf{x}_{it}'\delta + \delta_h h}$$

$$u_{it} \sim N^{+}(0, \sigma_{u,it}) \text{ with } \sigma_{u,it} = e^{\upsilon_o + \mathbf{z}_{it}'\upsilon + \upsilon_h h}, \tag{12.1}$$

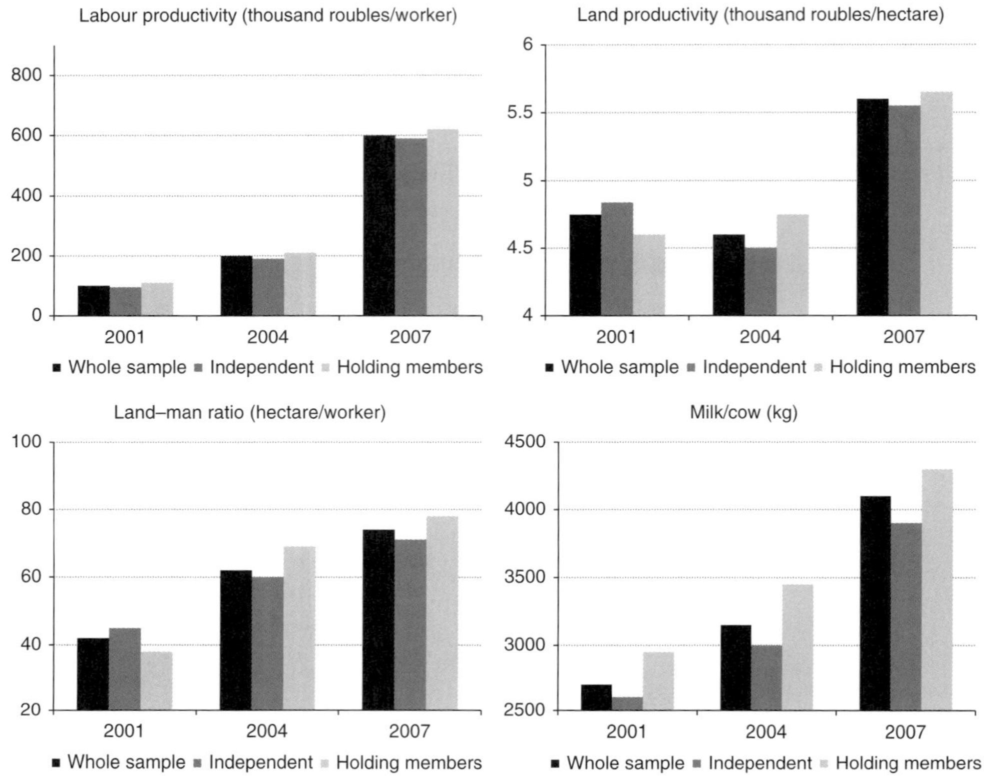

Fig. 12.3. Partial productivity by organizational form in Russia, 2001, 2004 and 2007. (From authors' calculations.)

where v and u are statistically independent. In Eqn 12.1, y_{it} is the output, $\mathbf{x}_{it}$ is the vector of inputs, t is a time trend representing technical change, h (or *Hol*) is a dummy variable which is 1 for agroholdings and zero otherwise, q_{it} is an index for factor quality and z_{it} is a vector of determinants affecting efficiency. The indices i and t represent farms and time, respectively.

This model was chosen because it allowed for the consideration as heteroscedasticity of v, which could be interpreted as the impact of factor input on risk and risk management on production. Using the power series expansion $e^{\ln y} = 1 + \ln y + \frac{1}{2!}(\ln y)^2 + \frac{1}{3!}(\ln y)^3 + ... \approx 1 + \ln y$ gives

$$\mathrm{var}(y_{it}) \approx \mathrm{var}(\ln y_{it}) = \sigma^2_{v,it} + \sigma^2_{u,it}$$

$$\frac{\partial\,\mathrm{var}(y_{it})}{\partial x_{j,it}} \approx \frac{\partial\,\mathrm{var}(\ln y_{it})}{\partial x_{j,it}} = 2\delta_j e^{\delta_o + \mathbf{x}_{it}'\boldsymbol{\delta} + \delta_h h} \gtrless 0, \qquad (12.2)$$

where the sign of Eqn 12.2 depends on whether the parameter estimate is positive or negative.

The production function $f(\ln \mathbf{x}_{it}, q_i, h, t)$ was approximated by a translog function in terms of effective inputs

$$f(\ln \mathbf{x}_{it}, q_i, h, t) = a_0 + a' \ln \mathbf{x}^*_{it} + \tfrac{1}{2} \ln \mathbf{x}^{*\prime}_{it} A \ln \mathbf{x}^*_{it}, \qquad (12.3)$$

$$\text{with } \ln \mathbf{x}^*_{it} = \ln \mathbf{x}_{it} + \boldsymbol{\tau} t + \boldsymbol{\eta} h + \boldsymbol{\theta} q_{it},$$

where $\mathbf{x}^*_{it}$ denotes a vector of effective inputs. Effective inputs are assumed to be functions of actual input x, organizational form h, technical change t and factor quality q.

Alternatively, the translog function could have been formulated using a parameter for technical change, organizational form and factor quality directly. This would have required first-order and second-order effects of the additional variables and their cross effects with the inputs to be parameterized. Expressing the function with effective inputs would have reduced the number of parameters and would have provided a more parsimonious formulation, which would have resulted in a more efficient estimation. In the empirical analysis, we used the four effective inputs above. Expressing the model in effective inputs requires the estimation of 23 parameters. The extensive form has 34 parameters, without being more informative than the form in effective inputs.

The SFA model Eqn 12.1 can be estimated using the standard SFA with some minor modifications. These concern the variance of the compound error terms that change to $\sigma_{it}^2 = \sigma_{v,it}^2 + \sigma_{u,it}^2 = (e^{\delta_o + \mathbf{x}_{it}'\delta + \delta_h h})^2 + (e^{\upsilon_o + \mathbf{z}_{it}'\upsilon + \upsilon_h h})^2$, and the skewness parameter that becomes $\lambda_{it} = \sigma_{v,it} / \sigma_{u,it} = (e^{\delta_o + \mathbf{x}_{it}'\delta + \delta_h h}) / (e^{\upsilon_o + \mathbf{z}_{it}'\upsilon + \upsilon_h h})$.

We estimated a pooled regression model. In principle, the panel structure could have been considered. However, most available panel data models assume that technology differences are considered sufficiently with differences in the constant term (a_0), which does not change over time. Alternatively, we could employ the Pitt and Lee (1981) estimator, which assumes that technology and efficiency are constant over time. Given the relatively long period, the use of a panel data estimator appears to be inappropriate. Thus, despite possible efficiency losses, we opted for a pooled estimation to allow for changes in firm-level technical efficiency and to account for the effects of group affiliation.

Data

Next, we used the same data set, but we considered only those that provided information for 2 or more years. The resulting data set consisted of an unbalanced panel of 191 agricultural enterprises, of which 126 agricultural enterprises belonged to an agroholding in 2001, 2004 and 2007. The total number of observations was 455.

The dependent variable was gross production y, an output variable that was constructed in three steps. In the first step, farm-specific producer prices for all production activities were calculated using the value and the amount of products sold. In the second step, gross production was weighted by the individual prices and summed up to total gross production value. This provided values in current prices. In the third step, the values of gross production were deflated by a Törnqvist-Theil index. The deflator was constructed from the price information obtained in the first step, along with the shares of individual products on gross production. We used the procedure proposed by Caves *et al.* (1982) to construct an index that allowed us to conduct multilateral consistent comparisons. This procedure allowed us to construct an output variable whose level and development was comparable among agricultural enterprises and over time.

We used four inputs $\boldsymbol{x}$: land *Lan*, labour *Lab*, capital *Cap* and material inputs *Mat* (Table 12.1). The land variable was comprised of hectares of arable land; labour was the number of workers in agricultural production; capital was represented by the costs of fixed assets; and material inputs was comprised of information on seed, fertilizer use, pesticide use and other variable inputs. Capital and material inputs were available only in current prices. However, we applied the deflation procedure as described for output. The only difference was that we could not use firm-level data, just regional-level data when calculating the index. This procedure provided enterprise deflators; however, the differences reflected only the firm-specific weight of the individual components in the aggregate and the firm-specific prices. Considering that we used depreciation as a proxy for the use of capital, it was highly probable that we underestimated the use of capital, especially with regard to agroholdings. This was due to the fact that leasing machinery was a common activity in agroholdings (Rylko and Jolly, 2005) that could not be calculated from the data set.

The variable land quality *Qua* included several soil indicators given by the European Commission (2012) and was calculated according to the procedure described in Schierhorn *et al.* (2011) that assessed the differences in yield potential on different locations resulting from external determinants. As such, land quality comprises a combination of soil data, climate data and the infrastructure related to arable land. The index was calculated based on square kilometres. The value of the land quality index was given as *Qua* at each firm's headquarters (Table 12.1).

The vector $\boldsymbol{z}$ consisted only of a variable measuring the availability of water at the different locations *Pet*; it was calculated following Hargreaves (2003) and was constructed from the mean temperature, the daily temperature range

Table 12.1. Parameter estimates. (From authors' estimates.)

		Variable	Parameter	Estimate	Standard error	*t*-Value
Production function parameters		Con	a_0	0.003	0.004	0.889
		Lab	a_1	0.255	0.031	8.060[c]
		Lan	a_2	0.148	0.029	5.093[c]
		Cap	a_3	0.124	0.016	7.818[c]
		Mat	a_4	0.481	0.027	17.310[c]
		Lab[b]	a_{11}	0.046	0.030	1.497
		Lan[b]	a_{22}	−0.399	0.050	−7.868[c]
		Cap[b]	a_{33}	−0.004	0.001	−2.399[c]
		Mat[b]	a_{44}	0.037	0.039	0.950
		LabLan	a_{12}	−0.009	0.069	−0.138
		LabCap	a_{13}	0.017	0.011	1.521
		LabMat	a_{14}	−0.057	0.021	−2.689[c]
		LanCap	a_{24}	0.097	0.013	7.059[c]
		LanMat	a_{24}	0.236	0.055	4.299[c]
		CapMat	a_{34}	−0.085	0.018	−4.712[c]
Effective input parameters	Time	Lab	τ_1	−0.201	0.007	−27.961[c]
		Lan	τ_2	0.128	0.026	4.822[c]
		Cap	τ_3	0.782	0.179	4.347[c]
		Mat	τ_4	−0.055	0.042	−1.286
	Group	Lab	η_1	−0.002	0.019	−0.130
		Lan	η_2	0.319	0.082	3.857[c]
		Cap	η_3	0.712	0.433	1.644[a]
		Mat	η_4	−0.363	0.102	−3.550[c]
	Land quality		θ_2	1.667	1.593	1.046
Heteroscedasticity in *v*	Con		δ_0	−2.830	0.396	−7.143[c]
	Lab		δ_1	1.149	0.717	1.601
	Lan		δ_2	−0.597	0.239	−2.496[c]
	Cap		δ_3	0.164	0.104	1.573
	Mat		δ_4	−1.372	0.335	−4.091[c]
	Hol		δ_h	0.059	0.074	0.796
Heteroscedasticity in *u*	Con		υ_0	−1.548	0.043	−35.812[c]
	Pet		υ_1	0.002	0.045	0.045
	Hol		υ_h	0.093	0.067	1.391

Notes: Lab = labour; Lan = land; Cap = capital; Mat = material inputs. [a]Significant at the 1% level; [b]significant at the 5% level; [c]significant at the 10% level.

and irradiation. *Pet* was constructed for each square kilometre using the ECA&D (2011) database. The farm-specific *Pets* were given by the *Pet* at the location of the firm's headquarters. The higher the *Pet*, the higher was the expected efficiency.

For estimation purposes, we normalized all variables by their mean (the inputs were expressed as logarithms, which implied that the geometric mean was chosen). This facilitated the interpretation of the estimation results since the first-order terms represented the production elasticities at the sample mean.

Estimation results

All first-order terms were positive and statistically significant (Table 12.1). Moreover, about 75% of the observations were consistent with the theoretical requirements (monotonically increasing in inputs, quasi-concave in inputs). The most critical input was land. Without land, the share of theoretically consistent observations increased to 95%. This result could be explained by the definition of the input: land input was approximated by the amount of arable land and not by the actual cultivated or harvested area.

This implied that our land input was overestimated, which explained the poor results regarding theoretical consistency for land.

The highest elasticity was estimated for material inputs: it amounted to about 50% of total production. The relatively high elasticity of labour (0.25) differred considerably from other studies (Bokusheva and Hockmann, 2006; Osborne and Trueblood, 2006). However, these studies employed other data sources or investigated other regions, so the results were not comparable. The sum of the four production elasticities was 1.008, suggesting that, on average, the firms were operating with constant returns to scale.

Both group membership and technical change affected production significantly. Their impact on production structures was calculated by expanding the production function (Eqn 12.3). This resulted in effects that were linear and quadratic, as well as effects that interacted with each other and interacted with the original inputs (Table 12.2).

The second column gives information about the influence of technical change. The calculations show a high rate of technical change (3.8%). Moreover, the rate is accelerating (0.005). These results correspond to the revival of the agricultural sector that started after 2000. Furthermore, technical change was material saving, but labour, land and capital using.

Table 12.2. Calculated additional parameters.[a] (From authors' calculations.)

Effect on	Effect of time	Holding	Quality
Constant	0.038	–0.594	0.246
Time	0.005	0.022	0.010
Holding	–	–	–0.239
Quality	–	–	–0.055
Labour	0.005	0.029	–0.016
Land	0.014	–0.143	–0.666
Capital	0.010	0.059	0.163
Materials	–0.027	0.001	0.394

Notes: [a] The values in the first row are given by $\mathbf{a}'\tau$, $\mathbf{a}'\eta$ and $\mathbf{a}'\theta$, respectively. The values in the second row are given by $\tau'\mathbf{A}\tau$, $\eta'\mathbf{A}\tau$ and $\theta'\mathbf{A}\tau$, the numbers in the third and fourth row are $\eta'\mathbf{A}\theta$ and $\theta'\mathbf{A}\theta$, respectively. The values given for the four inputs are given by $\tau'\mathbf{A}$, $\eta'\mathbf{A}$ and $\theta'\mathbf{A}$. The values represent percentage changes of production resulting from a change in the t, h or $\mathbf{q}$. Since Holding is a dummy variable, no squared term is shown separately; it is included in the constant.

Group affiliation had a negative impact on the original production possibilities (–0.594). This was in line with the observation that, according to the programme of the oblast government, the parent companies were required to contribute to the recovery of the agricultural sector by modernizing almost bankrupted agricultural enterprises (Wandel, 2007). The success of this strategy cannot be denied since agricultural enterprises belonging to an agroholding realized an extra technical change compared to private/smallholder farms (2.2%). Furthermore, group membership had a positive impact on labour and capital and, to a lesser extent, materials. The signs for materials and capital were reasonable, given the investment activities of agroholdings. At first glance, the sign for labour was questionable since the agroholdings closed unprofitable businesses or released unproductive labour on the agricultural enterprises (Hockmann *et al.*, 2009). However, one must consider that many agroholdings specialized in animal production and invested correspondingly in agricultural production. Since animal production is more labour-intensive than is crop production, the positive sign for labour appears reasonable.

The quality variable was not significant in the estimation, so the coefficients in the column 'Quality' in Table 12.2 should be regarded with care. The first term is the product of a_2 and θ_2, and the value of 0.246 indicates that land quality has a positive impact on production, given the high significance of a_2. Also, land quality appears to be a substitute for land (–0.666) and is complementary to the use of capital and materials.

The negative sign of land quality and group affiliation demonstrated that agricultural enterprises belonging to an agroholding had poorer land quality than did private/smallholder farms. This is in line with the observation that mainly bankrupted enterprises became part of an agroholding. In addition to economic reasons, the financial difficulties were due to conditions that prevented farms from being competitive.

Technical efficiency was estimated by the Jondow *et al.* (1982) procedure, and was shown for different business structures (Fig. 12.4). Technical efficiency is high, and there is no pronounced difference in the efficiency of agroholdings and

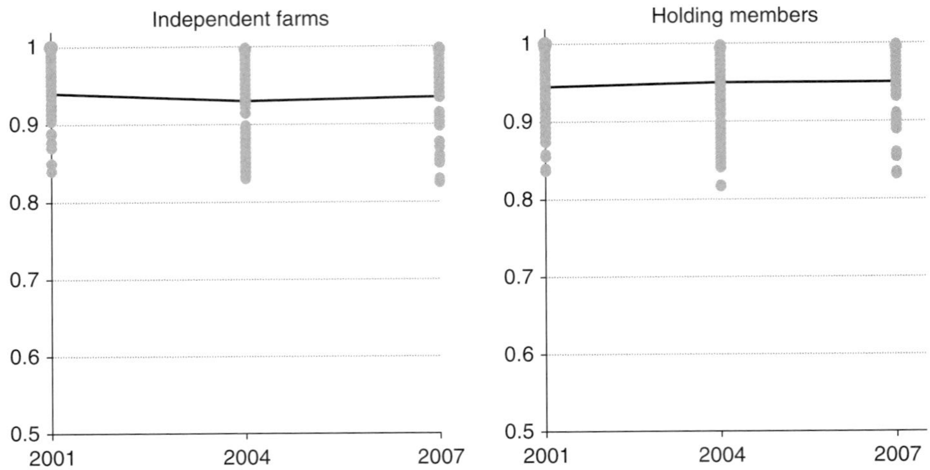

Fig. 12.4. Development of technical efficiency by organizational form in Russia, 2001, 2004 and 2007 (each dot represents one observation and the line complies with the annual mean). (From authors' calculations.)

private/smallholder farms. The relatively high efficiency scores and the small variance conflict with the efficiency scores of other studies conducted with farm-level data (Bokusheva and Hockmann, 2006). On average, given their available technology, the farms in the Belgorod Oblast produce almost on their production frontier.

Neither participation nor agroholding membership significantly impacted efficiency. Rather, the parameter for heteroscedasticity in v implied that active risk management must not be excluded since the parameters for land and capital were significantly different from zero and had a negative sign. This holds for both agroholdings and private/smallholder farms. Furthermore, the estimates showed that being a member of an agroholding had no effect on risk management practices.

Total factor productivity (TFP)

We defined TFP through the relationship of an output index and an input index. Since we represented production by a translog functional form, we applied the quadratic lemma (Diewert, 1976), and calculated a Törnqvist-Theil index for inputs. Because this index provided intransitive results in multilateral comparisons, we used a modified version introduced by Caves *et al.* (1982). This index was exact for the translog function and also allowed for consistent comparisons among individuals and over time

$$\ln(I) = \frac{1}{2}\sum_{j\neq 1}\left[\left(s_j+\bar{s}_j\right)\ln\frac{x^*_{j,it}}{\bar{x}^*_{j,it}} - \bar{s}_j\overline{\ln x^*_{j,it}} + s_j\overline{\ln x^*_{j,it}}\right]$$
$$\text{with } s_j = \frac{f(\ln \mathbf{x},\, h,\, t,\, q)}{\partial \ln x^*_{j,it}}. \tag{12.4}$$

Using Eqn 12.2, the index in Eqn 12.4 can be decomposed into an input, technical change, organizational form and a quality index. To calculate total productivity, the input index was taken with the assumption of a constant return to scale (Orea, 2002)

$$\ln\left(I^{Inp}\right) = \frac{1}{2}\sum_{j\neq 1}\left[\left(s_j+\bar{s}_j\right)\ln\frac{x_{j,it}}{\bar{x}_{j,it}} - \bar{s}_j\overline{\ln \bar{x}_{j,it}} + s_j\overline{\ln \bar{x}_{j,it}}\right], \tag{12.5}$$

with

$$s_j = \frac{f(\ln \mathbf{x},\, h,\, t,\, q)}{\partial \ln x^*_{j,it}}\left(\sum_h \frac{f(\ln \mathbf{x},\, h,\, t,\, q)}{\partial \ln x^*_{j,it}}\right)^{-1}$$

Since we only have one output, the index given by $\ln\left(I^{Out}\right) = \ln y_{it} - \ln \bar{y}_{it}$. Correspondingly, TFP is defined as $\ln(TFP) = \ln\left(I^{Out}\right) - \ln\left(I^{Inp}\right)$.

This indicator captures all the effects on production that result from exploiting economies of scale (SE), technical change (TC), quality (TQ), group membership (TG), and efficiency (TE) (e.g. TFP = SE*TC*TQ*TG*TE) (Coelli *et al.*, 2005). Since the data showed that: (i) almost constant returns to scale were present; (ii) agroholdings and private/smallholder farms had almost the same efficiency levels; and (iii) the estimate for quality was not significant, we restricted our analysis to the impact of technical change when discussing TFP development.

In 2001 and 2004, private/smallholder farms appeared to be better equipped from the technological point of view. In contrast, agroholdings were better equipped in 2007 (Fig. 12.5). These developments corresponded to the parameter estimates of the production function (Table 12.1). In the description of those results, the special role of technical change in agroholdings was already highlighted.

It can be concluded that agroholdings adopt technological changes faster and more intensely than do private/smallholder farms. Furthermore, this illustrates that agroholdings are defining the regional production frontier. This conclusion, at first glance, contradicts Hockmann *et al.* (2009), who have found that agroholdings are lagging behind private/smallholder farms. However, their data set captured developments from 1999 to 2003, a period for which we were unable to identify a superior role for agroholdings. Moreover, agroholdings and private/smallholder farms appear to have been similarly efficient and productive.

In addition, given the constant variation of technical efficiency among members of agroholdings and private/smallholder farms, it can be concluded that the positive developments detected for some members of an agroholding in general cannot be observed for all members. On the contrary, the developments detected from our data suggested that agroholding members evolved in diverse ways, as did other agricultural enterprises. Basically, the agroholdings benefited by including innovative farms that had supreme performance indicators. Thus, being a member of an agroholding was not a guarantee for success per se, but it allowed for a better exploitation of growth potential.

Conclusions

Group affiliation has an influence on both the structure and the performance of a farm. Agroholdings tend to focus on factory-style production units and private/smallholder farms focus on seasonal production. By looking at the partial productivities, agroholdings perform better overall, but the results are not straightforward or comprehensive. For the components of TFP, the findings from Hockmann *et al.* (2009) were confirmed. The result was twofold. On the one hand, the conclusion was verified that the technological and managerial innovations introduced by agroholdings did not necessarily increase the efficiency of farms belonging to agroholdings; on the other hand, we showed that, due to technical change, agroholdings were the driving force of the production frontier's shift.

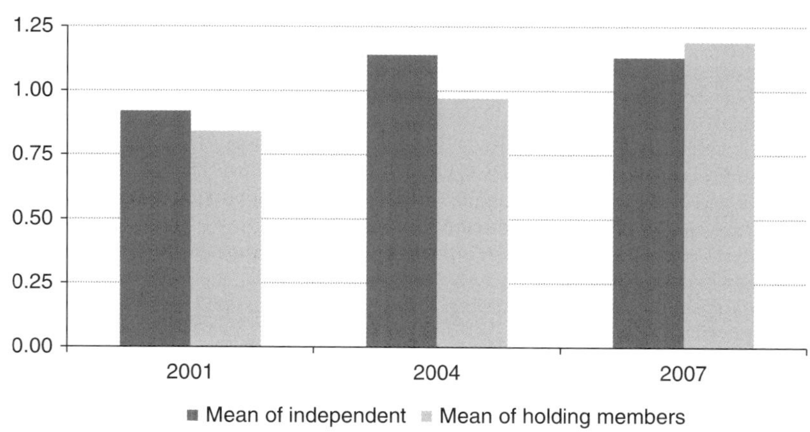

Fig. 12.5. TFP development by organizational form in Russia, 2001, 2004 and 2007. (From authors' calculations.)

Starting with a lower TFP level in 2001, the agroholdings outperformed the private/smallholder farms in the years leading up to 2007. This is true for some agroholdings, but not for all. Therefore, future research is needed on both efficiency and productivity within the agroholdings as a group.

References

Allen, D.W. and Lueck, D. (2003) *The Nature of the Farm: Contracts, Risk, and Organization in Agriculture*. MIT Press, Cambridge, Massachusetts.

Bokusheva, R. and Hockmann, H. (2006) Production risk and technical inefficiency in Russian agriculture. *European Review of Agricultural Economics* 33(1), 93–118.

Caves, D.W., Christensen, L.R. and Diewert, W.E. (1982) Multilateral comparisons of output, input, and productivity using superlative index numbers. *The Economic Journal* 92(1), 73–86.

Coelli, T.J., Rao, P.D.S., O'Donnell, C.J. and Battese, G.E. (2005) *An Introduction to Efficiency and Productivity Analysis*, 2nd edn. Springer, New York.

Diewert, W.E. (1976) Exact and superlative index numbers. *Journal of Econometrics* 4(2), 115–145.

Dries, L., Germenji, E., Germenji, E., Noev, N. and Swinnen, J.F.M. (2009) Farmers, vertical coordination, and the restructuring of dairy supply chains in Central and Eastern Europe. *World Development* 37(11), 1742–1758.

ECA&D (2011) E-Obs gridded dataset. European Climate Assessment and Dataset (ECA&D), The Netherlands. Available at: http://eca.knmi.nl/download/ensembles/ensembles.php (accessed 30 November 2011).

European Commission (2012) European Soil Database (ESDB). Available at: http://eusoils.jrc.ec.europa.eu/ESDB_Archive/ESDB/index.htm (accessed 3 September 2012).

FIRA-Database (2001–2007) FIRA Database. Available at: http://www.fira.co.uk/standards-plus-database (accessed 3 September 2012).

Gataulina, E.A., Uzun, V.Y., Petrikov, A.V. and Yanbykh, R.G. (2006) Vertical integration in an agroindustrial complex: agrofirms and agroholdings in Russia. In: Swinnen, J. (ed.) *The Dynamics of Vertical Coordination in Agrifood Chains in Eastern Europe and Central Asia: Case Studies*. World Bank, Washington, DC, pp. 45–71.

Hargreaves,G. (2003) History and evaluation of Hargreaves evapotranspiration equation. *Journal of Irrigation and Drainage Engineering* 129, S-53.

Hockmann, H. and Kopsidis, M. (2007) What kind of technological change for Russian agriculture? The transition crisis of 1991–2005 from the induced innovation theory perspective. *Post-Communist Economies* 1(19), 35–52.

Hockmann, H., Wandel, J. and Nedoborovskyy, A. (2005) Agroholdings in Russia: breaking the vicious circle. Paper prepared for presentation at the 94th EAAE Seminar, Ashford, UK.

Hockmann, H., Bokusheva, R. and Bezlepkina, I. (2009) Agroholding membership: does it make a difference in performance? *Quarterly Journal of International Agriculture* 48, 25–46.

Jondow, J., Lovell, C.A.K., Materov, I.S. and Schmidt, P. (1982) On the estimation of technical inefficiency in the stochastic frontier production function model. *Journal of Econometrics* 19, 233–238.

Kumbhakar, S. and Lovell, C.A.K (2000) *Stochastic Frontier Analysis*. Cambridge University Press, Cambridge, Massachusetts.

Liefert, W. (2007) Measuring the success of agricultural transition: an application to Russia. Paper prepared for presentation at the 104th EAAE-IAAE Seminar, Budapest.

Orea, L. (2002) Parametric decomposition of a generalized Malmquist productivity index. *Journal of Productivity Analysis* 18, 5–22.

Osborne, S. and Trueblood, M.A. (2006) An examination of economic efficiency of Russian crop production in the reform period. *Agricultural Economics* 34, 25–38.

Pitt, M.M. and Lee, L.-F. (1981) The measurement and sources of technical inefficiencies in the Indonesian weaving industry. *Journal of Development Economics* 9, 43–64.

Rylko, D. and Jolly, R.W. (2005) Russia's new agricultural operators. Their emergence, growth and impact. *Comparative Economic Studies* 47(1), 115–126.

Schierhorn, F., Hahlbrock, K. and Müller, D. (2011) The agricultural potential of European Russia. *IAMO 2011 (English Version)*, 13, S.11–27.

Swinnen, J.F.M., Berkum, S.V. and White, J. (2005) *The Dynamics of Vertical Coordination in Agrifood Chains in Eastern Europe and Central Asia, Implications for Policy and World Bank Operations*. World Bank, Washington, DC.

Wandel, J. (2007) Integrated structures, market forces and competition in Russia's agro-food sector: an assessment from the perspective of the Austrian School of Economics. Paper prepared for presentation at the 102th EAAE Seminar, Moscow, Russia.

13 Large-Scale Dairy and Poultry Production in Russia: Efficiency and External Environment

Vladimir N. Surovtsev, Evgeny Schedrin, Mikhail Ponomarev, Elena Chastikova and Uliana Saigusheva*
North-West Institute of Economics and Organization of Agriculture, St Petersburg, Russia

Abstract
This chapter analyses the production efficiency and the reasons behind the different development pace of dairy and poultry production in Russia. It also describes the reasons and consequences of growth in these two industries and the transition to addressing the environmental problems of these leading agricultural sectors. This chapter evaluates the development prospects for Russian dairy and poultry production in the global market based on rates of production concentration and intensification, the adoption of innovative production technologies, marketing and management and government support of the agrarian sector.

Cost-Efficiency of Dairy and Poultry Production

The cost-efficiency of dairy and poultry production is dependent on price fluctuations in the global market, seasonal fluctuations in the domestic market, changes in consumer preferences and government legislation. Domestic producers are becoming more dependent on fluctuations in prices of inputs for grain, breeding stock, technologies and fuel markets.

Owing to improvements in milk quality and reduced seasonality in dairy production (the main factors that determine dairy market prices and price fluctuations), milk selling prices (roubles per kilogram; RUB/kg) have increased much faster than milk prime costs in Russia. The profit margin on milk was 15.2% in 2011 (Table 13.1).

In Russia, there is a high level of milk purchase price differentiation throughout the country, and tough competition between domestic and foreign milk processing companies in the milk market. For example, in 2011, the raw purchase price for 1 tonne (t) of milk in the Central Federal District ranged between 12,400 RUB/t and 16,600 RUB/t; in the Siberian Federal District, it ranged between 13,000 RUB/t and 23,700 RUB/t; in the Volga Federal District, it ranged between 12,300 RUB/t and 14,500 RUB/t (Federal Service of State Statistics, 2003–2008). In addition, milk grading in Russia is subject to regulation under the federal law for technical regulations for milk and dairy products (Table 13.2).

Deviations between the producer price index and the consumer price index for milk are insignificant throughout the year (3%–5%), so fluctuations

*Corresponding author; e-mail: skarapejas@yandex.ru

Table 13.1. Cost-efficiency of milk production in Russian agricultural enterprises, RUB/kg and profit margin (%), selected years, 2005–2011. (From Federal Service of State Statistics, selected years, 2005, 2008, 2011.)

Indicators	Years		
	2005	2008	2011
Prime cost of products sold (including industrial processing), RUB/kg	5.97	9.45	12.88
Selling price, RUB/kg	6.78	11.15	14.84
Profit margin on products sold (including industrial processing) subsidy not included (%)	13.4	17.9	15.2

Table 13.2. Quality requirements for unpasteurized milk in Russia (GOST) and regulatory level for microorganisms and somatic cells in unpasteurized milk, 2003 and 2008. (From Federal Service of State Statistics, 2003, 2008.)

Quality indicators, unpasteurized milk	GOST 2003			Technical regulations of 2008		
	Premium quality	First quality	Second quality	Premium quality	First quality	Second quality
Somatic cells, cells/ml	$5*10^5$	$1*10^6$	$1*10^6$	$4*10^5$	$1*10^6$	$1*10^6$
Bacterial count, CFU/ml	$3*10^5$	$5*10^5$	$4*10^6$	$1*10^5$	$5*10^5$	$4*10^6$

in the purchase price of milk in Russia result from milk seasonality (Fig. 13.1).

Uneven calving and poorly organized year-round feeding lead to excessive milk production in summer and underproduction of milk in winter and autumn (Fig. 13.2). This also results in a shortage of the raw milk supply to the milk processing industry. In the Leningrad region, improvements in the management system, in conjunction with technical modernization of production, have helped to solve the issue of uneven milk yields (Fig. 13.3).

Traditionally, the calculation of milk production profitability in Russia does not take into account the costs for herd reproduction. These costs are charged to meat production. This results in increasing the profitability of milk production and increasing the losses of cattle meat sales. Thus, a more objective picture is obtained if we look at the profitability level of the industry in general, rather than milk and meat production individually.

In 2011/12, the average level of milk production profitability in the farming enterprises of the Leningrad region did not exceed 20%. The profitability of the industry in general (losses from selling cattle meat were taken into account) was close to zero, making it impossible for milk producers to secure herd reproduction using their own funds. Thus, government subsidy payments were needed to ensure technological modernization of the industry.

Low profitability of the industry is one of the key reasons why growth rates of total output remain low despite the industry's significant potential for development in both Russia and the Leningrad region. Even with increased livestock on highly efficient farms and massive budget support targeted to increase the herd throughout the region, dairy farming remains a low-income business in both Russia and the Leningrad region (Table 13.3).

Even with high rates of technological development, Russian poultry farming is observing a decline in cost effectiveness of production (Table 13.4). The two main reasons are: (i) insufficient growth rates of selling price resulting from increased competition, the market power of large retail chains and the slow growth of real income in Russia; and (ii) high growth rates of prices for inputs (such as breeding stock, electricity, feedstuff) connected with increased grain exports.

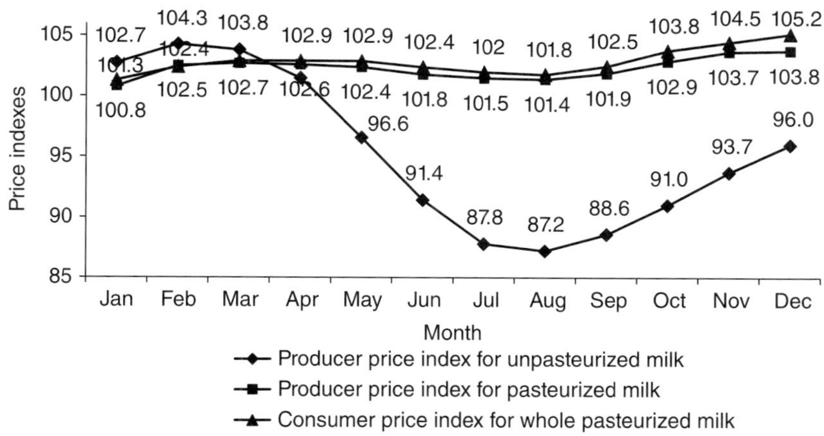

Fig. 13.1. Producer price index for unpasteurized and pasteurized milk, and the consumer price index for whole pasteurized milk in Russia, 2011. (From Federal Service of State Statistics, 2011.)

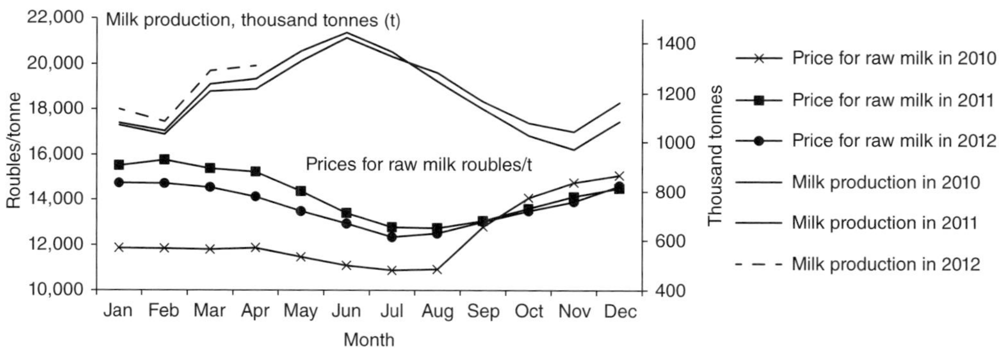

Fig. 13.2. Fluctuations in price for unpasteurized milk due to milk seasonality in Russia (RUB/t), 2010–2012. (From Federal Service of State Statistics, 2010–2012.)

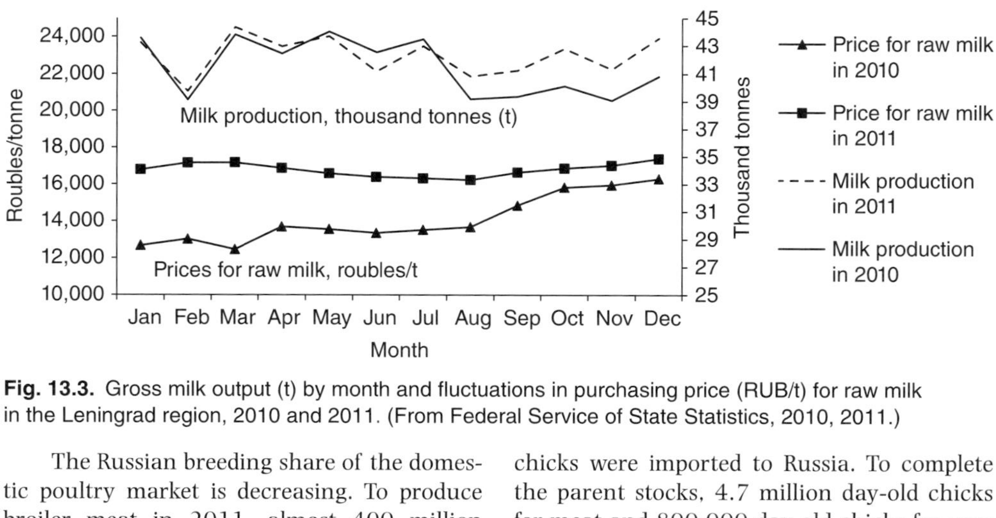

Fig. 13.3. Gross milk output (t) by month and fluctuations in purchasing price (RUB/t) for raw milk in the Leningrad region, 2010 and 2011. (From Federal Service of State Statistics, 2010, 2011.)

The Russian breeding share of the domestic poultry market is decreasing. To produce broiler meat in 2011, almost 400 million hatching eggs and 10 million hybrid day-old chicks were imported to Russia. To complete the parent stocks, 4.7 million day-old chicks for meat and 800,000 day-old chicks for eggs were imported (Ministry of Agriculture of the

Table 13.3. Percentage of profitability of dairy farming in Russia's Leningrad region (subsidies not included), 2008–2012. (From Federal Service of State Statistics, 2008–2012.)

	Years				
	2008	2009	2010	2011	2012
Milk production profitability (%)	26.7	14.8	17.2	15.7	14.3
Cattle meat (%)	–42.0	–41.3	–47.5	–44.7	–37.9
Total production of the industry (%)	7.2	–1.7	0.3	4.4	4.0

Table 13.4. Cost-efficiency of poultry meat production in Russia, RUB/kg, 2007–2011. (From Federal Service of State Statistics, 2007–2011.).

	Years				
	2007	2008	2009	2010	2011
Prime cost of products sold (RUB/kg)	39.52	46.96	47.66	49.04	57.21
Selling price (RUB/kg)	44.48	50.17	56.01	55.04	58.06
Profitability level of products sold, excluding subsidies (%)	12.5	6.8	17.5	12.2	10.2

Russian Federation, 2012). While the introduction of foreign poultry breeds and cross-breeds secured a significant growth in domestic poultry production, the breeding and raising of foreign poultry require veterinary preparations and feed additives of foreign origin, which increase prices. This contributes to a decrease in the industry's profit.

In 2007–2011, in the Leningrad region, the growth rates of the selling prices for poultry meat and eggs were lower than were the growth rates of the prime cost of production. In 2011, the cost-effectiveness of poultry meat production decreased by 2% and egg production decreased by 10.4% compared to 2007 (Table 13.5). Even so, the profitability of poultry meat production is much higher than that of beef and pork production, due to better production and management effectiveness.

The cost-efficiency of the livestock breeding industry is significantly dependent on compound feed prices. In 2012, compound feed production stood at 1.6 million tonnes (Mt). Producers with their own compound feed production facilities are more flexible and are better able to smooth price fluctuations in the grain and compound feed markets. In Russia, fodder produced by poultry farms comprises 70% of their total fodder requirements, securing improved fodder quality and reducing its price by 25%–30%.

Environmental Issues of Dairy and Poultry Production

Before 1990, the issues of environmental protection in Russia were addressed only by large-scale industrial poultry farms and pig complexes with large numbers of animals. Within this setting, traditional production technologies in dairy and crop production included a treatment system for solid manure that was a rich source of fertilizer but also a potentially dangerous environmental polluter when treated improperly. Traditional technologies involved high labour and energy costs to produce organic fertilizer (compost) and required expensive moisture-absorbing materials, such as peat.

Since 2000, there has been a push to increase agricultural output in Russia. Thus, agricultural producers have had to use available financial resources to modernize production and to improve the infrastructure instead of addressing production environmental safety (North-West Research Institute of Engineering and Electrification of Agriculture, 2012).

Table 13.5. Economic indicators of the profit (million RUB) and profitability (%) of poultry meat and egg production in Russia's Leningrad region, 2007–2012. (From Federal Service of State Statistics, 2007–2012.)

	Years					
	2007	2008	2009	2010	2011	2012
Profit from meat sales (million RUB)	784	527	1,506	1,392	1,152	4,346
Profitability of poultry meat (%)	17.1	8.1	23.2	19.8	12.4	32.8
Profit from eggs sales (million RUB)	614	934	791	843	632	868
Profitability level of egg production (%)	19.8	22.9	17.2	17.0	10.7	12.7

While the concentration and intensification of cattle and poultry production have contributed to the declining direct costs per production unit, they have improved the quality of products and enhanced logistics and marketing efficiency. However, they have also increased the environmental risks and the environmental load.

Addressing environmental issues has been difficult for many poultry producers due to the lack of efficient organic waste treatment technologies to process large volumes of organic waste into concentrated fertilizer or safe waste.

According to some estimates, the total amount of waste from livestock breeding enterprises and poultry farms in Russia is about 350 million m^3/year, with about 2 million hectares (Mha) of land being used for manure storage. At that, not more than 30% of manure is usually used as organic fertilizer (Table 13.6), which results in low application rates of organic fertilizers to crops (Table 13.7).

Industrial technologies in the livestock breeding industry have improved the share of liquid manure systems (keeping animals with no bedding), which comprised 60% of total cattle and poultry manure in 2011. According to statistical data, over 450,000 t of manure and poultry litter is produced every day in Russia, with almost one-half of it unused. This has caused the regions where industrial livestock breeding companies and poultry farms are located to become environmentally unfriendly regions.

To ensure environmental safety in agricultural areas in Russia, legislative standards have been introduced to regulate cattle and poultry manure management. Fines for the negative impact on the environment have been adopted by the federal law on the environmental protection for emissions of hazardous substances within the statutory limits, emissions of hazardous substances exceeding the limits or not permitted and waste disposal.

Table 13.6. Cattle and poultry manure application in Russia and the Leningrad region (Mt), 2006 and 2010. (From North-West Research Institute of Engineering and Electrification of Agriculture, 2012.)

	Produced	Applied	
Years	Million tonnes	Million tonnes	Per cent
Russia			
2006	300.0	47.8	15.9
2010	325.0	55.0	17.0
Leningrad region			
2006	2.85	1.17	41.1
2010	3.5	1.25	37.1

While many technologies for processing manure into organic fertilizer have been developed, none of these is functioning on the poultry farms in the Leningrad region, where the four largest poultry production enterprises account for 70% of total poultry manure produced per year in the region, or 780,000 t of the total 1.1 Mt (North-West Institute of Economics and Organization of Agriculture of the Russian Academy of Agricultural Sciences, 2012). The main constraining factors when introducing and developing manure management technologies in the Leningrad region are the limited opportunities to sell the end product, insufficient government incentives to change current practices, minimum penalties/fines and substantial subsidy payments for using mineral fertilizers in crop production.

Agroholdings are becoming popular in dairy production in Russia, wherein these commercial structures acquire or merge with agricultural

Table 13.7. Application of organic fertilizers before sowing at agricultural enterprises in Russia (Mt), selected years, 2005–2012. (From Federal Service of State Statistics selected years, 2005–2012.)

	Years			
	2005	2008	2011	2012
Applied organic fertilizers	Million tonnes			
Total	50.4	51.6	53.1	54.4
Agricultural crops (includes)	49.9	51.3	52.6	54.2
Grain and grain legume (maize excluded)	30.2	29.4	26.7	26.4
Forage crops, total	13.5	13.9	16.1	15.9
Per 1 ha of planting	Tonnes			
Agricultural crops (includes)	0.9	1.0	1.0	1.1
Grain and grain legume (maize excluded)	0.9	0.9	1.0	1.0
Forage crops, total	0.7	0.9	1.1	1.2
Area, where organic fertilizers are used, in percentage to the total planted area	3.4	6.2	7.3	7.6

enterprises to concentrate dairy production on megafarms. While this enables agroholdings to minimize costs, the resulting intensification and concentration have a negative environmental impact by increasing the environmental load and the number of point sources of pollution.

The majority of large dairy production agroholdings in the Leningrad region are moving to free-stall barns and milking parlours. These technologies create liquid manure (high-moisture manure) and greater volumes of organic wastes. The older concrete manure storage sites designed to store solid manure are inadequate for storing organic wastes.

Liquid manure fertilizers require the newer technologies for proper storage, transportation and application to the land. The older storage facilities do not meet environmental standards to protect against wastewater pollution leaching into groundwater, surface water and land reclamation canals. Because of these conditions, the procedures for organic fertilizer storage, transportation and application are often violated. These factors impact the environment negatively, leading to noticeable losses in agricultural output.

Innovative technologies for high-moisture manure storage and application are needed to meet the tough international environmental safety requirements. These technologies improve soil fertility and production and need more investment. Over 2009–2012, more than RUB 800 million were invested in agroindustrial environmental projects in the Leningrad region; as a result, facilities for processing 450,000 m^3 of cattle breeding waste per year have been built (Committee for Agroindustrial Complex of Leningrad Region, 2012).

In 2013, the Leningrad region promoted the use of organic waste from dairy and poultry production through subsidized credit interest rates and fuel prices for organic fertilizer storage facilities and application equipment.

The agricultural enterprises in the Leningrad region are converting to modern storage systems that can handle high liquid manure moisture contents of 93%–95%, which is necessary for sanitary dairy production. These include both in-ground and aboveground storage, either trench or round, that have a protective coating inside the tanks, which are equipped with manure mixing and discharging devices.

The efficiency of manure fertilizer application depends on several factors, including the technology, application rate, type of soil and distance from the storage facility to the field. The topographical peculiarities (e.g. irregular-shaped fields, farmlands scattered over vast territories and the prevalence of perennial grass) of the north-western and forest areas of Russia greatly increase the transportation distance to the manure fertilizer application areas, thereby contributing to rising fuel and labour costs.

Environmentally safe waste treatment systems require effort by both the agricultural enterprises and the government to ensure food safety. The government programme for 2013–2020, *Development of Agriculture and Regulation of Agricultural Commodities, Raw Materials and Food Markets*, issued ecological and biological initiatives based on innovative technologies in agriculture to increase food product safety (Ministry of Agriculture of the Russian Federation, 2012).

Market Development Prospects for Dairy and Poultry Production

The import market for dairy products is substantial in Russia. In 2011, milk sales of all agricultural enterprises in Russia were 19.2 Mt. Imports of milk and dairy products totalled 7.9 Mt in 2011 and 8.2 Mt in 2012 (Table 13.8).

The largest supplier to the Russian dairy market is Belarus. In 2011, its share of total dairy imports for dairy butter was 39%, cheese and curd was 30%, dry and condensed milk was 83% and whole milk was 87%. Producers from Belarus have the advantage in dairy imports due to established trading agreements between Russia, Belarus and Kazakhstan, as well as through large government subsidies to dairy producers in Belarus (Russian Union of Dairy Enterprises, 2013).

In 1990, the average level of per capita milk/dairy product consumption in Russia was 380 kg/person/year. By 2010, average per capita milk/dairy product consumption decreased to 250 kg/person/year, which was lower than the per capita consumption levels in pre-reform Russia and the developed countries of the world (Fig. 13.4).

Limited consumer demand has been one of the obstacles impeding the further development of the Russian dairy market (Ministry of Agriculture of the Russian Federation, 2008). Milk production in the Leningrad region satisfies only 27% of actual consumer demand in the region (Table 13.9). More than 70% of the milk and dairy products consumed are imported to the Leningrad region from other regions; this includes dry milk and dairy products with a long shelf life (Committee for Agroindustrial Complex of Leningrad Region, 2012).

Traditional types of dairy products, such as pasteurized milk, sour cream and curds, and buttermilk, are the major sellers in the Russian dairy market. Growing consumer demand for other dairy beverages, such as milkshakes and yogurt drinks, has increased the variety and output volume of dairy products and has increased the requirements for raw milk quality.

High growth rates of poultry production in 2012 allowed Russia to increase the share of Russian output in total poultry production resources up to 95%. In 2011, average per capita egg consumption in Russia was 273 eggs/person/year, which exceeded the recommended per capita dietary recommendation of 260 eggs (Fig. 13.5).

The share of imports in total meat and meat product resources (carry-over stocks are not included) dropped to 23.6% in 2012, which was 12% less than that in 2005 (Table 13.10). These results are reflected in the increased production of poultry and pork and in the decreased production of beef.

Table 13.8. Resources and use of milk and dairy products in Russia, selected years, 1990–2012. (From Russian Union of Dairy Enterprises, 2013.)

	Years						
	1990	1996	2000	2005	2010	2011	2012
Milk and dairy product resources							
Production (1,000 t)	55,716	35,819	32,259	30,826	31,847	31,646	31,917
Imports (1,000 t)	8,043	4,530	4,718	7,115	8,159	7,938	8,165
Share of Russian production in milk/dairy products (%)	82.9	83.7	84.2	77.8	76.1	76.3	75.9
Import share (%)	12.0	10.6	12.3	18.0	19.5	19.2	19.4
Milk and dairy product uses							
Exports (1,000 t)	335	490	507	493	460	614	593
Household consumption (1,000 t)	57,233	34,236	31,317	33,250	35,237	35,189	35,448
Self-sufficiency, milk/dairy products (%)	86.2	87.5	88.3	82.5	80.5	81.5	80.8

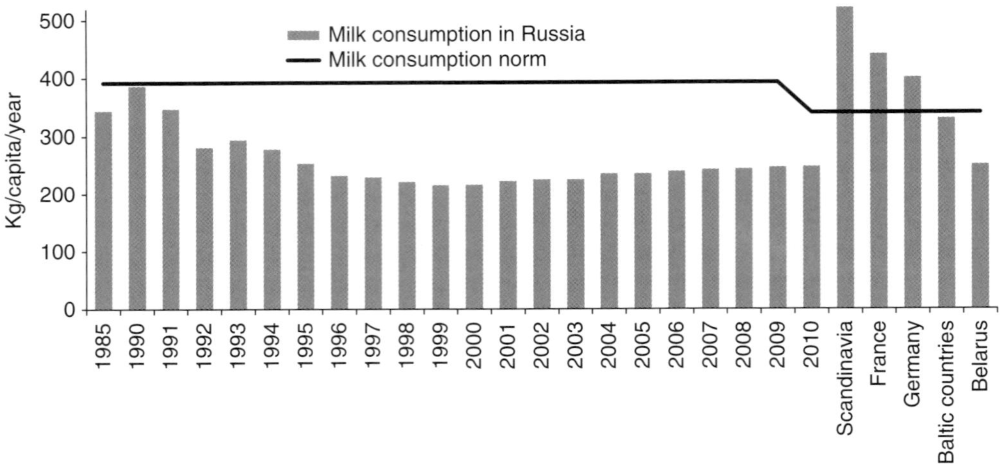

Fig. 13.4. Milk/dairy product consumption (kg/capita/year) in Russia, 1985–2010, and in other countries (Scandinavian countries, France, Germany, Baltic countries and Belarus), 2010. (From Federal Service of State Statistics, 1985–2010.)

Table 13.9. Per capita milk production and consumption in the Leningrad region (kg/person), 2006–2011, and 2011 to 2006 per cent share. (From Committee for Agroindustrial Complex of Leningrad Region, 2012.)

	Years						2011 to 2006, per cent share
	2006	2007	2008	2009	2010	2011	
Milk production (kg/person)							
Leningrad region and St Petersburg	91	89	90	90	83	84	91.6
Milk consumption (kg/person)							
Leningrad region and St Petersburg	287	292	288	304	308	310	108.0

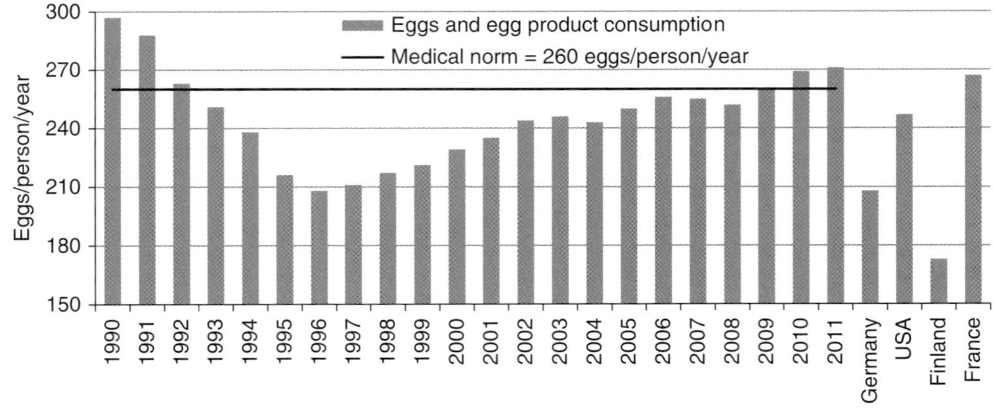

Fig. 13.5. Eggs and egg product consumpton in Russia, 1990–2011, and in other countries (Germany, USA, Finland and France), 2011. (From Federal Service of State Statistics, 1985–2011.)

Despite the growing self-sufficiency in poultry meat, per capita consumption of 25 kg in 2011 still remains lower than the recommended dietary standard of 30 kg per capita (Fig. 13.6).

The share of domestic production on the poultry meat market increased to 83% in 2010, compared with 39% in 2001 (Table 13.11).

Table 13.10. Percentage level of self-sufficiency in meat production (beef, pork and poultry) in Russia, selected years, 1990–2012. (From Federal Service of State Statistics, selected years, 1990–2012.)

	Years						
	1990	1996	2000	2005	2010	2011	2012
	Per cent						
Share in meat/meat products	80.0	65.0	63.0	57.0	66.0	68.0	70.0
Share in beef/pork/poultry	87.0	70.0	67.0	63.0	72.0	74.0	76.0
Share in poultry	97.6	35.5	52.7	50.9	81.3	88.1	95.0

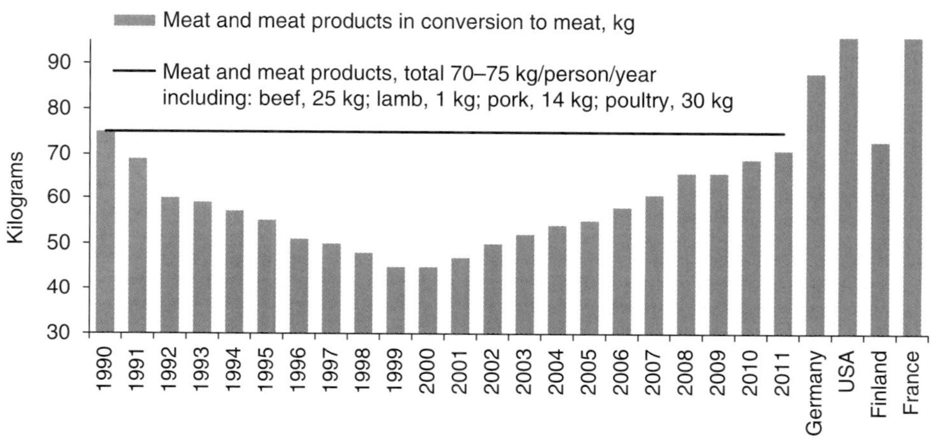

Fig. 13.6. Meat and meat product consumption (kg/capita/year) in Russia, 1990–2011, and in other countries (Germany, USA, Finland and France), 2011. (From Federal Service of State Statistics, 1990–2011.)

Table 13.11. Poultry meat resources in Russia, 1000 t, selected years, 1990–2010. (From Federal Service of State Statistics, selected years, 1990–2010.)

	1990	2001	2005	2010
	Thousand tonnes			
Domestic production carcass weight	1801	883.7	1379.8	2860
Imports, carcass weight	44.1	1390.7	1328.8	640
Total output, Russian market, carcass weight	1845.1	2274.4	2708.6	3500
Share of imports (%)	2.4	61.1	49.0	17.0
Per capita poultry meat consumption	Kilograms			
Total	12.4	15.3	18.8	24.5
Includes domestic production	12.0	5.8	9.6	20.4

Growing rates of import replacement contribute to the development of the industry's export possibilities. Rising competition forces agricultural producers to extend the variety of products and offers a wide price range to satisfy all consumer groups. Large poultry and egg enterprises were the first to use advanced processing. In 2012, 40% of the total poultry meat production was marketed as carcasses, with 53% of the total carcasses marketed as chilled meat and 40% marketed as processed food. Also in 2012,

9.8% of egg production was marketed as processed egg products, with 4% as liquid pasteurized egg products in aseptic packaging, 4.3% as dry egg products and 1.5% as ready-to-use egg products (Ministry of Agriculture of the Russian Federation, 2013).

In general, the state of dairy and poultry production markets in Russia is estimated as favourable for the future development of these industries. Government support and protectionism policies are additional drivers that allow local producers to remain competitive and to increase their agricultural output.

Government Support Given to Dairy and Poultry Production

Dairy and poultry production receives massive government support owing to the importance of these industries in Russia. The National Priority Project *Development of Agroindustrial Complex* (Ministry of Agriculture of the Russian Federation, 2005) announced livestock breeding as the priority direction of agricultural development in Russia. Since 2008, the main tool of the state agrarian policy at the federal level has been government programmes for the *Development of Agriculture and Regulation of Agricultural Commodities, Raw Materials, and Food Markets* for 2008–2012 and 2013–2020, and respective programmes in the regions (Committee for Agroindustrial Complex of Leningrad Region, 2012; Ministry of Agriculture of the Russian Federation, 2012).

Dairy farming activities included in the government programme for 2008–2012 were extended and included support of processing industries, stimulation of end demand and activities aimed at regulating seasonal fluctuations in prices for raw milk. In addition, several more government programmes were adopted as additional support, for example, the *Industry Programme Development of Dairy Farming and Increase in Milk Output in Russia for 2009–2012* (Ministry of Agriculture of the Russian Federation, 2008) and the *Industry Programme Development of Poultry Production in the Russian Federation for the Period 2012–2015* (Ministry of Agriculture of the Russian Federation, 2011).

The government protectionism policy has become stricter. Import rates for dairy products have been increased, with changes made to the balance of Russia and Belarus in terms of milk and dairy products to limit imports of cheap dairy products from Belarus (Russian Union of Dairy Enterprises, 2013).

Industry alliances have become more important (National Union of Milk Producers, Union of Milk Processing Companies), there are initiatives to identify and prevent counterfeiting on the dairy market and social commercials promoting milk and dairy product consumption are becoming more popular (Ministry of Agriculture of the Russian Federation, 2008).

The main activities of the government programme are transferred from the federal to the regional level, based on the co-financing of activities. The Ministry of Agriculture issues agreements with the regions of Russia annually. These agreements stipulate the mutual obligations of the parties, including the amount and conditions of subsidy payments provided by the Ministry of Agriculture and production output, number of animals, productivity of animals, reproduction indicators (e.g. since 2013, the dairy reproduction indicator is determined based on the number of calves per 100 cows) and milk quality is provided by a region (Ministry of Agriculture of the Russian Federation, 2012).

The main activity of the government programme to ensure the dairy industry's modernization is an increase in credit availability for agricultural producers. Industries receive credits from federal and regional budgets to reimburse part of the costs related to paying off credit interests. Reimbursement from the federal budget for investment credits in dairy farming is 100% of the Russian Central Bank rate. Co-financing from regional budgets should be not less than 3% above the Central Bank rate. State support given to dairy farming from the federal budget became 3.5 times higher in 2011 compared with 2008 (Russian Union of Dairy Enterprises, 2013).

One of the crucial directions of dairy production development is support of the livestock breeding sector. The value of incentives for dairy livestock breeding in 2011 was estimated at RUB 2 billion, and 20,000 head of pedigree young stock were imported.

Despite improvements in the financial situation of the dairy agricultural sector, the possibilities for industry modernization and innovative development still remain limited. The government programme subsidizing loans has covered just 19.5% of agricultural companies. This has affected negatively the pace of technical and technological modernization in the agricultural sector, inclusive of fodder production and forage grain production.

Taking into account social development programmes for rural areas, government incentives for the Agroindustrial Complex of Leningrad Region increased from RUB 1.4 billion in 2007 to RUB 5 billion in 2012. Over this period, federal budget support has increased from RUB 0.5 billion to RUB 1.9 billion; regional budget support has increased from RUB 0.9 billion to RUB 3 billion; and local budget support has increased from RUB 0.03 billion to RUB 0.12 billion. The major share of government support in the region falls on funding the interest loan rate (in 2011, 68% was from the federal budget and 32% was from the regional budget). In 2011, agricultural companies received subsidies valued at RUB 1021 million from the federal budget and RUB 250 million from the regional budget to pay investment credit interest. Additional government support has contributed to the increased profitability of the dairy industry. In 2011, milk production profitability was 10.5% without subsidies included and 14.7% with subsidies included. Government support promoted the growth of this indicator from 1.1 percentage points in 2006 to 4.2 percentage points in 2011 (Committee for Agroindustrial Complex of Leningrad Region, 2012).

Decaying market conditions for both dairy and poultry production in Russia resulted in Russia joining the World Trade Organization (WTO), and the establishment of the Customs Union has impacted negatively the financial situation of many agricultural producers, including those who had modernized production through the use of credits. Investment credit extension has created a situation in which the amount of state support allocated to subsidize investment credits is increasing, while the number of new investment projects is decreasing. Thus, in Russia's Leningrad region, the amount of regional budget subsidies to pay credit interest increased 69.8%, from RUB 200.4 million in 2007 to RUB 340.3 million in 2011; federal budget subsidies increased 3.9 times higher, from RUB 351.1 million in 2007 to RUB 1379 million in 2011. At the same time, the amount of investments in the Agroindustrial Complex of Leningrad Region increased 32.8%, from RUB 6.7 billion in 2007 to RUB 8.9 billion in 2011.

In the government programme for the *Development of Agriculture and Regulation of Agricultural Commodities, Raw Materials and Food Markets for 2013–2020,* the value of financing from the federal budget was estimated at RUB 1509 billion (in real 2012 terms) for the 2013–2020 period, with a growth in annual financing from RUB 159 billion in 2013 to RUB 218 billion in 2020 (Ministry of Agriculture of the Russian Federation, 2012).

The subprogramme for the *Development of the Livestock Breeding Industry, Processing, and Marketing of Livestock Breeding Commodities for 2013–2020* was allocated RUB 499 billion, or RUB 60 billion per annum (Ministry of Agriculture of the Russian Federation, 2013).

Beginning 1 January 2015, the government plans to cease subsidizing investment credits for the modernization of poultry production facilities, taking into account that poultry production output by 2015 should satisfy the country's internal demand (Ministry of Agriculture of the Russian Federation, 2012). The annual average growth rate of agricultural output until 2020 should be about 2.5%, while the growth rate of food production should be about 4%. Higher growth rates are expected for meat/meat products and milk/dairy products output.

To summarize, let us look at the conditions of agricultural development that can be both a challenge and a developmental opportunity for agricultural producers and agroindustrial complex management authorities. The external environment in Russian agriculture includes: economic globalization; rising global competition on products and resource markets; Russia's WTO accession; and the establishment of a Customs Union between Russia, Belarus and Kazakhstan. The internal environment in Russian agriculture includes recovery from the transformation crisis and the change in the structure of state support to agriculture.

Dairy and poultry production have good developmental perspectives, owing to significant domestic demand and massive government support. To take advantage of the development prospects, Russian dairy and poultry production should use economies of scale. Large dairy and poultry enterprises are better able to adopt the latest resource savings and high-performance technologies. Also, they are better able to accelerate their production output through new management approaches, to ensure that planned economic and production results are achieved while meeting environmental commitments.

Conclusions

A number of institutional, financial and production problems have been an obstacle in increasing production competitiveness for dairy production in Russia. Production efficiency has been hampered by limited access to government subsidies and investment credits. Adopting innovative technologies and attracting foreign investors, such as PepsiCo and Dannon, could improve the situation for dairy producers in Russia.

In contrast, there has been a high growth rate of poultry production in Russia due to poultry producers being able to take advantage of new institutional conditions, such as attracting domestic and foreign investments and acquiring subsidized investment credits for adopting global innovative technologies. These advantages have allowed Russian poultry production to become competitive on the global market in terms of resource intensity and product quality. At the same time, problems have arisen due to environmental and food safety issues, and because of the increasing monopoly power of the large-scale producers.

References

Committee for Agroindustrial Complex of Leningrad Region (2012) *Government Programme for Agricultural Development in Leningrad Region for 2013–2020*. St Petersburg, Russia.

Federal Service of State Statistics (1985–2013) *Major Indicators of Russian Agriculture*. Federal Service of State Statistics, Moscow.

Ministry of Agriculture of the Russian Federation (2005) *Government Programme for Development of Agroindustrial Complex*. Ministry of Agriculture of the Russian Federation, Moscow.

Ministry of Agriculture of the Russian Federation (2008) *Industry Programme Development of Dairy Farming and Increase in Milk Output in Russia for 2009–2012*. Ministry of Agriculture of the Russian Federation, Moscow.

Ministry of Agriculture of the Russian Federation (2011) *Industry Programme Development of Poultry Production in the Russian Federation for the Period 2012–2015*. Ministry of Agriculture of the Russian Federation, Moscow.

Ministry of Agriculture of the Russian Federation (2012) *Government Programme for Development of Agriculture and Regulation of Agricultural Commodities, Raw Materials and Food Markets for 2013–2020*. Ministry of Agriculture of the Russian Federation, Moscow.

Ministry of Agriculture of the Russian Federation (2013) *Government Programme for Development of the Livestock Breeding Industry, Processing, and Marketing of Livestock Breeding Commodities for 2013–2020*. Ministry of Agriculture of the Russian Federation, Moscow.

North-West Research Institute of Engineering and Electrification of Agriculture (2012) *Manure Treatment on Cattle Breeding Farms to Ensure Environmental Safety of Land, Surface and Ground Water in Leningrad Region*. North-West Research Institute of Engineering and Electrification of Agriculture, St Petersburg, Russia.

Russian Union of Dairy Enterprises (2013) *Dairy Industry in the World and in the Russian Federation*. Russian Union of Dairy Enterprises, Moscow.

14 International Crop Yield Comparisons: Selected KRU Regions

Charles B. Moss* and Andrew Schmitz
University of Florida, Gainesville, Florida, USA

Abstract

When studying Kazakhstan, Russia and Ukraine (KRU) regional crop yields, the effect of fertilizer choice on Russian grain yields is ultimately an empirical question requiring a point of comparison. In this chapter, we compare Russian input and output levels with those for the USA. Most of the shortfall in KRU regional grain yields can be explained by low yields in Kazakhstan. Extending these results, most of the shortfall in Kazakhstan is associated with lower fertilizer application rates.

Introduction

This chapter analyses the productivity and input use of wheat and other grains in Kazakhstan, Russia and Ukraine (KRU). We compare the KRU region with a selected US Great Plains region (i.e. the states of Colorado, Kansas, Nebraska and Oklahoma in the USA). We recognize that wheat production in these two regions is not exactly the same – the KRU region is colder and produces spring wheat, while the US Great Plains region is warmer and produces hard red winter wheat. We then develop an informational measure for the differences in output level per unit of input use of fertilizer. In general, the results indicate that fertilizer use is lower in the KRU region, but the difference among Russia, Ukraine and selected US Great Plains states is relatively small. This is not the case for Kazakhstan. We present implications for the lower resource use among these countries.

It is difficult to disentangle the question of input intensity in the Soviet republics before privatization, due to the impact of producer subsidy equivalents (PSE) (Johnson, 1991; Lerman *et al.*, 2003). In comparison, US farm policies included subsidies in commodity loans that functioned as price floors and transferred in the form of deficiency payments (Schmitz *et al.*, 2010).

Empirical Analysis

Table 14.1 presents a comparison of the 1991–2013 wheat yields in bushels/acre between the KRU region and the selected US Great Plains region. The results indicate that Ukraine's wheat productivity is somewhat higher than it is in the selected US Great Plains region, Russia's wheat yield averages that of the selected US Great Plains region and Kazakhstan's wheat

*Corresponding author; e-mail: cbmoss@ufl.edu

yield is much lower than that in the selected US Great Plains region.

The results in Table 14.1 are consistent with the regression results for a simple linear trend presented in Table 14.2. The trends are highest in Russia at 0.399, followed by the US Great Plains state of Nebraska at 0.377 and Kazakhstan at 0.294.

Another way of analysing the data involves comparing the informational inequality between wheat yields and acres planted. Specifically, as presented in Kullback (1997), the information measure ($I(.)$) is computed as

$$I\left(f(x,\theta_A),g(x,\theta_B)\right)=\int_{-\infty}^{\infty} f(x,\theta_A)\ln\left[\frac{f(x,\theta_A)}{g(x,\theta_B)}\right]dx, \quad (14.1)$$

where $f(x,\theta_A)$ is the prior distribution function for x, given the set of parameters θ_A, and $g(x,\theta_B)$ is a posterior distribution function for x, given the set of parameters θ_B. One application of the information measure presented in Eqn 14.1 is to provide a measure of how much information is contained in a sample. For example, if $f(x,\theta_A) \rightarrow g(x,\theta_B) \Rightarrow I(.) \rightarrow 0$. Alternatively, if $I(.) \rightarrow 0$, the sample provides little additional information. This formulation was popularized in economics as Theil's (1967) measure of income inequality

$$I(q,p)=\sum\nolimits_i q_i \ln\left(\frac{q_i}{p_i}\right), \quad q_i=\frac{Y_i}{\sum_j Y_j} \text{ and } p_i=\frac{P_i}{\sum_j P_j}, \quad (14.2)$$

where Y_i is the income in region i and P_i is the population in region i. $I(q,p)$ then measures the difference between the information in the income distribution and the population distribution – it measures income inequality. If income were distributed equally through the population, then the information inequality would approach zero, implying that information

Table 14.1. Wheat yields, KRU region and selected US Great Plains region, bushels/acre, 1991–2013. (From authors' computations based on data from USDA/NASS Quickstats, 1991–2013.)

	KRU region			Selected US Great Plains region			
Year	Kazakhstan	Russia	Ukraine	Colorado	Kansas	Nebraska	Oklahoma
				Bushels/acre			
1991	7.57	25.67	44.66	31.7	33.0	32.0	27.0
1992	19.59	29.08	45.70	30.9	34.0	30.0	28.5
1993	13.50	27.01	56.39	37.5	35.0	35.0	29.0
1994	10.68	22.70	45.55	30.8	38.0	34.0	27.0
1995	7.72	20.77	44.07	38.4	26.0	41.0	21.0
1996	9.35	23.00	34.13	33.3	29.0	35.0	19.0
1997	11.57	27.30	41.99	32.8	46.0	37.0	32.0
1998	7.72	20.03	39.32	39.6	49.0	46.0	39.0
1999	19.14	23.15	33.98	43.8	47.0	48.0	35.0
2000	13.35	24.04	29.38	29.8	37.0	36.0	34.0
2001	17.51	30.57	46.00	33.8	40.0	37.0	33.0
2002	16.17	30.72	45.26	22.8	33.0	33.0	28.0
2003	15.28	25.23	21.81	35.1	48.0	46.0	39.0
2004	12.46	29.38	47.04	27.4	37.0	37.0	35.0
2005	14.10	28.79	42.29	24.4	40.0	39.0	32.0
2006	16.77	29.08	37.54	21.6	32.0	36.0	24.0
2007	19.29	31.16	34.72	39.2	33.0	43.0	28.0
2008	14.39	36.21	54.46	30.8	40.0	44.0	37.0
2009	17.66	34.28	45.85	40.6	42.0	48.0	22.0
2010	10.83	28.34	39.77	45.5	45.0	43.0	31.0
2011	24.63	33.68	49.71	40.0	35.0	45.0	22.0
2012	11.72	26.26	41.55	34.3	42.0	41.0	36.0
2013	18.40	32.50	50.15	27.3	38.0	35.0	31.0

Table 14.2. Simple linear trend regression wheat yields, selected regions. (From authors' calculations.)

	Constant	Trend	R^2
Kazakhstan	–573.940[c]	0.294[c]	0.202
	(255.181)[a]	(0.127)	
Russia	–770.660[b]	0.399[b]	0.390
	(217.892)	(0.109)	
Ukraine	–35.851	0.039	0.001
	(507.973)	(0.254)	
Colorado	–14.730	0.024	0.014
	(413.929)	(0.207)	
Kansas	–402.935	0.220	0.059
	(383.733)	(0.192)	
Nebraska	–716.522[c]	0.377[c]	0.229
	(302.473)	(0.151)	
Oklahoma	–211.370	0.121	0.020
	(366.580)	(0.183)	

Notes: [a]Numbers in parentheses denote standard errors; [b]Denotes statistical significance at the 0.01 level of confidence; [c]Denotes statistical significance at the 0.05 level of confidence.

on the distribution of the population does not differ from information on the distribution of income. Foster (1983) demonstrates that the Theil measure of income inequality is superior to other measures of inequality.

In this study, we adapt Theil's formulation to focus on the distribution of yields

$$I(y,a)=\sum_i y_i \ln\left(\frac{y_i}{a_i}\right),\quad y_i=\frac{\widetilde{Y}_i}{\sum_j \widetilde{Y}_j} \text{ and } a_i=\frac{A_i}{\sum_j A_j}, \tag{14.3}$$

where $\widetilde{Y}_i$ is the crop yield (total bushels of wheat) in region i and A_i is the acres of wheat planted in region i. As the per acre yield becomes uniform, $I(y, a) \rightarrow 0$. In addition to the axiomatic advantages presented by Foster, Theil's measure has the advantage of exact decomposition

$$I(y,a)=I_R+\bar{I} \Leftarrow \left\{ \begin{array}{l} I_R=\sum_r y_r \ln\left(\frac{y_r}{a_r}\right),\ y_r=\sum_{i\in r} y_i \, and\ a_r=\sum_r a_r \\ \bar{I}=\sum_r y_r I_r,\ I_r=\sum_{i\in r}\left(\frac{y_i}{y_r}\right)\ln\left(\frac{y_i/y_r}{a_i/a_r}\right) \end{array} \right\}, \tag{14.4}$$

where I_R is the inequality between regions or regional inequality, I_r is the inequality within region r and $\bar{I}$ is the average inequality within each region.

Table 14.3 presents the information inequalities for wheat yields presented in Eqns 14.3 and 14.4 which indicate that the yield inequality within the KRU region is relatively large. The maximum value of the regional inequality is 0.18661 in 1991, followed by 0.16476 in 1995 and 0.13588 in 1998. Compared with these values, the inequality for yields in the selected US Great Plains region is very low, reaching a maximum of 0.03282 in 2009, followed by 0.02621 in 1995. The decomposition of the income inequality I_R indicates that most of the difference in yields can be explained by differences among Kazakhstan, Russia and Ukraine – the regional inequality that compares the information in the KRU region with the selected US Great Plains region is fairly small compared to the average inequality for the same two regions (i.e. $\bar{I}$ in Eqn 14.4).

To analyse the causes of the inequality in Russian agriculture, we present Table 14.4, which contains the geometric average wheat yields for the KRU region and the selected US Great Plains region. Computing the ratios of average wheat yields of the KRU region to the selected US Great Plains region, the results confirm the inequality results presented in Table 14.3. In Table 14.4, we present the impact of dropping the least productive state in each region. In the case of the KRU region, we drop Kazakhstan, while in the case of the selected US Great Plains region, we drop Oklahoma. Recalculating the geometric means, the average wheat yields for the KRU region are now much closer to those of the selected US Great Plains region. In fact, the average wheat yields for the KRU region are now higher in 2002 and 2008 than are the average wheat yields for the selected US Great Plains region.

Given these differences in wheat yields, the question is then: How much of these differences can be explained by differences in inputs – primarily, differences in fertilizer application rates? Because of the size of Table 14.5, we broke it down into two tables (Table 14.5a and b). Table 14.5a presents the total volume, measured in million tonnes (Mt), of fertilizer

Table 14.3. Information inequality for wheat yields, selected regions, 1991–2013. (From authors' calculations.)

	Information measure[a]					Information (%)	
Year	Russia	USA	Regional	Average	Total	Regional	Average
1991	0.18661	0.00335	0.00664	0.15704	0.16368	0.04054	0.95946
1992	0.03887	0.00301	0.00081	0.03307	0.03389	0.02402	0.97598
1993	0.10582	0.00391	0.00377	0.08868	0.09245	0.04082	0.95918
1994	0.10930	0.01010	0.01767	0.09094	0.10862	0.16272	0.83728
1995	0.16476	0.02621	0.00899	0.14024	0.14923	0.06024	0.93976
1996	0.09874	0.02327	0.00597	0.08723	0.09321	0.06408	0.93592
1997	0.09095	0.01393	0.01598	0.07807	0.09405	0.16990	0.83010
1998	0.13588	0.00552	0.05829	0.11161	0.16990	0.34307	0.65693
1999	0.01810	0.00744	0.02902	0.01627	0.04529	0.64071	0.35929
2000	0.03810	0.00262	0.01728	0.03230	0.04958	0.34857	0.65143
2001	0.05060	0.00365	0.00311	0.04427	0.04739	0.06564	0.93436
2002	0.05770	0.00717	0.00016	0.05139	0.05155	0.00307	0.99693
2003	0.02538	0.00675	0.04330	0.02197	0.06527	0.66342	0.33658
2004	0.09651	0.00375	0.00503	0.08326	0.08829	0.05694	0.94306
2005	0.06645	0.01345	0.00575	0.05896	0.06470	0.08881	0.91119
2006	0.03847	0.01358	0.00065	0.03500	0.03565	0.01817	0.98183
2007	0.02535	0.00909	0.00226	0.02314	0.02540	0.08914	0.91086
2008	0.09988	0.00414	0.00148	0.08743	0.08892	0.01669	0.98331
2009	0.05494	0.03282	0.00244	0.05225	0.05469	0.04468	0.95532
2010	0.10949	0.01200	0.01818	0.09641	0.11458	0.15863	0.84137
2011	0.02611	0.02453	0.00001	0.02593	0.02594	0.00051	0.99949
2012	0.09300	0.00339	0.01767	0.07980	0.09747	0.18132	0.81868
2013	0.05542	0.00687	0.00076	0.04950	0.05026	0.01516	0.98484

Note: [a]Information measure is the quantity of statistical information in the sample.

Table 14.4. Comparison of average wheat yields, selected regions, 1991–2013. (From authors' calculations.)

	Geometric average yield (weighted by acres)					
Year	USA	Russia	Russian/ US ratio	USA (w/o OK)	Russia (w/o Kz)	Russian/ US ratio
			Bushel/acre			
1991	31.18	19.17	0.615	32.66	29.28	0.897
1992	31.64	27.39	0.866	32.97	32.01	0.971
1993	33.64	24.22	0.720	35.40	31.15	0.880
1994	33.67	19.21	0.570	36.21	25.68	0.709
1995	27.15	16.84	0.620	29.54	24.19	0.819
1996	26.89	18.59	0.691	30.60	24.96	0.816
1997	39.33	23.08	0.587	42.19	29.93	0.709
1998	44.63	17.40	0.390	46.79	23.24	0.497
1999	43.40	23.56	0.543	46.52	25.29	0.544
2000	35.12	21.04	0.599	35.48	25.00	0.705
2001	37.07	28.25	0.762	38.45	33.61	0.874
2002	30.43	27.42	0.901	31.26	33.40	1.069
2003	43.75	21.12	0.483	45.42	24.83	0.547
2004	35.31	24.38	0.691	35.43	32.20	0.909
2005	35.61	25.09	0.705	36.75	31.22	0.849

Continued

Table 14.4. Continued.

	Geometric average yield (weighted by acres)					
Year	USA	Russia	Russian/ US ratio	USA (w/o OK)	Russia (w/o Kz)	Russian/ US ratio
2006	29.10	25.61	0.880	30.64	30.56	0.997
2007	33.71	27.39	0.812	35.45	31.85	0.898
2008	38.39	29.76	0.775	38.91	39.49	1.015
2009	36.87	29.29	0.794	42.42	36.35	0.857
2010	40.93	21.96	0.536	44.84	30.58	0.682
2011	33.02	32.44	0.982	37.01	36.57	0.988
2012	39.25	21.75	0.554	40.44	28.91	0.715
2013	34.69	29.38	0.847	35.89	35.70	0.995

Note: w/o OK = without Oklahoma; w/o Kz = without Kazkhstan.

used in the KRU region and the selected US Great Plains region. Table 14.5b presents the total volume (Mt) of nitrogen applied. In addition, Table 14.5a and b present the total area (per 1000 ha) planted to grains and oilseed crops. Finally, Table 14.5a and b presents the total output (Mt) of grains and oilseed crops. These data then provide two different comparisons. Slightly reworking the analysis above, we can examine if there are differences in the quantity of fertilizer applied per hectare and in the quantity of fertilizer applied per unit of output (see Figs 14.1 and 14.2).

Because of the size of Table 14.6, we broke it down into two tables (Table 14.6a and b). The informational measures for fertilizer applied to crop area are presented, with Table 14.6a giving total area (in hectares) and Table 14.6b giving grain area (in hectares). Returning for a moment to the informational formulation in Eqn 14.3, the natural logarithm of the share ratio

$$z_i = \ln\left(\frac{y_i}{a_i}\right) \Rightarrow I(y,a) = E[z] = \sum_i y_i z_i, \quad (14.5)$$

where z_i provides a measure of how different the prior distribution is from the posterior distribution. The information inequality is the expected value of this geometric difference. Table 14.6a and b present this distance along with the overall informational inequality. The largest discrepancy between yields and fertilizer application rates occurs in Kazakhstan for all fertilizer and output measures. These results are consistent with the differences in the geometric means presented in Table 14.4. In addition, the positive number implies that the share of hectares seeded to grain in Kazakhstan is larger than is the share of fertilizer consumed. This provides evidence that the fertilizer application rate for Kazakhstan is significantly lower than for the other countries.

Because of the size of Table 14.7, we broke it down into two tables (Table 14.7a and b). Table 14.7a and b present the same analysis, focusing on crop yields and fertilizer applications. Again, the logarithmic distance is higher for Kazakhstan, but its magnitude is much smaller than that in Table 14.6a and b. A comparison between the tables is informative in that Table 14.6a and b measure differences in application rates, while Table 14.7a and b measure differences in average physical product. The results indicate that the impact of the differences in application rates on yields is much larger than the differences in average physical products, which implies that the average physical product curves are fairly flat around the observed point. Furthermore, the results presented in Table 14.7a and b indicate that the average physical products for Russia and Ukraine in the KRU region are particularly close to those of the selected US Great Plains region. To examine this hypothesis, Table 14.8 computes the inequality of the geometric average between Russia and Ukraine in the KRU region and the selected US Great Plains region, as well as the information between Russia and Ukraine; between the selected US Great Plains region and Russia/Ukraine; and between Kazakhstan and Russia/Ukraine/the selected US Great Plains region.

Table 14.5a. Aggregate fertilizer (nitrogen, phosphorus and potassium, NPK) use for grains and oilseeds yield data (Mt). (From IFA, 2013.)

Crop year	Russia	Ukraine	Kazakhstan	USA
		NPK application rates (Mt)		
2000/01	2.589	1.264	0.069	20.436
2001/02	2.365	0.930	0.065	19.725
2002/03	2.377	0.967	0.060	18.880
2003/04	2.263	1.106	0.058	16.046
2004/05	1.922	0.889	0.053	19.455
2005/06	1.657	0.860	0.048	20.771
2006/07	1.528	0.675	0.045	19.273
2007/08	1.517	0.590	0.064	20.091
2008/09	1.378	0.500	0.064	21.203
2009/10	1.423	0.535	0.064	19.353
2010/11	1.602	0.474	0.049	19.614
2011/12	1.420	0.440	0.037	18.798
		Total area (1000 ha)		
2000/01	47.99	14.67	12.47	93.93
2001/02	48.45	16.19	12.77	90.93
2002/03	49.53	16.34	13.66	88.37
2003/04	45.70	14.30	13.47	92.32
2004/05	47.11	18.10	13.87	91.95
2005/06	47.90	18.08	13.77	91.76
2006/07	47.96	18.32	14.57	88.43
2007/08	47.84	17.67	15.08	92.50
2008/09	51.64	21.09	15.98	94.75
2009/10	52.15	20.64	16.96	92.02
2010/11	38.73	21.22	16.20	92.97
2011/12	48.41	22.41	15.51	90.60
		Yield (Mt)		
2000/01	67.06	26.84	11.18	418.78
2001/02	85.24	40.71	15.56	404.95
2002/03	88.17	41.13	15.44	371.37
2003/04	70.04	23.74	13.60	415.51
2004/05	80.63	44.06	11.97	474.10
2005/06	82.74	42.44	13.02	451.57
2006/07	83.18	39.97	15.92	426.04
2007/08	85.66	34.44	19.47	488.13
2008/09	113.25	62.82	14.87	483.09
2009/10	101.64	54.29	20.35	508.03
2010/11	65.17	49.83	11.61	490.68
2011/12	101.75	70.00	26.13	470.49

Table 14.5b. Aggregate fertilizer use with crop and production data, nitrogen (Mt). (From IFA, 2013.)

Crop year	Russia	Ukraine	Kazakhstan	USA
		Nitrogen application (Mt)		
2000/01	1.678	0.900	0.058	12.055
2001/02	1.483	0.650	0.055	11.645
2002/03	1.494	0.700	0.050	11.117
2003/04	1.423	0.800	0.050	10.398
2004/05	1.166	0.600	0.045	11.396
2005/06	0.967	0.550	0.043	11.966
2006/07	0.918	0.500	0.041	10.926

Continued

Table 14.5b. Continued.

Crop year	Russia	Ukraine	Kazakhstan	USA
2007/08	0.897	0.470	0.039	11.191
2008/09	0.838	0.430	0.050	11.819
2009/10	0.907	0.450	0.050	10.970
2010/11	1.090	0.380	0.041	10.895
2011/12	0.960	0.350	0.030	10.467
		Grain area (1000 ha)		
2000/01	42.80	11.61	12.47	57.12
2001/02	44.40	13.61	12.77	53.61
2002/03	45.25	13.45	13.66	52.09
2003/04	40.25	10.25	13.47	56.28
2004/05	41.65	14.34	13.87	55.14
2005/06	41.60	13.77	13.77	55.17
2006/07	40.80	13.32	14.57	51.47
2007/08	41.60	12.74	15.08	60.53
2008/09	44.30	14.64	15.98	59.48
2009/10	45.20	14.81	16.96	56.51
2010/11	31.53	13.92	16.20	55.79
2011/12	39.19	14.67	15.51	55.46
		Grain yield (Mt)		
2000/01	62.65	23.19	11.18	333.89
2001/02	82.05	38.27	15.56	315.12
2002/03	83.95	37.67	15.44	287.43
2003/04	64.60	19.20	13.60	338.91
2004/05	75.00	40.50	11.97	378.16
2005/06	75.30	36.84	13.02	356.04
2006/07	75.10	33.18	15.92	329.43
2007/08	78.73	28.49	19.47	405.68
2008/09	104.40	52.12	14.87	393.89
2009/10	93.61	45.01	20.35	409.13
2010/11	57.93	38.28	11.61	390.30
2011/12	89.32	55.80	26.13	378.14

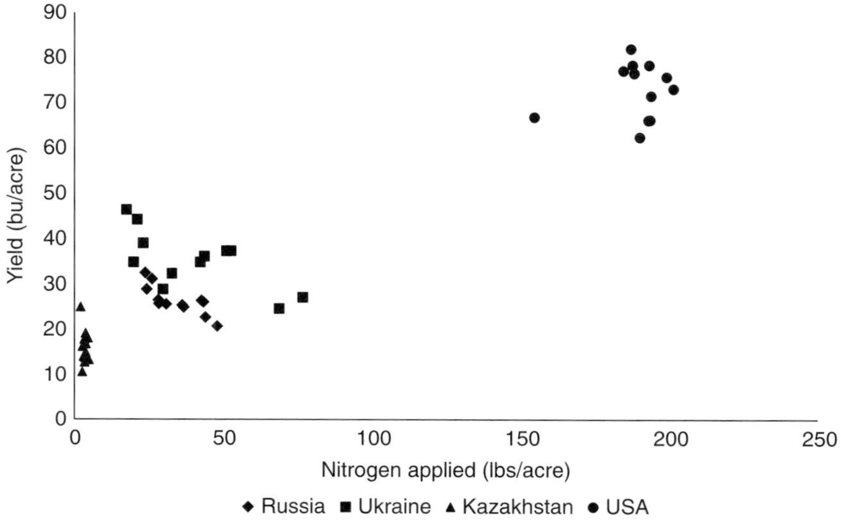

Fig. 14.1. Total physical product for wheat as a function of nitrogen applied. (From authors' calculations.)

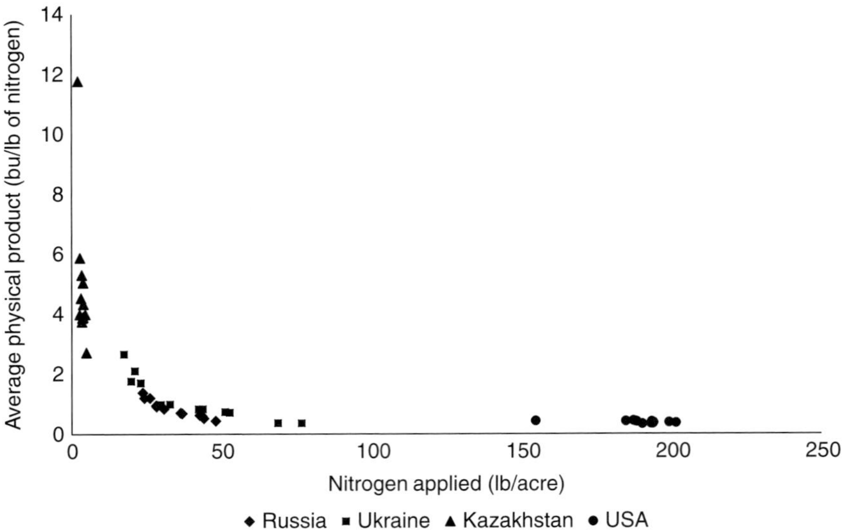

Fig. 14.2. Average physical product of nitrogen in the production of wheat. (From authors' calculations.)

Table 14.6a. Information measures for relationship between fertilizer consumed and total area. (From authors' calculations.)

	Logarithmic ratio of shares				
Crop year	Russia	Ukraine	Kazakhstan	USA	Inequality
		Total area (hectares/NPK consumption)			
2000/01	0.983	0.514	3.262	–0.412	0.335
2001/02	1.033	0.870	3.292	–0.459	0.383
2002/03	1.017	0.808	3.417	–0.476	0.406
2003/04	0.864	0.418	3.303	–0.392	0.324
2004/05	1.163	0.977	3.523	–0.483	0.450
2005/06	1.369	1.051	3.664	–0.509	0.515
2006/07	1.384	1.239	3.713	–0.539	0.564
2007/08	1.401	1.349	3.405	–0.524	0.541
2008/09	1.553	1.672	3.448	–0.573	0.634
2009/10	1.461	1.512	3.438	–0.581	0.617
2010/11	1.134	1.750	3.758	–0.495	0.567
2011/12	1.383	1.785	3.892	–0.573	0.652
		Total area (hectares/nitrogen consumption)			
2000/01	0.910	0.348	2.928	–0.390	0.288
2001/02	0.987	0.716	2.949	–0.444	0.337
2002/03	0.970	0.619	3.079	–0.458	0.356
2003/04	0.898	0.312	3.025	–0.388	0.304
2004/05	1.138	0.846	3.161	–0.473	0.405
2005/06	1.363	0.953	3.229	–0.503	0.471
2006/07	1.341	0.986	3.270	–0.524	0.494
2007/08	1.357	1.006	3.335	–0.508	0.497
2008/09	1.485	1.256	3.130	–0.555	0.548
2009/10	1.365	1.139	3.140	–0.560	0.530
2010/11	0.958	1.410	3.367	–0.468	0.461
2011/12	1.213	1.452	3.541	–0.549	0.545

Table 14.6b. Information measures for relationship between fertilizer consumed and grain area. (From authors' calculations.)

Crop year	Logarithmic ratio of shares				Inequality
	Russia	Ukraine	Kazakhstan	USA	
	Grain area (hectares/NPK consumption)				
2000/01	1.178	0.590	3.572	–0.600	0.545
2001/02	1.248	0.999	3.595	–0.684	0.629
2002/03	1.226	0.912	3.716	–0.705	0.657
2003/04	1.058	0.406	3.624	–0.566	0.530
2004/05	1.353	1.058	3.837	–0.681	0.698
2005/06	1.550	1.101	3.986	–0.696	0.773
2006/07	1.565	1.263	4.056	–0.738	0.847
2007/08	1.547	1.308	3.692	–0.661	0.744
2008/09	1.711	1.618	3.760	–0.728	0.865
2009/10	1.627	1.489	3.746	–0.760	0.870
2010/11	1.293	1.693	4.122	–0.641	0.812
2011/12	1.521	1.710	4.241	–0.715	0.888
	Grain area (hectares/nitrogen consumption)				
2000/01	1.106	0.424	3.238	–0.577	0.481
2001/02	1.203	0.845	3.251	–0.669	0.567
2002/03	1.179	0.724	3.379	–0.687	0.590
2003/04	1.092	0.300	3.346	–0.561	0.503
2004/05	1.328	0.926	3.474	–0.671	0.638
2005/06	1.543	1.002	3.551	–0.690	0.715
2006/07	1.522	1.010	3.613	–0.723	0.757
2007/08	1.504	0.966	3.621	–0.646	0.696
2008/09	1.642	1.202	3.442	–0.709	0.768
2009/10	1.531	1.116	3.448	–0.739	0.767
2010/11	1.117	1.353	3.731	–0.614	0.683
2011/12	1.351	1.377	3.890	–0.691	0.762

Table 14.7a. Information measures for relationship between fertilizer consumed and total production. (From authors' calculations.)

Crop year	Logarithmic ratio of shares				Inequality
	Russia	Ukraine	Kazakhstan	USA	
	Total production (Mt/NPK consumption)				
2000/01	0.186	–0.013	2.022	–0.048	0.028
2001/02	0.421	0.615	2.312	–0.142	0.072
2002/03	0.471	0.608	2.416	–0.163	0.084
2003/04	0.142	–0.223	2.164	–0.036	0.036
2004/05	0.427	0.594	2.103	–0.116	0.050
2005/06	0.681	0.669	2.373	–0.151	0.081
2006/07	0.729	0.813	2.596	–0.172	0.108
2007/08	0.695	0.728	2.372	–0.149	0.093
2008/09	1.037	1.462	2.075	–0.245	0.180
2009/10	0.802	1.154	2.294	–0.198	0.132
2010/11	0.359	1.309	2.130	–0.127	0.083
2011/12	0.797	1.595	3.085	–0.255	0.229

Continued

Table 14.7a. Continued.

Crop year	Logarithmic ratio of shares				Inequality
	Russia	Ukraine	Kazakhstan	USA	
	Total production (Mt/nitrogen consumption)				
2000/01	0.114	–0.179	1.687	–0.026	0.021
2001/02	0.375	0.461	1.969	–0.128	0.054
2002/03	0.424	0.419	2.079	–0.145	0.063
2003/04	0.176	–0.330	1.886	–0.032	0.032
2004/05	0.402	0.462	1.741	–0.106	0.038
2005/06	0.674	0.571	1.938	–0.144	0.068
2006/07	0.686	0.561	2.153	–0.157	0.083
2007/08	0.651	0.386	2.302	–0.133	0.078
2008/09	0.969	1.046	1.757	–0.227	0.136
2009/10	0.706	0.780	1.996	–0.177	0.095
2010/11	0.184	0.969	1.739	–0.100	0.051
2011/12	0.627	1.262	2.733	–0.231	0.172

Table 14.7b. Information measures for relationship between fertilizer consumed and grain production. (From authors' calculations.)

Crop year	Logarithmic ratio of shares				Inequality
	Russia	Ukraine	Kazakhstan	USA	
	Grain production (Mt/NPK consumption)				
2000/01	0.313	0.036	2.218	–0.080	0.043
2001/02	0.574	0.745	2.504	–0.201	0.114
2002/03	0.617	0.715	2.612	–0.224	0.129
2003/04	0.242	–0.255	2.345	–0.059	0.052
2004/05	0.544	0.699	2.292	–0.153	0.076
2005/06	0.790	0.731	2.577	–0.185	0.113
2006/07	0.847	0.847	2.816	–0.210	0.149
2007/08	0.775	0.703	2.537	–0.169	0.116
2008/09	1.132	1.451	2.251	–0.274	0.211
2009/10	0.906	1.152	2.480	–0.229	0.165
2010/11	0.456	1.260	2.344	–0.141	0.094
2011/12	0.863	1.564	3.281	–0.277	0.264
	Grain production (Mt/nitrogen consumption)				
2000/01	0.241	–0.130	1.883	–0.057	0.033
2001/02	0.529	0.591	2.161	–0.186	0.091
2002/03	0.570	0.527	2.274	–0.206	0.103
2003/04	0.276	–0.361	2.067	–0.055	0.047
2004/05	0.519	0.567	1.930	–0.143	0.061
2005/06	0.783	0.633	2.141	–0.179	0.097
2006/07	0.804	0.594	2.373	–0.195	0.118
2007/08	0.731	0.361	2.467	–0.153	0.101
2008/09	1.063	1.036	1.933	–0.256	0.165
2009/10	0.810	0.779	2.182	–0.208	0.124
2010/11	0.280	0.920	1.953	–0.114	0.059
2011/12	0.693	1.231	2.930	–0.253	0.203

Table 14.8. Information measures for relationship between nitrogen consumption and grain yield. (From authors' calculations.)

Crop year	Russia/ Ukraine	USA/ (Russia and Ukraine)	Kazakhstan with others	Total
2000/01	0.0143	0.0027	0.0271	0.0325
2001/02	0.0004	0.0477	0.0445	0.0907
2002/03	0.0002	0.0540	0.0506	0.1027
2003/04	0.0398	0.0016	0.0376	0.0468
2004/05	0.0003	0.0363	0.0257	0.0611
2005/06	0.0025	0.0635	0.0344	0.0968
2006/07	0.0048	0.0677	0.0520	0.1185
2007/08	0.0141	0.0421	0.0573	0.1007
2008/09	0.0001	0.1397	0.0286	0.1646
2009/10	0.0001	0.0798	0.0469	0.1239
2010/11	0.0462	0.0254	0.0257	0.0595
2011/12	0.0326	0.1039	0.0953	0.2029

Conclusions

By definition, economies in transition from central planning to open markets are experiencing radical changes. Some evidence suggested that agricultural input usage during the later years of the Soviet Union were similar to that observed in the rest of the world. However, several studies concluded that input use deviated significantly from optimal in the 1990s, particularly with the use of fertilizers. In this study, we examined the empirical evidence for this suboptimality from 1991 through 2012. Specifically, we compared fertilizer use and wheat yields between the KRU region and the selected US Great Plains region. In general, output levels are lower for the KRU region than for the selected US Great Plains region, but the variation within each region is smaller than are the differences within each region. Furthermore, the differences in each region are explained largely by the smaller production yields of Kazakhstan and Oklahoma. The results of relative input use indicate that fertilizer use is somewhat lower in the KRU region than it is in the selected US Great Plains region.

Finally, it is important to remember that the standard economic model of agricultural production is significantly simplified. Inputs are transformed into outputs in a single period by a mechanical process. Even in the mathematical development of the production function, we often make assumptions such as strong and weak necessity (i.e. whether or not at least one variable is necessary for production or whether or not all inputs must occur at a non-zero level). Agronomically, the decay of the parent material for soils provides some of the necessary nutrients for plant production. This background source of nutrients is often augmented by the decay of past crops and carry-over fertilizer that has not been used by past crops. Of course, nutrients are often lost through erosion. These considerations are important, because the current level of production in the KRU region may be due in part to soil mechanics. KRU cropland may be fallowed as part of a normal crop rotation and may augment the latent level of soil productivity. In addition, the parent materials of KRU soils may be higher than in the selected US Great Plains region.

References

Foster, J.E. (1983) An axiomatic characterization of the Theil measure of income inequality. *Journal of Economic Theory* 31(1), 105–121.

IFA (2013) *IFA Statistics*. International Fertilizer Industry Association, Paris. Available at: http://www.fertilizer.org/En/Statistics/IFADATA.aspx (accessed 24 April 2014).

Johnson, P. (1991) *Modern Times: The World from the Twenties to the Nineties*, Revised Edn. HarperCollins Publishers, New York.

Kullback, S. (1997) *Information Theory and Statistics*. Dover Publications, Mineola, New York.

Lerman, Z., Kislev, Y., Biton, D. and Kriss, A. (2003) Agricultural output and productivity in the Former Soviet Republics. *Economic Development and Cultural Change* 51(4), 999–1018.

Schmitz, A., Moss, C.B., Schmitz, T.G., Furtan, H.W. and Schmitz, H.C. (2010) *Agricultural Policy, Agribusiness, and Rent-Seeking Behavior*, Second Edn. University of Toronto Press, Toronto, Canada.

Theil, H. (1967). *Economics and Information Theory*. North Holland, Amsterdam.

USDA/NASS (1991–2013) *Quickstats*. US Department of Agriculture, National Agricultural Statistics Service, Washington, DC.

15 Modelling Fertilizer Demand: The Three Bads

Andrew Schmitz* and Charles B. Moss
University of Florida, Gainesville, Florida, USA

Abstract

Theoretically, optimal fertilizer use depends on many factors, including price and input subsidies, weather, credit constraints and land quality differences. In this context, it is possible that even though fertilizer applications are low in all three countries, especially in Kazakhstan, fertilizer rates may be at the optimum level. This may be the case even though the fertilizers applied in these regions are from one-third to one-half the amount applied in the USA.

Introduction

This chapter examines the theoretical foundation for optimal fertilizer use in Kazakhstan, Russia and Ukraine (KRU region). Theoretically, optimal fertilizer use depends on many factors, such as input prices (and subsidies), weather, credit availability and land quality. It is possible that the low fertilizer use in all three countries may be optimal even though the amount of fertilizer applied is from one-third to one-half the amount applied in the USA (Chapter 14, this volume). The optimal level of fertilizer selection has a long history, beginning with work by Heady (1952). Empirical applications and extension prescriptions for fertilizer use in even the most advanced agricultural systems are controversial, especially for fertilizer use in the KRU region.

Several studies have examined the implications for agricultural input use in the post-Soviet era, but most of the focus is on Russia. Liefert (2002) evaluated the comparative advantage of Russian agriculture using the social cost–benefit (SCB) ratio, defined as

$$SCB_i = \frac{P_i^d + (s_i - t_i) + \Sigma_j a_{ij}(P_j^b - P_j^d)}{P_i^b}, \quad (15.1)$$

where P_i^d is the price of output i in the domestic market, s_i is the level of subsidy, t_i is the tax imposed on the output, a_{ij} is the amount of input j used to produce output i, P_i^b is the border price of input j and P_i^d is the domestic price of input j. Intuitively, if $SCB_i \leq 1$, a country has a comparative advantage in the production of good i. Based on this formulation, Liefert (2002) found that Russia had a comparative disadvantage in the production of agricultural commodities relative to agricultural inputs, especially in the production of meat.

Russia is also a major exporter of fertilizer. In fact, in 1996 and 1997, Russia exported over 80% of its total fertilizer output, mostly to European Union (EU) countries. Yet the SCB calculations do not appear to show a strong CA

*Corresponding author; e-mail: aschmitz@ufl.edu

(comparative advantage) for fertilizer vis-à-vis crops. Nevertheless, the SCB (social cost–benefit) calculations are consistent with the country's fertilizer trade (Liefert, 2002, p. 766).

The pressure to export fertilizer instead of exporting crop products may have been to raise hard currency (Liefert, 2002). Note that the SCB ratio for fertilizer from 1996 to 1997 was 7.25 compared with an SCB of 3.30 for crude oil over the same period. Hence, from 1996 to 1997, Russia had a relative comparative advantage in the exportation of nitrogen over that of crude oil.

Liefert *et al.* (2003) focused on the application of fertilizer on wheat, using production elasticities estimated by Osborne and Trueblood (2003) and Sedik *et al.* (1999). In their analysis, Liefert *et al.* (2003) computed

$$\lambda = \frac{MP_g^f \times P^g}{P^f} \Leftarrow VMP_g \equiv MP_g^f \times P^g = P^f, \quad (15.2)$$

where $VMP_g^f = MP_g^f \times P^g$ is the value of the marginal product of input f used to produce output g, which is equal to the marginal product of input f in the production of output g (MP_g^f) times the price of output $g(P^g)$. In equilibrium, this value of marginal product should equal P^f, which is the price of input f. They found that $\lambda = 1.16$ for firms with low marginal products in 2000, so these farms were close to optimality (i.e. implying that $\lambda = 1$). However, for farms with higher marginal products, they found that $\lambda = 3.67$. Hence, their evidence suggests that the higher marginal product farms are significantly under-applying fertilizer. Liefert *et al.* (2003) gives several reasons for this underutilization of inputs. First, farms may have significant constraints allocated for operating capital. Second, a somewhat related problem is the market structure for inputs and outputs in Russia, in which low interest credit is extended by the authorities in return for the first option to buy the production output (Liefert *et al.*, 2003). This basically amounts to a mercantilist form of economic policy (Schmoller, 2008).

Bokusheva and Hockmann (2006) examined the production risk and technical inefficiency in Russian agriculture using a nested production function based on the formulation proposed by Kumbhakar (2002). Bokusheva and Hockmann's results indicated that Russian agricultural production was subject to considerable risk that was dependent on input use. In addition, differences in technical inefficiency increased the variability of output in Russian agriculture. However, neither of these conclusions appears unique to Russian agriculture. The most unique finding of their study was the conclusion that the relative level of risk and technical inefficiency might be dependent on the stage of agricultural production in Russia. Specifically, modern production technologies and practices reduced output volatility and allowed more flexibility (Bokusheva and Hockmann, 2006).

Fertilizer Industry: An Overview

Fertilizers include nitrogen, phosphorus and potassium (NPK), which help produce better crop yields and quality. Nitrogen aids chlorophyll and plant proteins in plant growth; phosphorus helps store energy in plants; and potassium aids in disease resistance and photosynthesis. Almost one-half of the fertilizer use in the KRU region is for cereal crops such as wheat and maize. Between 1960 and 2005, KRU fertilizer use increased from 30 million tonnes (Mt) to 154 Mt. There was a 3-year decline of fertilizer use in the KRU region after the collapse of the Soviet Union in 1991 (Deutsche UFG, 2006). During the 1991–1994 period, KRU fertilizer application rates and crop production declined significantly.

The Potash Corporation of Saskatchewan, Canada, is the largest global regulator and producer of potash (fertilizer) that produces almost 25% of the world's potash and holds 75% of the world's unused capacity. Russia has 35% of the global potash reserves. In the early 1990s, Russia redirected its production volumes to the export market to increase its market share of potash.

The post-Soviet economic reforms in the KRU region brought several changes to crop production, livestock production and fertilizer use when the state funding of its agricultural enterprises decreased (Deutsche UFG, 2006). As an example, fertilizer use decreased to only 4 kg/ha, because its price surpassed agricultural product prices.

The basic choice of fertilizer level is a simple application of economic optimization, but uncertainty complicates the choice of inputs.

As described in Bokusheva and Hockmann (2006), the choice of inputs may increase the impact of weather risk, credit constraint and bad policy (the three *bads*). We discuss each of these factors in turn.

The first bad: weather risk

To develop the implications for weather on input decisions, we start with the Kumbhakar (2002) production function as presented by Bokusheva and Hockmann (2006)

$$y = f(x;\alpha)(1-u) + g(x;\beta)v, \qquad (15.3)$$

where y is the level of output, $f(x;\alpha)$ is the deterministic portion of the production function, x is the level of inputs, α are parameters, u is the level of technical inefficiency, $g(x;\beta)$ is the stochastic portion of the production relating the choice of inputs x with the risk of production and v is a random error (e.g. independent and identically normal). The optimal level of inputs without risk aversion then becomes

$$max\,E[\pi] = E\left[p_y\left(f(x;\alpha) + g(x;\beta)v\right) - p_c x\right]$$
$$\Rightarrow \frac{\partial f(x;\alpha)}{\partial x} + \frac{\partial g(x;\beta)}{\partial x}E[v] = \frac{p_c}{p_y}, \qquad (15.4)$$

where p_y is the output price and p_c is the price of the input. Under standard assumptions, $E[v] = 0$, so that the optimal use of inputs is the same as that presented in Eqn 15.2.

The production function presented in Eqn 15.5 assumes that all decisions are made before ascertaining the weather information. A slight modification involves assuming that some information about the weather becomes available before all inputs are applied. For example, in the case of hard red winter wheat in the US Great Plains region, some fertilizer is applied when the crop is planted. Wheat can be top dressed with additional fertilizer in February or March before its flag leaf stage of growth. Under this assumption, input use becomes dependent on the state of nature (i.e., amount of rainfall).

In Fig. 15.1, there is a different production function for each state of nature (i.e., amount of rainfall).

$$f(x_1, x_2; \alpha_1, \alpha_2) \Rightarrow \begin{Bmatrix} f(x_1, x_{21}; \alpha_1) \\ f(x_1, x_{22}; \alpha_2) \end{Bmatrix}, \qquad (15.5)$$

where x_1 is the level of input applied when the crop is planted, x_{21} is the amount of fertilizer added at the second period when event 1 occurs (i.e. a high rainfall event), x_{22} is the amount of fertilizer applied when event 2 occurs (i.e. a low rainfall event) and α_1 and α_2 are the different production function parameters given each state of nature (see Moss and Schmitz (2014) for a more complete discussion of state dependent production functions).

Again, consider Fig. 15.1 where $f(x_1, x_{22}; \alpha)$ is the fertilizer response function under certainty for the impact of nitrogen x_1 on wheat yields Y (following the expectation of the production function in Eqn 15.5) and x_{22} is applied when event 2 occurs. Following the results of Eqn 15.5, the optimal level of fertilizer is depicted by the point

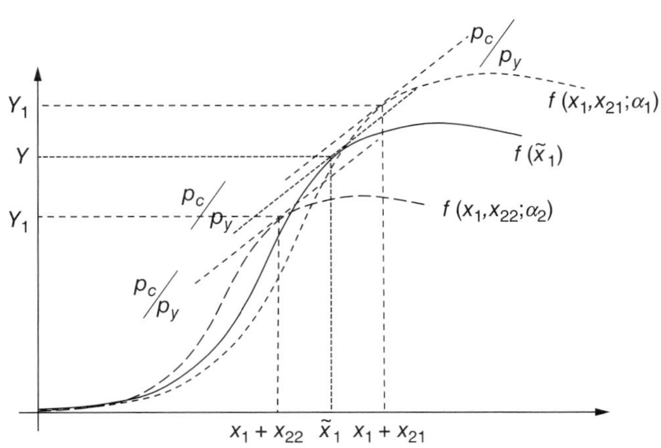

Fig. 15.1. Fertilizer use and weather impacts. (From authors' illustration.)

$$\frac{\partial f(x_1, x_{22}; \alpha)}{\partial x_1} = \frac{p_c}{p_y}, \tag{15.6}$$

where the slope of the production function is equal to the price line.

Now we introduce weather variability. In the years of high rainfall, the yield response function is $f(x_1, x_{21}; \alpha_1)$, but in a drought period the function is $f(x_1, x_{22}; \alpha_2)$. Given a fixed relative price ratio p_c / p_y, not enough fertilizer is used in the wet period (the optimal amount is $x_1 + x_{21}$, but only $\tilde{x}_1$ is applied). However, in the drought period, too much fertilizer is used, as the optimum amount is $x_1 + x_{22}$, which is less than $\tilde{x}_1$. If we compare the standard production formulation in Eqn 15.5 with that of Eqn 15.6, the profit under the formulation in Eqn 15.6 will be higher if the signal is observed without cost (Moss, 2010, pp. 173–182, pp. 227–249).

Consider now the product price impact of variable weather conditions. Suppose the price of wheat falls in the wet period (high rainfall) due to increased production. The price ratio now is p_c/p'_y ($>p_c/p_y$) (Fig. 15.2). In this case, the optimal fertilizer use is $\tilde{x}'_1$, which is below the level $\tilde{x}_1$. Unlike in the previous result, too much fertilizer has been used. However, the amount of fertilizer cannot be adjusted to the optimal level (i.e. one cannot remove fertilizer once it is applied).

In general, the introduction of risk does not affect the profit-maximizing level of input use in the standard stochastic production function formulation presented in Eqn 15.5. The optimal level of applied fertilizer declines due to risk aversion, especially when the input is risk increasing (assumed in the case of fertilizer). If we assume that the ability to integrate information into the input decision as depicted in Eqn 15.6, the net impact of weather on input use is a function of relative concavity of each production surface. The expected amount of fertilizer applied is dependent on weather events.

The second bad: credit constraint

At least for less developed countries, arguments are made that imply input use is constrained because of lack of capital. To develop this scenario, consider the case where inputs are purchased at the beginning of the production cycle on credit. Assume that the farmer pays an interest rate of $r(\delta)$, where δ represents some aspect of the borrower's credit worthiness. In the current example, we assume that $\partial r(\delta)/\partial\delta > 0$ with $\lim_{\delta\to 0} r(\delta) > 0$. The farmer's optimization problem can then be expressed as

$$\pi = p_y f(\tilde{x}_1) - p_c(1+r(\delta))\tilde{x}_1$$
$$\Rightarrow \frac{\partial f(\tilde{x}_1)}{\partial \tilde{x}_1} = \frac{p_c(1+r(\delta))}{p_y} = \frac{p_c(\delta)}{p_y}. \tag{15.7}$$

In Fig. 15.3, as the financial condition of farmers becomes more binding, the optimal quantity of fertilizer applied declines from $\tilde{x}_1$ to $\tilde{x}''_1$, while the level of output declines from Y_1 to Y''.

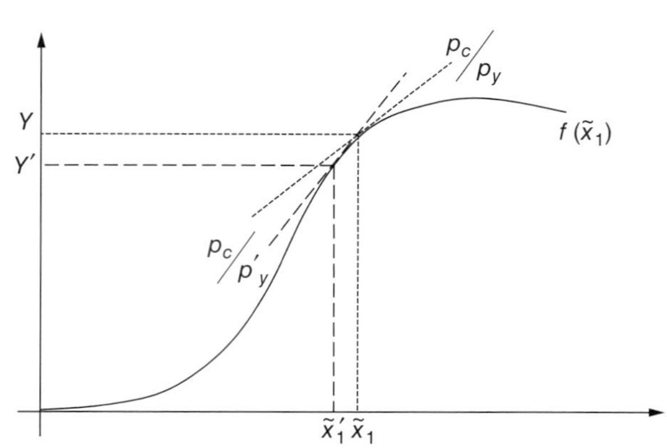

Fig. 15.2. Effect of low price and bad policy. (From authors' illustration.)

The third bad: bad policy

As discussed in Liefert *et al.* (2003), one of the potential distortions in the Russian fertilizer market involved the potential gains from the exportation of fertilizer rather than from the exportation of crops. Liefert *et al.* (2003) found that the ratio between the value of marginal product and the cost of the input was 3.67 for productive Russian farms. Earlier, Liefert (2002) noted that Russia was a net exporter of fertilizer (8.31 Mt of nitrogen in 1996 and 613 Mt of nitrogen in 1997), while at the same time it imported wheat (1.70 Mt in 1996 and 1.60 Mt in 1997). At the most basic economic level, these results would appear inconsistent – replacing one dollar of export revenue from the export of nitrogen would yield a monetary increase of US$3.67 from wheat. Several possible explanations for this disequilibrium are suggested that fall in the realm of policy distortions, such as taxes on agricultural production or implicit and explicit subsidies on exports.

The potential consequences of these distortions are presented in Fig. 15.4. The level of fertilizer is determined by the tangency between the production function $f(\tilde{x}_1)$ and the relative input/output price ratio p_c / p_y, resulting in $\tilde{x}_1$ units of fertilizer. Next, we rotate this input/output price ratio by considering the imposition of taxes on agricultural output τ or increasing the price of fertilizer from export subsidies or enhancements ψ. Either of these policies results in lower fertilizer use (a shift from $\tilde{x}_1$ to $\tilde{x}_1'$).

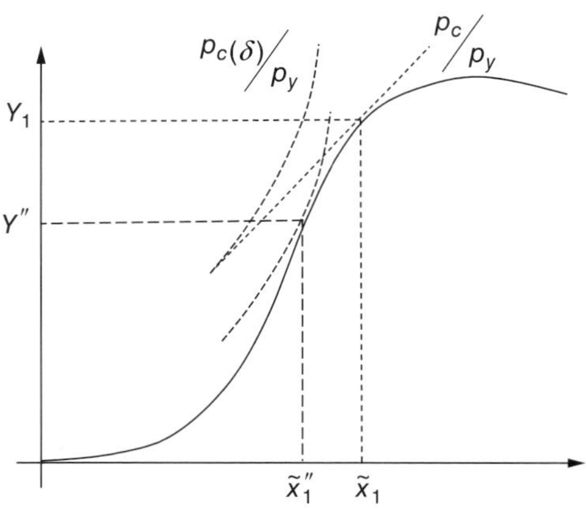

Fig. 15.3. Fertilizer use and credit constraint. (From authors' illustration.)

Combined Policy, Uncertainty and Credit Constraint

Figure 15.5 presents the combined effects of bad policy, uncertainty and credit constraints. The price distortions of bad policy imply an input/output price ratio of $p_c + \psi / p_y - \tau$. This results in an optimal level of fertilizer of x_1, which is lower than that of the open market (undistorted price) equilibrium. By using less fertilizer, a lower level of output Y is produced. Next, we consider the effect of risk and uncertainty that yields an optimal level of fertilizer of x_1' ($x_1' \leq x_1$). This reduction implies a lower output level ($Y' \leq Y$). Finally, we consider the effects of credit

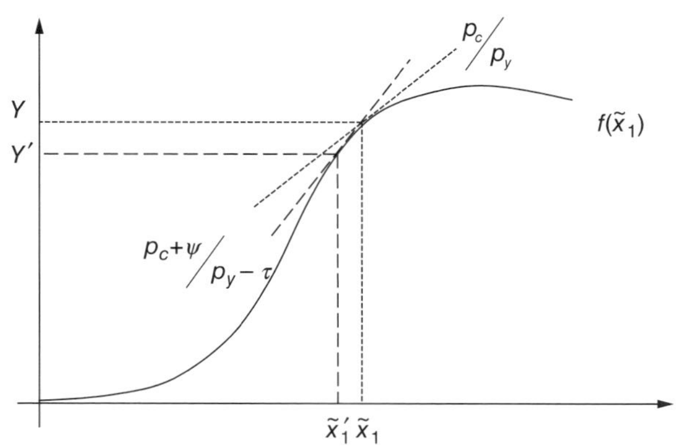

Fig. 15.4. Fertilizer use and bad policy. (From authors' illustration.)

constraints that change the input/output price ratio to $p_c(\delta, \psi)/(p_y - \tau)$. Intuitively, the consideration of credit constraints shifts the optimal level of fertilizer applied further to the left (from $\tilde{x}_1'$ to $\tilde{x}_1''$), with an additional decline in output (from Y' to Y'').

It is argued that Ukraine has the largest percentage of the most productive soils in the world. Consider Fig. 15.6, where P_F' is the wheat fertilizer response function for Ukraine I, while P_F is a region of poorer soil quality II. With X_2 of fertilizer, the yield in I is Y_2, while in II it is Y_1. In this formulation, only an amount of fertilizer X_1 in I is needed to generate the same yield from X_2 applied in II (Fig. 15.6).

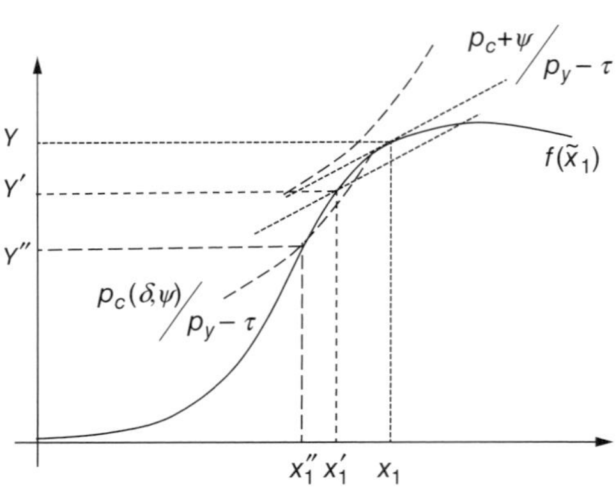

Fig. 15.5. Combined policy uncertainty and credit constraint. (From authors' illustration.)

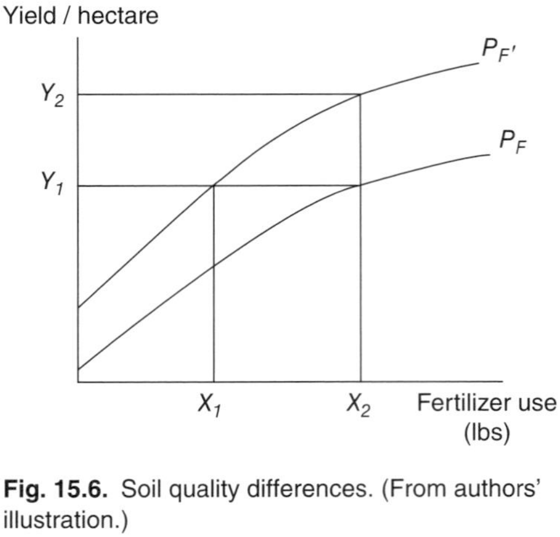

Fig. 15.6. Soil quality differences. (From authors' illustration.)

Discussion and Conclusions

Fertilizer use in the KRU region is lower than that observed in many of the developed countries (of the three KRU countries, per hectare fertilizer use in Kazakhstan is the lowest). We develop three different market imperfections (or bads) that may explain these discrepancies. First, we examined the impact of uncertainty. Intuitively, risk alone is insufficient to result in lower input use. To generate lower fertilizer use, we must either assume risk aversion or we must assume some state-dependent structure with an asymmetric yield function. A significant conjecture is that risk aversion cannot be used to explain differences in fertilizer decisions across countries. Input use in the USA is subject to risk, just as input use is subject to risk in Russia. Of course, agricultural policies may affect the consequences of those risks. In the USA, federally subsidized crop insurance may imply differences in fertilizer decisions. Next, we turned to the effect of credit constraints on fertilizer decisions. Unlike risk, differences in credit availability in transition economies may explain differences in fertilizer use in Russia and the USA. Last, we investigated the effect of a variety of agricultural policies on input use. Unlike in the KRU, the USA is accustomed to agricultural policies that increase input use such as loan deficiency payments. Often, policies in the KRU are put in place to accomplish different policy objectives such as the generation of foreign exchange.

This paper has focused on optimal fertilizer use where the production function is unknown. Our analysis has assumed that a production function exists that is affected by weather and other sources of uncertainty. However, additional empirical information on the complementarity of inputs, including fertilizer, seeds and mechanical technology, may shed additional light on crop production in the KRU. It is possible that the observed fertilizer application rates are optimal, given the equipment and seeds available to agriculture in the KRU.

Improved seed genetics may be insufficient to improve output regardless of the amount of fertilizer applied, unless mechanical technology (e.g. zero-tillage technology) is also available.

References

Bokusheva, R. and Hockmann, H. (2006) Production risk and technical inefficiency in Russian agriculture. *European Review of Agricultural Economics* 33(1), 93–118.

Deutsche UFG (2006) *The Global Fertilizer Industry: An Overview*. Deutsche Bank AG, London.

Heady, E.O. (1952) *Economics of Agricultural Production and Resource Use*. Prentice-Hall, Upper Saddle River, New Jersey.

Kumbhakar, S.C. (2002) Specification and estimation of production risk, risk preferences, and technical efficiency. *American Journal of Agricultural Economics* 84, 8–22.

Liefert, W.M. (2002) Comparative (dis?) advantage in Russian agriculture. *American Journal of Agricultural Economics* 84(3), 762–767.

Liefert, W.M., Gardner, B. and Serova, E. (2003) Allocative efficiency in Russian agriculture: the case of fertilizer and grain. *American Journal of Agricultural Economics* 85(5), 1228–1233.

Moss, C.B. (2010) *Risk, Uncertainty, and the Agricultural Firm*. World Scientific, Hackensack, New Jersey.

Moss, C.B. and Schmitz, A. (2014) Choosing between increased means and reduced variance: Implications for genetically modified crops. *AgBioforum* 17(2), 161–165.

Osborne, S. and Trueblood, M. (2003) *An Examination of Economic Efficiency of Russian Crop Production in the Reform Period*. Economic Research Service, US Department of Agriculture, Washington, DC.

Schmoller, G. (2008) *The Mercantile System and its Historical Significance*. Evergreen Review, Inc, New York City, New York.

Sedik, D., Trueblood, M. and Arnade, C. (1999) Corporate farm performance in Russia, 1991–1995: an efficiency analysis. *Journal of Comparative Economics* 27, 514–533.

16 Russian Trade in Agricultural Products: Current State and Influences of Trade Integration

Vasily Erokhin*
Moscow University of Finance and Law, Moscow, Russia

Abstract

Trade in agricultural products has become more and more internationalized and globalized. A global trading system is now both more free and fair than ever before, which will boost global prosperity and contribute significantly to global economic development. Favourable conditions for international trade in agricultural products include trade liberalization, trade unions and free trade areas. Most of the regulations on global market agricultural products are implemented by the World Trade Organization (WTO), which unites the majority of countries, including the main agricultural producers, exporters and importers. The WTO oversees issues such as trade regulation, sanitary control rules, custom duties and tariff quotas on imported agricultural production. It also oversees the structure, direction and dynamics of future trade in agricultural products. This chapter analyses the state of Russian foreign trade in agricultural products from 2002 to 2012 and investigates WTO threats to and opportunities for Russian agriculture and trade in the global market. This chapter examines state support of agricultural production in Russia and its influence on volume, direction, structure and effectiveness of international agricultural trade.

Introduction

Trade in agricultural products in the 21st century has become more and more globalized. Most of the regulatory functions of the global agricultural market are implemented by the World Trade Organization (WTO). While one of the objectives of the WTO is to unite agricultural producers, exporters and importers, agricultural international trade is far from being fully liberalized. Issues such as the state regulation of trade, sanitary and phyosanitary standards and the administration of customs duties and tariff quotas affect the dynamics of global trade in agricultural products.

Russia joined the WTO in 2012, which has brought advantages and disadvantages to Russia's agricultural industries. The main disadvantage for Russia is its dependence on food imports. Because of this situation, some economists anticipate a decrease in the share of local agricultural producers, which would affect employment in related industries negatively. Food processing industries, especially meat and dairy, are expected to be the most impacted (Erokhin and Ivolga, 2011). The main advantage to trade integration is that WTO membership provides easier access to foreign markets for Russian agricultural producers, which may lessen its future dependence on food imports.

*E-mail: basilic@list.ru

Russia's Trade in Agricultural Commodities

Agriculture counts for a relatively small portion of Russia's gross domestic product (GDP) (about 3%–4% in 2002–2012). According to VLANT (2012), the volume of Russia's GDP (agriculture) exceeded US$111 billion in 2012 (Fig. 16.1).

Compared to other industries that comprise the Russian economy, such as oil production, gas production and mining, agriculture plays second fiddle considering the volume of value added and its share of total exports. The Russian agriculture share of total exports is small – only 1.5%. However, this industry is of great importance for Russia in the context of the employment of rural workers and the overall structural sustainability of the country.

The main arable acreage crops produced in Russia are wheat, barley, sunflowers, oats, potatoes and rye. In 2012, wheat dominated Russia's crop rotation system with 26.6 million hectares (Mha). It produced 20% of the world's sunflower oil, 11.2% of the barley, 5.9% of the wheat, 4% of the milk and dairy products and 3% each of the sugar and poultry (VLANT, 2012).

Despite the high volume of domestic agricultural production, Russia remains a net importer of agricultural commodities. Moreover, its dependency on imports is increasing (Fig. 16.2).

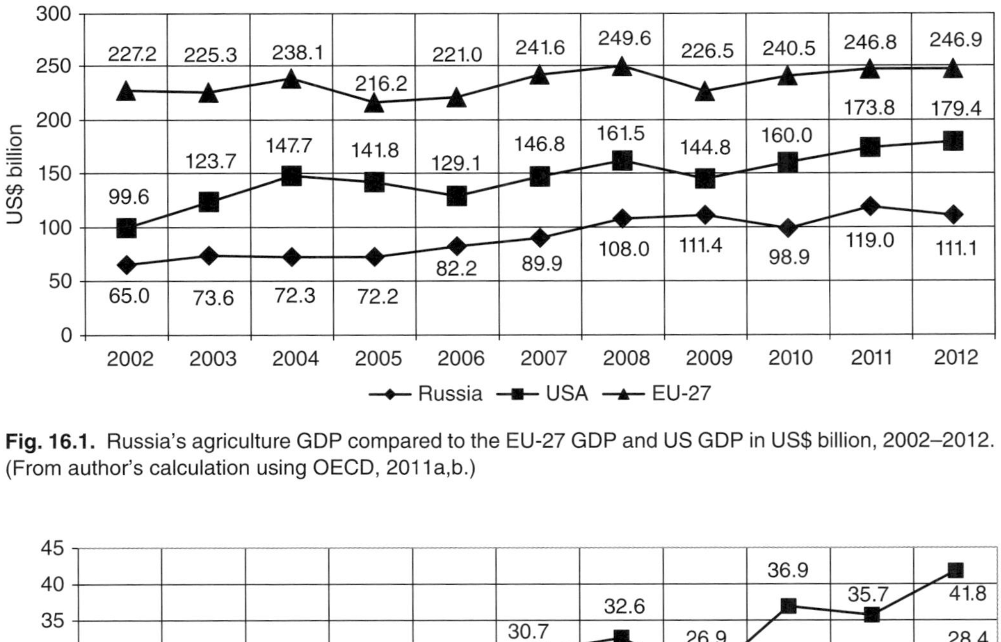

Fig. 16.1. Russia's agriculture GDP compared to the EU-27 GDP and US GDP in US$ billion, 2002–2012. (From author's calculation using OECD, 2011a,b.)

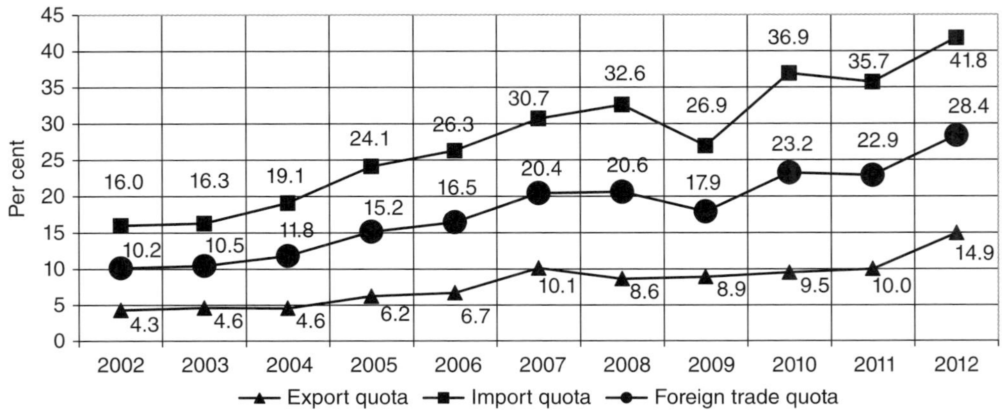

Fig. 16.2. Percentage of export, import and foreign trade quotas of Russia's trade in agricultural commodities, 2002–2012. (From author's calculation using OECD, 2011a,b.)

Import and export quotas show correspondingly the share of imports and exports in GDP, while the foreign trade quota results in the importance of external trade activities for the domestic economy. Agricultural import quota in Russia exceeded 41% of Russia's agricultural GDP in 2012 (Fig. 16.2). The European Union (EU) and the USA accounted for 20% of the agricultural imports to Russia (FAO, 2012d). While agricultural export quotas grew as well (14.2% in 2012), they were quite low. Russia's exports of agricultural commodities, as calculated in US dollars, resulted in US$16.6 billion in 2012 and imports reached US$46.4 billion (passive trade balance of US$29.8 billion) (Nilson, 2011). Passive trade balance was a consistent trend of Russia's foreign trade in agricultural commodities between 2002 and 2012 (Fig. 16.3).

The largest import items in 2002–2012 were meat, milk, dairy products, beverages and sugar. Beef, pork and poultry together made up over 19% of the agricultural imports in 2012, followed by alcoholic and non-alcoholic beverages at 6.3%, cheese at 5.1% and tobacco at 4.2% (VLANT, 2012). Import deliveries of meat (beef, pork and poultry combined) increased fourfold over the 2002–2012 period, while beverages and cheese increased almost fivefold and sugar increased threefold (Table 16.1).

Russia's agricultural exports are more homogeneous than its imports, and are essentially made up of wheat, sunflower oil and barley. Since 2002, Russia has become one of the world's top wheat exporters, with an over 14% share of the world wheat market (positioned fifth in the world after USA, Australia, Canada and the EU) (Nilson, 2011). World trade in wheat in 2012 grew 11% in comparison with 2011 (the maximum growth among the goods

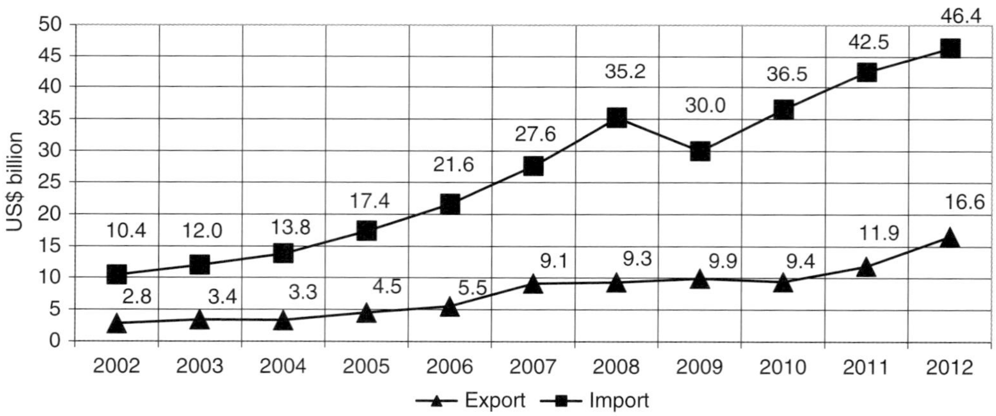

Fig. 16.3. Export and import agricultural products for Russia in US$ billion, 2002–2012. (From author's calculation using OECD, 2011a,b.)

Table 16.1. Imports of selected agricultural commodities for Russia in US$ billion, 2002–2012. (From author's calculation using FAOSTAT, 2013.)

Item	2002	2003	2004	2005	2006	2007	2008	2009	2010	2011	2012
Beef	0.42	0.48	0.42	0.57	0.93	1.03	1.96	1.63	1.63	1.73	1.82
Beverages	0.24	0.37	0.52	0.72	0.71	1.13	1.28	0.81	1.05	1.11	1.13
Pork	0.68	0.65	0.63	0.82	1.40	1.64	2.20	1.91	1.92	1.98	2.03
Milk	0.01	0.03	0.05	0.06	0.02	0.05	0.06	0.04	0.24	0.32	0.36
Tobacco	0.66	0.68	0.70	0.82	0.77	0.94	1.04	1.08	1.08	1.09	1.11
Sugar/honey	0.05	0.06	0.08	0.06	0.07	0.09	0.06	0.08	0.12	0.13	0.14
Poultry	0.81	0.69	0.66	0.84	0.91	1.04	1.32	1.07	0.84	1.05	1.08
Cheese	0.23	0.36	0.42	0.61	0.54	0.78	0.99	0.83	1.26	1.31	1.34

considered), or 14 Mt. The considerable growth of trade in wheat was supported by large crop yields in Russia and other former USSR countries (in Russia and Ukraine, long-term maxima of wheat crops were observed in 2008–2012).

In 2008–2012, wheat provided one-third of the total volume of Russia's agricultural and food exports, and became the most important crop in relation to both export earnings and farm income. Wheat is easy to produce and demand is growing, but such a high dependency on one crop is a serious threat to the sustainability of Russia's agriculture.

Alongside wheat, Russia is one of the leading barley suppliers, with over 16% of the world's barley market share in 2012. Sunflower oil is becoming an essential export item for Russia (VLANT, 2012); its export volume increased 12-fold over the 2002–2012 period (US$0.51 billion in 2012) (Table 16.2).

Russia's main export destinations are Asia, the Middle East and the EU (Table 16.3). China is the largest consumer of Russian agricultural and food commodities (34.4% in Russia's agricultural exports in 2012) (FAO, 2012b). Its share was even higher in 2005–2008. Turkey is the second largest consumer of Russia's agricultural commodities, primarily of grains (13.7% of Russia's agricultural exports in 2012). Turkey has become an important trade partner for Russia, trading its fruit and vegetables in exchange for Russia's wheat, sunflower oil and barley.

The top ten suppliers to Russia include the countries of Latin America, Europe and Asia (Nilson, 2011). The largest agricultural supplier to Russia is Turkey. Its volume of deliveries (4%) has grown rapidly from US$71.1 million in 2002 to US$1212.8 million in 2012 (Table 16.4).

Table 16.2. Exports of selected agricultural commodities for Russia in US$ billion, 2002–2012. (From author's calculations using FAOSTAT, 2013.)

Item	2002	2003	2004	2005	2006	2007	2008	2009	2010	2011	2012
Wheat	0.77	0.78	0.54	1.13	1.37	3.61	2.86	2.75	2.07	1.98	2.13
Sunflower oil	0.04	0.05	0.09	0.19	0.39	0.47	0.68	0.58	0.38	0.44	0.51
Tobacco	0.04	0.05	0.05	0.09	0.08	0.10	0.15	0.14	0.10	0.11	0.13
Beverage	0.08	0.09	0.07	0.10	0.15	0.22	0.25	0.22	0.22	0.21	0.25
Barley	0.21	0.30	0.11	0.20	0.16	0.42	0.32	0.44	0.20	0.25	0.28
Milk	0.03	0.03	0.04	0.05	0.05	0.05	0.07	0.05	0.03	0.05	0.06
Sunflower cake	0.01	0.02	0.05	0.05	0.05	0.12	0.11	0.11	0.11	0.12	0.11
Sugar/honey	0.05	0.03	0.04	0.04	0.07	0.12	0.03	0.06	0.02	0.03	0.04

Table 16.3. Top ten importers of Russian agricultural commodities in US$ million (USMD) and percentage share, 2002–2012. (From author's calculations using FAOSTAT, 2013.)

		2002		2005		2008		2012	
No	Country	Volume	Share	Volume	Share	Volume	Share	Volume	Share
		USMD	Per cent	USMD	Per cent	USMD	Per cent	USMD	Per cent
1	China	906.6	32.4	2290.1	50.9	4,665.9	50.2	3,404.3	34.4
2	Turkey	302.0	10.8	241.3	5.4	1,468.4	15.8	1,356.8	13.7
3	Korea	220.7	7.9	538.4	12.0	383.7	4.1	341.4	3.4
4	Finland	582.3	20.8	792.4	17.6	1,261.0	13.6	326.1	3.3
5	Italy	41.5	1.5	85.5	1.9	410.8	4.4	222.8	2.3
6	India	48.4	1.7	7.5	0.2	121.0	1.3	148.9	1.5
7	Saudi Arabia	8.0	0.3	61.8	1.4	90.3	1.0	140.7	1.4
8	Indonesia	8.9	0.3	0.4	0.0	45.3	0.5	111.7	1.1
9	Germany	375.2	13.4	139.3	3.1	203.7	2.2	102.9	1.0
10	USA	109.1	3.9	63.8	1.4	87.7	0.9	62.1	0.6

Table 16.4. Top ten exporters of agricultural commodities to Russia in US$ million (USMD) and percentage share, 2002–2012. (From author's calculations using FAOSTAT, 2013.)

		2002		2005		2008		2012	
No	Country	Volume	Share	Volume	Share	Volume	Share	Volume	Share
		USMD	Per cent	USMD	Per cent	USMD	Per cent	USMD	Per cent
1	Turkey	71.1	0.7	479.5	2.8	952.7	2.7	1,212.8	4.0
2	Netherlands	94.9	0.9	242.1	1.4	870.9	2.5	1,102.7	3.7
3	Brazil	5.7	0.1	171.8	1.0	605.5	1.7	867.3	2.9
4	Poland	73.4	0.7	399.2	2.3	255.6	0.7	788.6	2.6
5	Norway	0.9	0.1	154.4	0.9	369.8	1.1	621.2	2.1
6	Argentina	14.1	0.1	225.5	1.3	413.8	1.2	582.2	1.9
7	China	47.9	0.5	280.5	1.6	709.9	2.0	515.3	1.7
8	France	25.6	0.2	164.0	0.9	396.2	1.1	471.0	1.6
9	USA	101.4	1.0	115.7	0.7	226.4	0.6	451.8	1.5
10	India	429.3	4.1	317.4	1.8	347.9	1.0	429.7	1.4

However, in spite of the recent successes and solid positions in a number of world agricultural markets, there are still many problems in Russia's agriculture industry. Joining the WTO in 2012 has created trade problems based on WTO veterinary, sanitary and phytosanitary measures or restrictions. As a WTO member, countries must meet the WTO Agreement on Sanitary and Phytosanitary Measures based on the scientifically proven principles of phytosanitary risk.

Principal WTO Obligations of Russia

We analyse the principal obligations undertaken by Russia and examine how they affect its domestic agricultural production and Russia's foreign trade in agricultural commodities. Each WTO member country has to adjust its foreign trade policy in accordance with the WTO's international regulations that are focused on the liberalization of international trade and that have access to the domestic markets of member states. The main disadvantages of such liberalization are related to the competitiveness of domestic products on the domestic market, and also are related to the adaptation of producers to the changing regulations. Each member country is allowed a transition period of 2–7 years based on classification status. For Russia, it is estimated that once the transition period expires, almost one-half of all tariff rates will remain above or at least equal to the current common customs tariff of the Customs Union.

According to the WTO standard approach, each member country binds the aggregate volume of support that distorts trade during the 3-year period preceding Russia's 2012 WTO membership. The allowed level of support for Russia equalled US$9 billion, which was twice that of the standard rules. This allowed level will be decreased gradually, and after 2018 it will be bound on the existing level at that time.

According to the Research Institute of Agricultural Economics of the Russian Academy of Agricultural Sciences (2013), the average weighted tariff rate for agricultural and food commodities in Russia will decrease by one-third (from 15.6% to 11.3% by the end of the transition period). The major changes will take place in 11 commodity groups (Table 16.5).

Tariff rate quotas were excluded from the Russia–WTO agricultural agreement, although they proved effective as protective measures for Russia's domestic meat market. Consequently, Russia lost its capability to decrease meat quotas, which are fixed at the 2013 level. Moreover, quotas for pork will be repealed altogether after 2020. They will be replaced by a 25% flat tariff rate quota.

Membership in the WTO obviously limits opportunities in the independent regulation of external economic activities. Particularly, bound import tariffs limit the manoeuvrability and the flexibility of state customs and tariff regulations. Agriculture in Russia is somewhat protected through subsidies to domestic producers that distort trade and production.

State Support of Agriculture: Challenges of Trade Integration for Russia

The importance of state support of agriculture cannot be overstated, as it distorts the character of international trade in agricultural products. Developed countries, primarily the USA and the EU, have implemented agricultural policies that affect the competitiveness of international trade both directly and indirectly (FAO, 2012a). In 2013, the agricultural GDP share of state support was 39% in the EU and 36% in the USA (Table 16.6).

WTO affiliation has limited Russia's capabilities to regulate its foreign trade activities. In particular, the binding of import customs tariffs restricts the flexibility of the state administration of customs tariff measures (Fig. 16.4).

Although Russia has declared US$9.9 billion as the maximum level of support of domestic agriculture in 2012 and its gradual reduction to US$4.4 billion by 2018, the problem is that Russia does not spend even one-half of this maximum, and no alterations are in sight. Rapid market liberalization and state support of agriculture have been replaced by import substitution and food security provisions in Russia (Erokhin and Ivolga, 2011). The drastic reduction of state support of agriculture in the early 1990s can be illustrated by the dynamics of the producer support estimate (PSE) share in the total revenues of Russia's agriculture (Fig. 16.5).

In 2011, the PSE in Russia reached 21.7%, which was more than the OECD average (18.8%) (Erokhin and Ivolga, 2012). The remarkable fact is that PSE growth in Russia contradicts the situation in the OECD states where PSE has decreased. The bulk of support in Russia is provided through customs tariff regulations. Before its WTO accession, access to the domestic agricultural market was affected by both high tariffs and import quotas (meat is a major import item). About 62% of Russia's imports of agricultural and food commodities are affected by tariffs that are greater than or equal to 10% (Table 16.7).

Table 16.5. Per cent changes in tariffs for agricultural commodities introduced by protocol on Russia's accession to the WTO. (From Research Institute of Agricultural Economics of the Russian Academy of Agricultural Sciences, 2013.)

Item	WTO tariff rate	Common customs tariff rate
	Change (%)	Change (%)
Pigs	5.0	40.0
Pork (fresh, chilled or frozen):		
Non-quota pork	65.0	75.0
Quota pork	0.0	15.0
Edible bypass	15.0	25.0
Other meat and edible meat bypass	15.0	25.0
Milk and concentrated cream	15.0	25.0
Milk whey	15.0	38.0
Cheeses and curd cheese	9.5	19.0
Tea with flavoured additives	12.5	20.0
Rice	10.0	29.0
Palm oil	3.0	0.4
Sausages	9.0	25.0

Table 16.6. Cumulative support of agriculture by WTO member states in 2012, in US$ million (USMD) and GDP per cent of share. (From Erokhin and Ivolga, 2012.)

WTO member countries	Amber box (AMS)		*De minimis*		Blue box		Green box		Cumulative support	
	USMD	Per cent	USMD	Per cent	USMD	Per cent	USMD	Per cent	USMD	Per cent
EU	39,758	58.2	678	6.6	20,239	92.1	19,895	19.1	80,570	39.4
USA	16,803	24.6	7,341	71.4	0	0.0	50,057	48.1	74,200	36.3
Japan	6,461	9.5	0	0.0	845	3.8	23,367	22.5	30,673	15.0
Others	5,272	7.7	2,263	22.0	895	4.1	10,694	10.3	19,124	9.3
Total	68,294	100.0	10,282	100.0	21,979	100.0	104,013	100.0	204,567	100.0

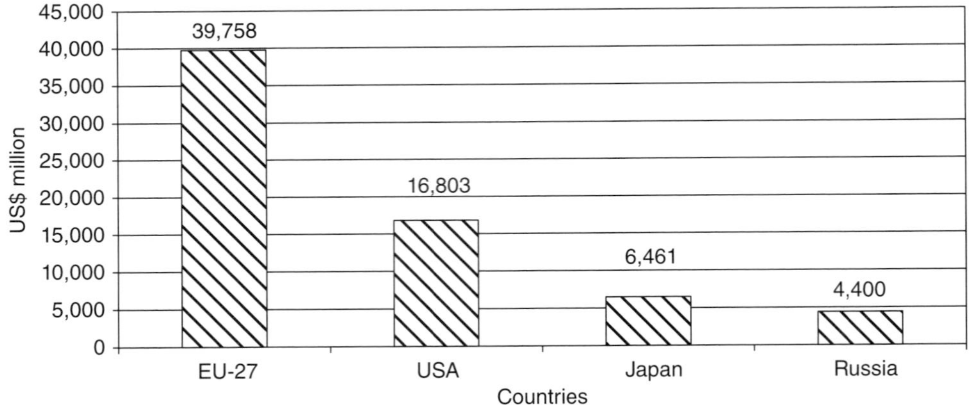

Fig. 16.4. Volume of amber box support in particular counties in 2012, US$ million. (From Erokhin and Ivolga, 2012.)

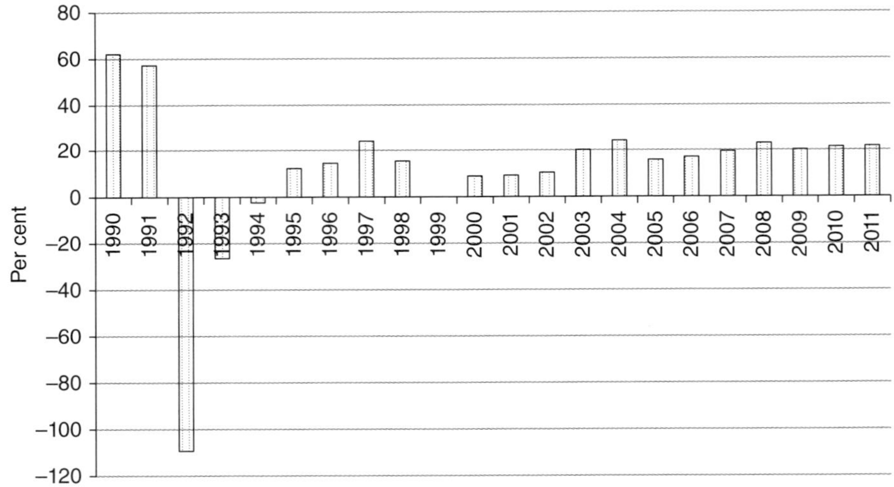

Fig. 16.5. Producer support estimate (PSE) per cent share in total revenues of Russia's agriculture, 1990–2011. (From OECD, 2011a.)

Table 16.7. Percentage distribution of import tariffs for agricultural and food commodities in Russia in 2012. (From FAO, 2012c.)

Distribution	No duties	0 ≤ 5	5 ≤ 10	10 ≤ 15	15 ≤ 25	25 ≤ 50	50 ≤ 100	>100
				Per cent				
No of tariffs	7.8	34.8	8.2	31.8	10.5	3.5	2.9	0.3
Import value	9.9	22.3	6.1	28.7	18.9	8.2	5.9	0.1

Global Agricultural Market: Influences of Trade Integration for Russia

Food and agricultural products are essential commodities, and generally developed countries achieve food security by means of domestic production. To enter those markets, foreign producers must have a substantial competitive advantage, which is unusual in developing countries. Moreover, there are high custom barriers (either tariffs or sanitary regulations) on trade of foreign agricultural commodities (Josling *et al.*, 2010).

The principles of competition and agricultural trade liberalization that underlie WTO activities are difficult for developing countries to achieve (Liapis, 2011). For Russia, this has not been as difficult, since Russia is able to expand its trade with its traditional trade partners through developing its production, processing, storage and transportation infrastructures by refocusing its state support on green box measures.

Conclusions

Russia's accession to the WTO and its involvement in international trade liberalization threatens its food security and increases the country's dependence on foreign agricultural and food commodities. To provide for food security within the context of trade liberalization, Russia needs to receive qualitative state support for its domestic agriculture and needs to maximize its trade effect.

Assessing the effect of trade integration for Russia is complicated. In spite of the negative forecasts about the state of Russia's future agriculture, there are opportunities to protect the domestic market and to support Russia's domestic farmers. The major challenge to protect Russia's agriculture is to enhance its long-term growth and to improve its global competitiveness in relation to its WTO obligations.

References

Erokhin, V.L. and Ivolga, A.G. (2011) Entrepreneurship in agriculture: new challenges of international trade integration. *Contemporary Agriculture: The Serbian Journal of Agricultural Sciences* 60(3–4), 398–402.

Erokhin, V.L. and Ivolga, A.G. (2012) How to ensure sustainable development of agribusiness in the conditions of trade integration: Russian approach. *International Journal of Sustainable Economies Management (IJSEM)* 1(2), 12–23.

FAO (2012a) *Review of Trade and Agricultural Policy*. European Union. Food and Agriculture Organization of the United Nations (FAO), Rome. Available at: http://www.fao.org/fileadmin/templates/est/meetings/wto_comm/RU/Trade_Policy_Brief_EU_RU_final.pdf (accessed 31 July 2013).

FAO (2012b) *Review of Trade and Agricultural Policy, People's Republic of China*. FAO, Rome. Available at: http://www.fao.org/fileadmin/templates/est/meetings/wto_comm/RU/Trade_policy_brief_China_RUS_final.pdf (accessed 31 July 2013).

FAO (2012c) *Review of Trade and Agricultural Policy, Russian Federation*. FAO, Rome. Available at: http://www.fao.org/fileadmin/templates/est/meetings/wto_comm/RU/Trade_Policy_Brief_Russia_Rus_final.pdf (accessed 31 July 2013).

FAO (2012d) *Review of Trade and Agricultural Policy, United States of America*. FAO, Rome. Available at: http://www.fao.org/fileadmin/templates/est/meetings/wto_comm/RU/Trade_policy_brief_US_RUS_final.pdf (accessed 31 July 2013).

FAOSTAT (2013) FAOSTAT website. Available at: http://faostat3.fao.org/home/E (accessed 31 July 2013).

Josling, T., Anderson, K., Schmitz, A. and Tangerman, S. (2010) Understanding international trade in agricultural products. One hundred years of contributions by agricultural economists. *American Journal of Agricultural Economics* 92(2), 424–446.

Liapis, P. (2011) *Changing Patterns of Trade in Processed Agricultural Products*. OECD Food, Agriculture and Fisheries Paper No 47. Organisation for Economic and Co-operation Development (OECD), Paris.

Nilson, A. (2011) Commodity trade between EU-27 and CIS countries, 2000–2010. Eurostat, Statistics in Focus 40/2011. Available at: http://ec.europa.eu/eurostat/documents/3433488/5579324/KS-SF-11-040-EN.PDF/d11e8459-b828-449e-b769-5f0640816689?version=1.0 (accessed 31 July 2013).

OECD (2011a) *Evaluation of Agricultural Policy Reforms in the European Union*. Organisation for Economic and Co-operation Development (OECD), Paris. Available at: http://dx.doi.org/10.1787/9789264112124-en (accessed 31 July 2013).

OECD (2011b) *Agriculture Policy Monitoring and Evaluation 2011*. OECD Countries and Emerging Economies. OECD, Paris.

Research Institute of Agricultural Economics of the Russian Academy of Agricultural Sciences (2013) RIAE website. Available at: http://www.vniiesh.ru/news/9651.html (accessed 31 July 2013).

VLANT (2012) *International Trade and International Cargo Flows in 2011* (final report). Available at: http://vlant-consult.ru (accessed 31 July 2013).

17 Farm Support in Ukraine and Russia under the Rules of the WTO

Lars Brink*
Independent Advisor, Chelsea, Quebec, Canada

Abstract

Ukraine acceded to the WTO in 2008 and Russia acceded in 2012. As WTO members, Ukraine and Russia have binding commitments not to exceed given levels of farm support. The ceilings apply to the sum of certain types of domestic support, calculated in a particular way. This chapter reviews the policies and support measurements in the base periods for the two countries' WTO commitment levels. Also, it examines the farm policy evolution in Ukraine and Russia since WTO accession in terms of compatibility with their WTO domestic support commitments. It gauges the potential for using budgetary support data from OECD policy monitoring to preview the classification and measurement of support for WTO reporting. This chapter assesses the latitude the two countries will have in the future to provide farm support that is limited by WTO commitments or that is exempt from those limits.

Acknowledgements

The author thanks Tim Josling, Iryna Kobuta, William Liefert, David Sedik and Stephan von Cramon-Taubadel for helpful comments on earlier drafts; the author is responsible for any remaining errors and omissions. The author also thanks the Global Issues Initiative at Virginia Tech (Virginia Polytechnic Institute and State University, USA) for financial support.

Introduction

In 1993, Ukraine and the Russian Federation (Russia) applied for accession to the General Agreement on Tariffs and Trade (GATT). After the establishment of the World Trade Organization (WTO) in 1995, the accession processes for both countries continued under WTO rules. Ukraine became a member on 16 May 2008 and Russia became a member on 22 August 2012. The accession process of the Republic of Kazakhstan started in 1996 and was still in negotiations in 2014. In agriculture, the WTO Agreement on Agriculture (Agreement) governs members' policies concerning market access, domestic support, export competition and export prohibitions and restrictions (WTO, 1995). The rules of the Agreement apply, along with the commitment levels on, for example, tariffs, tariff rate quotas, export subsidies and certain domestic support inscribed in the member's schedule of concessions and commitments. In domestic support, the scheduled commitment level or the bound

*E-mail: lars.brink@hotmail.com

total aggregate measurement of support (AMS), is the legally binding ceiling for the yearly amount of support provided through certain kinds of policies and calculated in the prescribed way.

This chapter reviews the classification of policies and measurement of support of Ukraine and Russia in the base periods for their Total AMS calculations that gave rise to their WTO commitment levels in domestic support. It interprets the subsequent evolution of agricultural policy support in Ukraine and Russia to show how classification of policies and measurement of support influence their ability to comply with their bound total AMS: Ukraine's compliance in 2009, apparent non-compliance in 2010 and possible non-compliance in later years, and Russia's secure compliance in its first few years as a WTO member. The policy identification and budgetary support measurements of the Organisation for Economic Co-operation and Development (OECD) complement the information provided to the WTO, but allow only a very guarded expectation of the potential use of OECD data to preview elements of budgetary support for years not yet available through the WTO for Ukraine and Russia. The focus on budgetary support follows from the differences between the OECD and WTO methods for calculating what in both cases is called market price support. Finally, this chapter discusses the latitude the two countries may have in the future to provide farm support of different kinds under the WTO rules.

WTO Rules on Domestic Support

The Agreement's provisions in domestic support concern policies that are domestic; that is, they are not applied at the border. Based on the Agreement's rules for classifying policies and measuring support, a country calculates for each year its current total AMS. This measurement must not exceed the bound total AMS. Policies are excluded from the measurement of support if they meet the requirements and criteria of the Agreement's Annex 2, designed to allow the exclusion of policies with no or at most minimal effects on production or trade (the green box). Support under a second set of policies is excluded from current total AMS if they meet certain other conditions and criteria: direct payments under production-limiting programmes where the basis for payment is fixed or partial, as per the Agreement's Article 6.5 (the blue box). A developing country may exclude support under a third set of policies if they meet conditions and criteria relating to, for example, general availability, low-income or resource-poor producers, or diversification from growing certain crops, as per Article 6.2 (the development box).

Support through policies that do not meet the criteria of Annex 2, Article 6.5 or Article 6.2 is measured through a number of aggregate measurements of support (AMSs). An AMS is calculated for each basic agricultural product (these are product-specific AMSs), and one non-product-specific AMS is calculated for support provided in favour of producers in general. An AMS aggregates the support provided through different policies, such as administered prices, direct payments and input subsidies. The current total AMS is calculated by summing all the AMSs while observing the so-called *de minimis* rules, which allow any AMS that is small enough to be exempted from the summation. An AMS can be exempted if it is no larger than 5% of the product's value of production or, for the non-product-specific AMS, of the value of production of the agriculture sector. Developing countries use 10% (China 8.5%) instead of 5%.

When a country accedes to the WTO, its bound total AMS is established on the basis of the support provided in an agreed base period. As the accession negotiations proceed over time, a succession of base periods can be considered. Ukraine's eventual base period was 2004–2006 and Russia's was 2006–2008. Ukraine and Russia classified their policies and measured their support in these base periods essentially in line with the Agreement's rules and presented these data in the prescribed formats (WTO, 1996). The measurements of AMS support made it possible to calculate current total AMS for the base period years, net of *de minimis* AMSs.

The 3-year average of current total AMS in the base period became the basis for the bound total AMS. Ukraine's bound total AMS of UAH 3043 million is simply that average. As a result of negotiations, Russia's bound total AMS was set at US$9 billion for 2012 and 2013, then declining in equal steps to settle in 2018 at US$4.4 billion, which was the average current total AMS in the 2006–2008 base period. (Russia showed its data in both RUB (Russian roubles) and US$ (US dollars) and scheduled its commitment level in US$). Russia also undertook to keep the sum of its product-specific AMSs at a level no higher than 30% of its non-product-specific AMS through 2017. This makes it impossible for Russia to shift rapidly into a pattern of large product-specific, relative to non-product-specific, AMS support.

The classification of policies and the measurement of support under the WTO deviate in some ways from economics-based practices. The following analysis uses information provided by Ukraine and Russia in their WTO accession and information reported by the OECD in its policy monitoring (OECD, 2013a). The OECD database identifies agricultural policies and measures support through a family of indicators, including the PSE (producer support estimate) and the GSSE (general services support estimate). The Food and Agriculture Organization (FAO, 2013a,b) and von Cramon-Taubadel *et al.* (2008) also review the underlying policies.

A particular issue in OECD and WTO measurements of support is the different meaning of market price support (MPS). The OECD MPS measures the gap between the domestic price and the international reference price in the same year and multiplies it by total production of the commodity. The WTO MPS is calculated only if an administered price is applied. It multiplies the gap between the applied administered price and a fixed external reference price by eligible production, which can be less than total production. While the Agreement requires the fixed external reference price to be based on the years 1986–1988, many countries acceding to the WTO have used the average external reference price for the 3 years of the country's own base period for domestic support in the accession. The OECD PSE measures support provided to producers through agricultural policies. The OECD GSSE measures support to general services for agriculture. The WTO AMS measures support provided in favour of agricultural producers through a set of policies that cannot be excluded on grounds of conforming to Annex 2 or Articles 6.2 or 6.5. There is neither general concordance between GSSE and green box support, nor between PSE and the sum of AMSs. Still, the policies in PSE in the base periods of Ukraine and Russia overlap considerably with those in AMSs. Both countries' green box policies comprised mainly general services and not payments, which would have been classified as PSE support.

Domestic Support in Ukraine

In its 2004–2006 base period, Ukraine indicated sets of policies that varied across the 3 years. Some earlier policies were being phased out in 2004 or 2005, and others were introduced in 2005 or 2006, or applied only in one year. One large support component introduced in 2005 related to the lower rate at which agricultural enterprises contributed to the mandatory pension insurance scheme, with the shortfall being made up from the government budget. The major components of Ukraine's base data and notifications are shown in Table 17.1.

Ukraine reported MPS only for sugar in the base period, thus suggesting that an administered price did not apply to any other product. The reference price for each year was derived from a European export price for that year and was thus not the average price for the 3-year period. The calculation of MPS only for sugar contrasts against the reporting of MPS in PSE, where MPS is calculated for 11 commodities or groups of commodities. However, the OECD MPS is negative for several commodities in one or more of the years from 2004 through 2006, resulting from the domestic price being below the border price (OECD, 2013d). Nevertheless, for sugar, poultry meat and eggs, the OECD MPS is positive in all 3 years.

A significant item in Ukraine's 2004–2006 PSE is the *benefits from fixed agricultural tax* that replaced about a dozen earlier taxes, fees and contributions. Ukraine does not report this support in AMS. While Annex 3 of the Agreement requires the AMS to include the revenue forgone by governments (i.e. support through tax concessions), the practice among most WTO members, and also accession candidates, has been to ignore almost all tax concessions when measuring AMS support. A different kind of benefit provided through the tax system in Ukraine concerns the value-added tax (VAT). In principle, a VAT scheme operates when the agricultural producer pays VAT on the value of inputs purchased (an expense) and charges VAT on the value of the products

Table 17.1. Composition of AMS support in Ukraine, by year (UAH millions) for selected years (2004–2010). (From WTO, 2007, 2011a, 2012b.)

	2004	2005	2006	2009	2010
Market price support for sugar	1,135	676	240	414	1,850
Per hectare payments for crops	–	–	949	–	–
Payments for livestock sold to processing and increased number of head	350	552	590	319	–
Livestock (not milk) subsidies using VAT	409	594	594	293	381
Milk subsidies using VAT	731	1,023	1,023	981	1,448
Interest subsidy (short and long term)	144	415	320	374	622
Benefit/reduced pension contribution rate	–	1,207	1,670	626	–
Input subsidy based on VAT accumulation	1,580	1,713	1,713	2,825	7,099
Sum of all other AMS support	724	618	553	492	1,139
Sum of all AMS support	5,073	6,798	7,652	6,324	12,539
Total ag production value (UAH billion)	83.9	92.5	94.3	1,54.2	192.7
AMS support as share of production value	6.0%	7.3%	8.1%	4.1%	6.5%
Current total AMS	2,558	3,018	3,555	–	–

Note: Current total AMS is the sum of all AMS support less *de minimis* AMSs. Bound total AMS is the 2006–2008 average of yearly current total AMS.

sold to the buyers of agricultural products, such as processors (a revenue). With the VAT revenue larger than the VAT expense, the producer delivers to the government the difference between the two VAT amounts. Complications arise when different products or transactions are subject to different VAT rates or to exemptions. In 2004–2006, Ukraine reported AMS support generated by two special features of the VAT scheme as applied in agriculture: (i) accumulation of VAT for purchasing agricultural production inputs; and (ii) subsidies using VAT.

Accumulation of VAT for purchasing agricultural production inputs is a large item in Ukraine's base data, as reported in the non-product-specific AMS. Under Ukraine's VAT provisions for agriculture, the difference between the producer's VAT revenue and VAT expense is not delivered to the government, but is placed in a special account held by the producer. The producer may use the funds accumulated in this account to purchase only inputs for agricultural production. Beginning in 2009, some of the funds can be used only to pay VAT on purchased inputs and other funds can be used for any other production purpose (OECD, 2013b). Other changes have been proposed (Berlin Economics, 2010; Kobuta, 2013). Sales of milk and meat are subject to their own special VAT provisions. The support in the WTO base data coincides with that in PSE (*Input subsidy based on VAT accumulation*) in 2004 and 2005, and the WTO entry for 2006 is the same as for 2005. While the PSE entry is attributed explicitly to all commodities except milk and meat, the AMS entry does not make it explicit that the non-product-specific nature of the subsidy is compromised by excluding the sales of milk and meat.

Subsidies using VAT is the other significant VAT-related item in Ukraine's base data. These subsidies are reported for milk, cattle, pigs and poultry (i.e. products that do not enjoy the accumulation of VAT for input purchases). Processing enterprises that are buyers of milk, cattle, pigs and poultry determine the amount of VAT payable to the government on their sales of processed milk and meat products. However, instead of paying VAT to the government, the processing enterprises accumulate the funds in separate bank accounts and transfer the funds to producers of milk, cattle, pigs and poultry as price top-ups on their sales to the processors. The scheme also has been called redirection of processors' VAT (OECD, 2013b). The amounts reported in 2005 and 2006 AMS data coincide with those reported in PSE as per tonne payments for livestock, broken down by milk, beef and veal, pig meat and poultry.

Matching the policies and amounts of support in Ukraine's 2004–2006 AMS data with those in PSE is difficult, even though the underlying budget-sourced data are the same.

The different names used for the same policy is only a minor issue. Different breakdowns of support under different headings require arithmetic to establish the matching entries and create complexity. For example, one entry in the AMS data can be the sum of two PSE entries (the interest subsidy in AMS is broken down by long-term and short-term credit in PSE), and the sum of two or more entries in AMS, even in separate supporting tables, appears as one entry in PSE (payments for young cattle sold to processing enterprises, payments for number of beef cattle, and subsidies using VAT). Still, it is possible to match most AMS entries with most PSE entries. Also, it is possible to single out PSE items that are not in the AMS data, such as the benefit from fixed agricultural tax, and vice versa, such as the benefit from reduced pension fund contributions. Matching the amounts of several 2004–2006 green box policies with similar entries in the OECD GSSE requires calculation for some items. In sum, the matching exercise reveals that numerous policies and their associated support in the WTO AMS and green box can be inferred from data on OECD PSE and GSSE, but the differences in data selection, aggregation and presentation would make it difficult to generate a WTO notification using only OECD data.

In 2011 and 2012, Ukraine notified the WTO Committee on Agriculture of its 2009 and 2010 support. Three of the largest support items in 2004–2006 were the input subsidy based on VAT accumulation, the livestock subsidies for milk using VAT and the livestock subsidies for meat (cattle, pigs, poultry) using VAT (Table 17.1). The reduced rate of pension contribution was the second largest support item in 2005 and 2006. MPS for sugar was also large, but it declined through 2006. By 2009 and 2010, the livestock subsidies for meat using VAT had dropped significantly in value, and the benefit from the reduced rate of pension contribution was shown as nil in 2010. While the livestock subsidies for meat using VAT fell from the 2004 to 2006 period, the livestock subsidies for milk using VAT increased considerably. The most spectacular increase, however, was that of the input subsidy based on VAT accumulation, which rose fourfold between 2006 and 2010 to become the largest single support entry. Most support amounts in the 2009 and 2010 notifications did not match those in PSE, although some were reasonably close. Similar to 2004–2006, this seems to be due to differences in how data were selected, aggregated and presented, which reinforces the doubt about the feasibility of developing a preview of notifications using only PSE and GSSE estimates for 2011 and 2012.

In 2009, MPS for sugar remained approximately at the levels of 2005 and 2006: the administered price was raised significantly but the eligible production was cut sharply. In 2010, however, MPS for sugar soared, to become the second largest AMS item in the notification. This was mainly the result of a further 36% increase in the administered price between 2009 and 2010 (Table 17.2). Nevertheless, the MPS for sugar would have been even larger both in 2009 and 2010 if Ukraine had followed the rules of the Agreement. Ukraine's MPS calculations in 2009 and 2010 did not use the fixed external reference price required by Annex 3 of the Agreement. Instead, Ukraine increased the average external reference price from 2004 to 2006 by a factor corresponding to the increase in the consumer price index since 2006, which reduced the price gap considerably (Table 17.2). If Ukraine had used the fixed external reference price without adjustment, the country would still have complied with its bound total AMS commitment in 2009. However, in 2010, the MPS for sugar, and hence the AMS for sugarbeet, would have been so large that Ukraine would have exceeded its bound total AMS.

The Agreement's Article 18.4 requires members in the review process (i.e. the Committee on Agriculture) to *give due consideration to the influence of excessive rates of inflation on the ability of a member to abide* by its commitment. This is not the same as enabling a notifying member to adjust its fixed external reference price unilaterally (as Ukraine did for 2009 and 2010), even if it does so to appear to stay within its bound total AMS (as Ukraine did for 2010). There is no definition of excessive rates of inflation, which leaves the determination of what is *excessive* to the Committee on a case-by-case basis. There is no definition of due consideration – it may mean that the Committee can allow a violation of the bound total AMS commitment if the violating member argues that this is due to excessive inflation. In the case of Ukraine, some members argued strongly that Ukraine should submit a revised

Table 17.2. Current total AMS with alternative methods for sugar market price support in Ukraine. (From author's calculations.)

			2009		2010	
Sugar produced from sugarbeet:						
Administered price	a	UAH/t	3125	3125	4250	4250
Fixed external reference price (FERP)	b	UAH/t	1747	–	1747	–
Adjusted external reference price (AERP)	c	UAH/t	–	2798	–	3053
Price gap using FERP	d=a–b	UAH/t	1378	–	2503	–
Price gap using AERP	e=a–c	UAH/t	–	327	–	1197
Eligible production	f	1000 t	1267	1267	1545	1545
MPS using FERP	g=f*d	UAH million	1746	–	3867	–
MPS using AERP	h=f*e	UAH million	–	414	–	1850
Sugarbeet:						
Payments	i	UAH million	0	0	198	198
AMS using FERP	j=i+g	UAH million	1746	–	4065	–
AMS using AERP	k=i+h	UAH million	–	414	–	2048
Current total AMS	l	UAH million	2445	1113	4198	2181
Bound total AMS		UAH million	3043	3043	3043	3043

Notes: a = minimum price for sugar produced from sugarbeet (WTO, 2011a, 2012b); b = average of external reference prices 2004–2006 (WTO, 2007). Ukraine's notifications use the currency symbol Hrv, not the ISO symbol UAH in Ukraine's base data and schedule; c = external reference prices as increased by yearly inflation rates in 2007, 2008, 2009 and 2010 of 16.6%, 22.3%, 12.3% and 9.1%, respectively (WTO, 2011a, 2012b); f = production of sugar refined from domestic sugarbeet (WTO, 2011a, 2012b); i = per hectare payments for sugarbeet (WTO, 2011a, 2012b); l = includes non-*de minimis* AMSs for other products (WTO, 2011a, 2012b).

notification without inflating the fixed external reference price (perhaps similar to Table 17.2). Even if it did not change the facts, this would increase transparency significantly by showing the difference between MPS for sugar with and without adjusting the fixed external reference price for all inflation. This would also clarify what the Committee considers to be excessive rates of inflation, as opposed to Ukraine's implied claim that any inflation is excessive.

Using only eligible production instead of total production deviates from economics-based indicators of support. However, calculating MPS based on the administered price and eligible production would use only variables controlled by the government in combination with the fixed external reference price. The WTO MPS is designed as a policy variable the government controls; thus, it is not a measurement of support. It is effectively a penalty for using an administered price: applying that price generates a WTO MPS that enters the product's AMS, which, if it is large enough, enters the current total AMS, and the larger current total AMS may or may not fit within the bound total AMS. Rejecting the WTO MPS because it is not an economic measurement of support is therefore not a convincing argument in the WTO context. If there is no administered price, the domestic price can still be kept higher than the international price with the help of border measures.

The large increase in the administered price for sugar would have caused Ukraine to exceed its bound total AMS in 2010 if it had used a fixed, rather than an adjusted, external reference price. Most of the other products fell below their *de minimis* AMS thresholds of 5% of the value of production. The non-product-specific AMS of UAH 8.2 billion in 2010 was much larger than in 2009, a result of the increase in support through special VAT regimes. The non-product-specific *de minimis* threshold was UAH 9.6 billion in 2010 (Table 17.1).

If Ukraine reports a large increase in some item, such as support through the VAT regime, in the non-product-specific AMS, it will eventually notify for 2011 and later years that non-product-specific AMS may exceed its *de minimis* threshold. The whole non-product-specific AMS of more than UAH 10–12 billion would then be included in the current total AMS, greatly violating the bound total AMS of about UAH 3 billion. The excess in 2011 or later would be even larger if the administered price for sugar was also raised, generating a still larger AMS for sugarbeet. In April 2013, a draft law was approved, which would have abolished the minimum prices for sugar and sugarbeet for 2013/14. If such an initiative were actually to become law in the future, the absence of an administered price for sugar would eliminate the need to calculate market price support in AMS for sugarbeet, and make Ukraine's current total AMS smaller.

In the notifications Ukraine will eventually submit for 2011 and later years, it might seek to invoke Article 18.4 for the scenarios contemplated here and argue it exceeded the bound commitment level because of excessive inflation. The reaction of the Committee to such a move would test the strength or lack thereof of the whole set of provisions in the Agreement designed to constrain large amounts of distorting support. If the Committee were to accept the argument that bound total AMS was exceeded only because of excessive inflation, the door would be open for any number of members to argue along similar lines. Article 18.4 would effectively make it possible to provide large amounts of ongoing AMS support without consideration of the rules and commitment intended to curb such support. Instead of being used only sparingly, invocation of Article 18.4 would become the standard for many members wishing to bypass the constraint on some types of distorting domestic support.

Domestic Support in Russia

In its 2006–2008 base period, Russia reported a stable set of green box policies and a more variable set of AMS policies, generating an average current total AMS of US$4.4 billion. Annual green box support was US$2 billion or more, mainly for agricultural science and research, training and education, pest and disease control measures, inspection services and infrastructural services. Direct payments in the green box were minor. Most of the product-specific AMSs, including milk and meat, were *de minimis*. The AMSs for flax, hemp, reindeer and horses were not *de minimis*. Per head payments for sheep, reindeer and horses were delivered as production subsidies, and per unit payments for milk, meat, eggs and wool were output subsidies. Russia did not calculate market price support in 2006 or 2007, but it did so for wheat, rye, barley and maize in 2008. The 2008 external reference price was the average export price for the 2006–2008 period. The administered price for each of these crops was below the external reference price, which generated negative market price support. Because there were no product-specific AMS payments for these crops, their AMSs were nil.

In the 2006–2008 period, Russia reported 91%–94% of its AMS support as non-product-specific AMS; the data included no tax concessions. A variety of input subsidies dominated the non-product-specific AMS in 2006–2008 (Table 17.3). They included soft credit (mainly interest subsidies) and subsidies for fertilizer, chemicals, fuel, lubricants and livestock feed. Smaller amounts of production subsidies and output subsidies were also reported as non-product-specific AMS, broken down into two groups of subsidies: crops and livestock. The classification of these subsidies as non-product-specific (i.e. support in favour of agricultural producers in general) was discussed in the accession process (WTO, 2011b). Russia explained the non-product-specific classification as covering a wide range of products.

Russia's 2006–2008 data showed both a federal and a regional component of budgetary support. MPS was calculated only at the federal level. The federal and regional components need to be added to compare it to the OECD's PSE data. Matching the support items in Russia's base data with the entries in the PSE data for 2006–2008 is complicated, because very few of the policies have the same names. Many of the 2006–2008 policies in the PSE data were not identified in the AMS data. While the PSE data identify about 50 policies in 2006–2008, the AMS data identify only 27 policies.

On the other hand, some policies showing no support in the 2006–2008 PSE data show support in the 2006–2008 AMSs. For example, the PSE data reported support in the form of an electricity price subsidy up through 2002 but not thereafter, since the electricity prices for agricultural and industrial producers were

Table 17.3. Components of non-product-specific AMS for Russia (RUB billion), 2006–2012. (From WTO, 2012a; OECD, 2013c.)

Component	2006	2007	2008	2009	2010	2011	2012
Fertilizer and chemical subsidies	3.2	3.5	14.8	13.3	14.5	8.6	10.0
Fuel and lubricant subsidies	21.4	9.7	13.8	11.0	7.0	20.1	24.2
Feed subsidies	0	0	11.8	0.7	0.8	6.3	6.7
Machinery and livestock leasing	5.3	3.0	4.1	1.3	3.5	4.8	13.5
Soft lending	15.9	13.9	39.6	52.2	56.2	55.9	58.2
Interest and investment subsidies	15.1	23.2	10.6	–	–	–	–
Other subsidies/debt write-off	6.8	21.9	13.5	–	–	–	–
Regional support for fixed costs	–	–	–	–	–	1.5	24.7
Sum of all other NPS AMS items	34.0	22.7	30.7	30.7	30.7	30.7	30.7
Non-product-specific AMS	101.8	97.9	138.8	109.1	112.8	127.9	168.1
De minimis threshold, NPS AMS	80.9	96.6	123.1	125.8	129.4	163.1	159.5

Notes: 'Sum of all other NPS AMS items' assumed the same as in 2008. *De minimis* threshold 2009–2012: 5% of value of production in Table 15.1 in Federal State Statistics Service (2013). The 2009–2012 support is the author's selection from OECD (2013c). NPS = non-product-specific.

aligned (OECD, 2011). The AMS data include such subsidies for 2006–2008. With regard to the reporting of green box support for 2006–2008, it is generally not possible to match the entries in the accession database with corresponding entries in the OECD GSSE data. However, the sum of the federal and regional components of the Federal Task Programme 'Fertility' in the accession database does match the entry for the Federal Programme Improvement of Soil Fertility: Land Amelioration in the OECD GSSE data in 2008.

To the extent possible, Table 17.3 uses OECD PSE data to estimate a few large support items in Russia's non-product-specific AMS for 2009–2012. Soft lending – a form of interest subsidy – saw a large increase from the 2006 to 2008 period. While some kinds of investment-related support were phased out, new forms were introduced, particularly involving regional governments. The *de minimis* threshold for non-product-specific AMS increased only slowly from 2008, followed by a large jump in 2011. While the projections for 2012 are highly tentative and include the assumption that many support items remain unchanged from 2008, the 2012 non-product-specific AMS may well have been larger than its *de minimis* threshold.

The State Programme for Development of Agriculture and Regulation of Agricultural Commodities Markets 2013–2020, or State Programme 2013–2020, reveals some of the structure and level of support after the completion in 2018 of Russia's reduction of its bound total AMS to US$4.4 billion (OECD, 2013a; USDA, 2012, 2013b). While soft lending will be extended to investors downstream of farming, credit support is not slated for a further increase. From 2013, support started to be shifted toward direct payments to producers rather than the traditional subsidies for interest, seeds, fuel, fertilizer and chemicals. The direct payments would be paid per hectare of sown area for crop production and per litre of milk for milk production, with some adjustment to take account of animal numbers. Such a shift would result in a smaller non-product-specific AMS and larger product-specific AMSs. The payments have been referred to as decoupled by Russian officials, and it appears they might be viewed as not being AMS support. It is unclear on what basis the payments would meet the criteria for the green or blue boxes in general and the green box criteria for decoupled income support in particular. The State Programme 2013–2020 introduced a premium subsidy for livestock insurance, complementing the existing premium subsidy for crop insurance.

The federal budgetary support envisaged in the State Programme 2013–2020 would start at a budgeted 2013 level only slightly above the 2012 level (USDA, 2013a). Expenditure would rise from RUB 159 billion in 2013 to RUB 218 billion in 2020. Much of the support is delivered by regional governments, often with a co-financing requirement. Assuming an average co-financing level of 50% of the federal budgetary support, the sum of federal and regional budgetary support would be RUB 238 billion in 2013 and RUB 327 billion in 2020. In the 2006–2008 base period, green box support accounted for between 40% and 48% of the sum of green box and AMS support. Assuming that in the 2013–2020 period, green box (and any blue box) support would make up 40% of each year's budgetary support, the budgetary AMS support (federal and regional) would rise from RUB 143 billion in 2013 to RUB 196 billion in 2020. At an exchange rate of 31 roubles per US dollar, the 2013 budgetary AMS support would amount to US$4.6 billion, while the corresponding 2020 amounts would be US$6.3 billion. With somewhat different assumptions, Sedik *et al.* (2013) project similar levels of budgetary AMS support.

Under the *de minimis* rules, the current total AMS in each year would be smaller than the full budgetary AMS support, particularly if the shift away from interest subsidies and subsidies for various inputs makes the non-product-specific AMS less than 5% of value of production in agriculture. There is no evidence of a trend over time to replace AMS support with green box support. In the Single Economic Space, Russia has agreed with Kazakhstan and Belarus to keep certain support to producers of agricultural commodities at no more than 10% of the value of production in agriculture (Brink, 2014).

Higher administered prices in the future might be applied for important crops, and the resulting MPS might increase the product's AMS. Increased direct payments could raise some product-specific AMSs such that they might become larger than their *de minimis* thresholds and thus increase current total AMS (Brink *et al.*, 2013). Keeping the current total AMS below the bound total AMS of US$9 billion in 2013 was not a problem. However, staying below the future bound total AMS of US$4.4 billion may depend on how much AMS support is exempted from the current total AMS on *de minimis* grounds. If the non-product-specific AMS was *de minimis*, there would be plenty of room to increase future product-specific AMSs. These considerations underscore the crucial roles played by the potential market price support and by the *de minimis* exemptions in projecting future compliance with the bound total AMS, roles often overlooked when focusing on the amount of budgetary support alone.

Conclusions

The domestic support data for the base periods of the WTO accessions of Ukraine and Russia confirm the pattern of policy support apparent from other data sources. The 2004–2006 policy instability in Ukraine continued in 2009 and 2010, the two notified years. The dominant evolution in 2009 and 2010 was the spectacular rise in input subsidies based on VAT accumulation enjoyed by livestock producers, a policy that has come under review. Ukraine's calculation of MPS for sugar has encountered severe scrutiny in the WTO Committee on Agriculture. In Russia, a stable set of policies in 2006–2008 provided most of the AMS support, but support under other policies was introduced or eliminated in that period. Soft lending increased rapidly until 2012.

OECD support data are usually available one or more years before notified WTO support data are circulated, although much of the data likely come from the same sources. Attempting to match GSSE and PSE data with green box and AMS data met with limited success. Differences in naming the policies and in various breakdowns (short term, long term, federal, regional, etc.), along with a lack of explanation as to why a policy was or was not included in the reporting to the WTO, made the matching of GSSE and PSE data with green box and AMS data difficult. The conclusion about the feasibility of preparing pre-notifications using OECD data is therefore less sanguine with regard to Ukraine and Russia than is suggested more generally by Josling and Mittenzwei (2013).

Looking to the future, Ukraine may face problems staying within its bound total AMS if its input subsidies through VAT accumulation continue to increase. At some point, this support may become large enough for the non-product-specific AMS to exceed its *de minimis* threshold, making the current total AMS several times larger than the bound total AMS. Calculating MPS for sugar in line with WTO methodology (i.e. without Ukraine's proposed indexing) would by itself also generate a current total AMS larger than the bound total AMS. The legal consequences of violating the bound total AMS commitment remain untested. In the case of Russia, the increases in support from 2008 appear moderate enough for the current total AMS to stay well below the bound total AMS. A key variable will be the value of production in agriculture. If it is large enough in 2018 and beyond, Russia's non-product-specific AMS may become *de minimis*, which would allow product-specific AMSs to increase within the bound total AMS.

References

Berlin Economics (2010) *VAT Refund Arrears in Ukraine.* Policy paper on behalf of the American Chamber of Commerce in Ukraine. BE Berlin Economics GmbH, Berlin. Available at: http://www.berlin-economics.com/download/studien/VAT%20Refund%20Arreas%20UKR%20BE_en.pdf (accessed 15 November 2013).

Brink, L. (2014) Countries in the Commonwealth of Independent States: Agricultural policy issues in the context of the World Trade Organization. Food and Agriculture Organization, Rome. Available at: http://www.fao.org/3/a-i3883e.pdf (accessed 17 July 2014).

Brink, L. Orden, D. and Datz, G. (2013) BRIC agricultural policies through a WTO lens. *Journal of Agricultural Economics* 64(1), 197–216, doi:10.1111/1477-9552.12008.

FAO (2013a) Agriculture and trade policy background note: Russia. Food and Agriculture Organization (FAO), Rome. Available at: http://www.fao.org/fileadmin/templates/est/meetings/wto_comm/Trade_Policy_Brief_Russia_final.pdf (accessed 15 November 2013).

FAO (2013b) Agriculture and trade policy background note: Ukraine. FAO, Rome. Available at: http://www.fao.org/fileadmin/templates/est/meetings/wto_comm/Trade_Policy_Brief_Ukraine_final.pdf(accessed 15 November 2013).

Federal State Statistics Service (2013) Agricultural production by types of farms. Moscow, Russia. Available at: http://www.gks.ru/bgd/regl/b12_12/IssWWW.exe/stg/d01/15-01.htm (accessed 15 November 2013).

Josling, T. and Mittenzwei, K. (2013) Transparency and timeliness: the monitoring of agricultural policies in the WTO using OECD data. *World Trade Review* 12, 533–547, doi:10.1017/S1474745612000535.

Kobuta, I. (2013) Ukraine: Domestic support in agriculture and its membership in the WTO. Presented at *FAO Workshop on WTO Commitments and Support to Russian Agriculture: Issues and Possible Solutions*, January, Kazan, Russia.

OECD (2011) Definitions and sources. Russia: estimates of support to agriculture. Organisation for Economic Co-operation and Development (OECD), Paris. Available at: http://www.oecd.org/tad/agricultural-policies/48708508.pdf (accessed 15 November 2013).

OECD (2013a) *Agricultural Policy Monitoring and Evaluation 2013: OECD Countries and Emerging Economies.* OECD, Paris.

OECD (2013b) Definitions and sources. Ukraine: Estimates of Support to Agriculture. OECD, Paris. Available at: http://www.oecd.org/agriculture/agricultural-policies/48708610.pdf (accessed 15 November 2013).

OECD (2013c) Russia. Agricultural policies and support – Producer and consumer support estimates database. OECD, Paris. Available at: http://www.oecd.org/agriculture/agricultural-policies/producerandconsumersupportestimatesdatabase.htm (accessed 15 November 2013).

OECD (2013d) Ukraine. Agricultural policies and support – Producer and consumer support estimates database. OECD, Paris. Available at: http://www.oecd.org/agriculture/agricultural-policies/producerandconsumersupportestimatesdatabase.htm (accessed 15 November 2013).

Sedik, D., Lerman, Z. and Uzun, V. (2013) Agricultural policy in Russia and WTO accession. *Post-Soviet Affairs* 29, 500–527, doi:10.1080/1060586X.2013.817160.

USDA (2012) *Agriculture Development Programme 2013–2020*. GAIN Report Number RS1270. Foreign Agricultural Service, US Department of Agriculture (USDA), Washington, DC.

USDA (2013a) *Federal Agricultural Budget in 2013*. GAIN Report Number RS1314. Foreign Agricultural Service, USDA, Washington, DC.

USDA (2013b) *GOR Resolutions on Distribution of Agricultural Subsidies*. GAIN Report Number RS1303. Foreign Agricultural Service, USDA, Washington, DC.

von Cramon-Taubadel, S., Nivyevskiy, O., von der Malsburg, E. and Movchan, V. (2008) Ukraine. In: Anderson, K. and Swinnen, J. (eds) *Distortions to Agricultural Incentives in Europe's Transition Economies*. The World Bank, Washington, DC, pp. 175–217.

WTO (1995) Agreement on Agriculture. In: *The Results of the Uruguay Round of Multilateral Trade Negotiations. The Legal Texts. World Trade Organization*. Cambridge University Press, Cambridge, UK.
WTO (1996) Information to be provided on domestic support and export subsidies in agriculture. WT/ACC/4. Technical note by the Secretariat. World Trade Organization (WTO), Geneva, Switzerland.
WTO (2007) Domestic support and export subsidies in the agricultural sector – Revision. WT/ACC/SPEC/UKR/1/Rev.12. Working Party on the Accession of Ukraine, WTO, Geneva, Switzerland.
WTO (2011a) Notification. G/AG/N/UKR/9. Committee on Agriculture, WTO, Geneva, Switzerland.
WTO (2011b) Report of the working party on the accession of the Russian Federation to the World Trade Organization. WT/ACC/RUS/70. Working Party on the Accession of the Russian Federation, WTO, Geneva, Switzerland.
WTO (2012a) Domestic support and export subsidies in the agricultural sector. WT/ACC/SPEC/RUS/39. Working Party on the Accession of the Russian Federation, WTO, Geneva, Switzerland.
WTO (2012b) Notification. G/AG/N/UKR/13. Committee on Agriculture, WTO, Geneva, Switzerland.

18 Accession of KRU to the WTO: The Effect of Tariff Reductions on KRU and International Wheat Markets

Saule Burkitbayeva[1]* and William A. Kerr[2]

[1]*LICOS KU Leuven, Belgium;* [2]*University of Saskatchewan, Saskatoon, Canada*

Abstract

Wheat trade accounts for one-third of the world grain trade, and it is expected to double by 2050. The KRU (Kazakhstan, Russia and Ukraine) countries account for approximately one-quarter of world wheat exports and are collectively considered one of the key wheat exporting regions. Ukraine became a member of the World Trade Organization (WTO) in 2008, Russia in 2012, and Kazakhstan is expected to reach accession soon. WTO accession entitles member countries to most-favoured-nation (MFN) tariffs and access to important markets largely inaccessible due to very high tariffs that can be charged on imports from WTO non-member countries. The KRU region's increased market accessibility as a result of WTO membership has the potential to foster important re-alignments in world wheat trade flows and prices and changes in welfare economics among the major wheat trading countries. Therefore, we examine world wheat trade liberalization effects using the global simulation model (GSIM).

Introduction

Kazakhstan, Russia and Ukraine (KRU) have long been recognized as having the potential to be major contributors to international trade in wheat (Falkus, 1966; Schmitz and Bawden, 1973). Major events in the 20th century have thwarted the realization of the KRU countries' international trade potential: the First World War; the Russian Revolution; the establishment of a command economy under communism; agricultural collectivization under Stalin; the Second World War; the fostering of a domestic grain-based livestock industry in the latter part of the communist era; the end of the command economy and the Soviet Union in 1991; and the chaotic early years of the transition reform to a market-based economy (Hobbs *et al.*, 1997; MacKay and Kerr, 2007). It is only during the 21st century that sufficient stability has existed for the creation of new investments in grain production and supply chains (Hobbs and Boyd, 2007; Burkitbayeva, 2013). The result has been that the KRU export share in global wheat trade increased from 2% in 1991 to 23% in 2008 (USDA, 2009).

Factors inhibiting the full integration of the KRU countries into the global trade in wheat include both domestic and external constraints. As non-members of the WTO for many years, the KRU countries were unable to benefit from the low most-favoured-nation (MFN) tariff rates of WTO member states. The tariff rates of trade barriers that are applied to WTO non-members are often very high and are not subject to WTO

*Corresponding author; e-mail: Sab868@mail.usask.ca

disciplinary actions (Kerr, 2010). The accession negotiations for Ukraine (Gaisford and Kerr, 2004; Antonyuk and Kerr, 2005), Russia (Gaisford and Iourkova, 2007), and Kazakhstan (Burkitbayeva, 2013) have been long and difficult, with Ukraine joining the WTO in 2008, Russia in 2012 (Kerr, 2012) and Kazakhstan still in negotiations in 2014 (Burkitbayeva, 2013). As stated, WTO membership allows member countries MFN tariff rates and provides access to importers' markets on an equal footing with other WTO members. Reductions in tariffs can be substantial. For example, pre-accession and post-accession MFN tariffs for wheat are as follows: China, 180% pre-accession versus 65% MFN; Turkey, 130% pre-accession versus 65% MFN; Switzerland, 118% pre-accession versus 36% MFN; and Kenya, 35% pre-accession versus 0% MFN. This chapter assesses the effect of the movement to MFN rates for the KRU countries and other major players in the international trade of wheat.

Accession Process of Russia, Ukraine and Kazakhstan to the WTO

One of the underlying premises of the multilateral system of trade rules is that countries would join to receive preferred access to member state markets (Kerr, 2010). The number of member state countries has expanded over the years (159 as of June 2014).

While the rules of trade impose market-based disciplines on acceding countries, they also provide considerable benefits. Firms that wish to make investments in international trading activities need to know that their investments are not at risk from government actions that could make their investments unprofitable. For example, governments can put profitable investments at risk through trade barriers. Liberalizing trade flow is at the core of the WTO trading system. Having confidence in being treated equally and fairly is important for countries and firms that invest in international trade activities. Albeit trade disputes will always remain, the WTO trading system with a set of rules and conditions reached by consensus allows handling trade disputes in a transparent and structured manner.

While there is considerable incentive to join the multilateral system and membership is encouraged, the process of WTO accession is demanding and can require years of relatively one-sided negotiations. Each applicant country is required to go through a five-stage accession process. First, after gaining observer status, the country has to submit a memorandum outlining how its trade and economic policies align, or do not align, with the WTO's rules. The memorandum is examined by a General Council Working Party (Working Party) that includes all member countries expressing an interest in being part of the Working Party. Second, applicant countries conduct bilateral negotiations with individual members of the Working Party. Commitments made in bilateral negotiations are applied equally to all member states following the WTO principle of non-discrimination. Third, after the bilateral negotiations are completed, the applicant's trade regime is re-examined for compliance with all the WTO rules. Fourth, the final accession terms are contained in a formal protocol of accession and tariff schedules. The fifth and final step is the submission of a final package that consists of the protocol and tariff schedules. This final package is presented to the WTO General Council or a Ministerial Conference. The observer (applicant country) is then free to join the organization if two-thirds of the WTO members vote in the affirmative (WTO, 2011).

The duration of the process of accession to the WTO differs from one case to another, depending on the number of bilateral negotiations that must be completed to achieve consensus. Political matters can also slow down the accession process. For example, while it took 19 years of negotiations for Russia to become a WTO member state, it took less than 3 years for Kyrgyzstan to join the WTO.

After the dissolution of the Soviet Union in 1991 and the achievement of independent country status, Russia, Ukraine and Kazakhstan applied for WTO membership. Russia was the first of the three countries to express its willingness to join the multilateral system. In June 1993, Russia submitted its official application to the Secretariat of the General Agreement on Tariffs and Trade (GATT). (Note: GATT, 1947–1995, was the forerunner of the WTO, 1995 and onward.) The General Council Working Party was established that same year and consisted of 58 members. A memorandum on Russia's foreign trade regime was circulated by the GATT Secretariat in 1994.

The process proceeded with questions and feedback to Russian authorities. The meetings of the Working Party commenced in 1995 (WTO replaced GATT that year). There were, in total, 31 formal Working Party meetings in addition to numerous informal meetings. Russia's most difficult bilateral negotiations were with the European Union (EU), the USA and China (Babkin *et al.*, 2012). After 2000, negotiations covered all aspects of Russia's accession to the WTO, including its commitments concerning the schedule of tariffs, market access for goods and services, and agriculture. Finally, 19 years of difficult negotiations by Russia yielded positive results. Russia's accession package, the biggest accession package in the history of the WTO, was approved by the Eighth Ministerial Conference in 2011. On 22 August 2012, Russia notified the WTO Secretariat of the ratification of the package and officially became a member of the WTO.

Among the KRU countries, Ukraine was the first to accede to the WTO. Ukraine submitted its application in November 1993 to the WTO, after Russia. A General Council Working Party was established in December 1993. The memorandum on Ukraine's trade regime was circulated in 1994. The process of Ukraine's accession can essentially be divided into three periods. The first period, 1993–1997, was dedicated to the analysis and monitoring of Ukraine's economy to determine if it reflected GATT/WTO norms and regulations. The second period, 1998–2003, entailed the process of reformulating laws deemed a priority for accession in order to cast regulations in accordance with the WTO norms. The third period, 2003–2008, can be described as a process of identifying Ukraine's commitments regarding its WTO membership. After a standstill, the 2005–2007 negotiations produced a breakthrough. During that period, Ukraine was able to reach an agreement with 52 WTO member states and was able to ratify 55 laws in accordance with the WTO norms (Pugachev, 2012). In 2008, after 15 years of negotiation, Ukraine was officially welcomed into the WTO.

The negotiation process for Kazakhstan's accession to the WTO started in January 1996, with submission of the official application to the WTO Secretariat. In February 1996, Kazakhstan was assigned status as an observer country to the WTO. In April 1996, the General Council of the WTO established a Working Party to examine the Kazakhstani application and to provide recommendations, including comments on a draft of the Protocol of Admission. Members of the Working Party were then encouraged to submit questions on the Kazakhstani memorandum on its foreign trade regime that was circulated by the WTO Secretariat in November 1996. Feedback and questions were then transmitted to the Kazakhstani authorities for further examination and comment. The memorandum reflected on the Kazakhstani economy, economic policies and foreign trade regimes at that time. Government officials acknowledged that the Kazakhstani key trade legislation met the most important WTO principles and committed Kazakhstan to bringing the country into full compliance with the WTO rules. All the ministries of the Kazakhstani government were urged to ensure full compliance of their legislation with the WTO standards (WTO, 1996). In 2014, Kazakhstan was still in negotiations for WTO membership.

Modelling International Trade in Wheat

Much of the research related to modelling world wheat markets dates back to the 1970s. Early studies accounted for the Soviet Union as a single market. In the 1990s, the KRU countries were not yet considered important exporting countries, and therefore were often ignored in the model design. Furthermore, the KRU data before the transition were not considered reliable (Hobbs *et al.*, 1993; Gaisford *et al.*, 2001). Therefore, our model uses only the more recent data to assess the effect of WTO accession by Russia and Ukraine on international trade in wheat.

Since the subject of interest is the effect of tariff reduction on world wheat trade and on the resulting welfare changes, the global simulation model (GSIM) model by Francois and Hall (2003) was adapted to assess the effect of WTO accession. The GSIM model is capable of addressing the issue of trade liberalization on a single industry level and has resource requirements that are manageable. It is a static, multi-regional, partial equilibrium, Armington-type product differentiation model that solves for equilibrium prices by satisfying global market clearing conditions (i.e. global imports must equal exports).

Armington-type product differentiation means that the products originating from different countries are imperfect substitutes. The heterogeneity of the products among countries comes from the different bilaterally imposed tariffs (e.g. the USA and Russia faced different tariffs imposed by China before Russia's accession to the WTO). These tariffs determine the relative price of goods. Therefore, changes in tariffs translate into import changes by source.

The GSIM model requires a bilateral trade matrix at world prices, an initial and final matrix of bilateral import tariffs, export supply elasticities, aggregate import demand elasticities and elasticities of substitution. Trade liberalization effects are reflected in terms of bilateral trade changes, welfare effects (producer surplus, consumer surplus and change in tariff revenues), price and change in output. Although fairly recently developed, the GSIM model has found application in several recent studies of trade liberalization, including: Vanzetti *et al.* (2005); Mutambatsere (2006); Holzner (2008); Serletis and Fetzer (2008); Worz *et al.* (2008); Holzner and Peci (2012); and Leudjou (2012).

A full description of the GSIM model can be found in studies by Francois and Hall (2003) and Jammes and Olarreaga (2005). The study in this chapter includes a total of 24 regions, including major wheat exporting and wheat-importing countries. The data required for the empirical analysis using the GSIM model are trade values by origin and destination; bilateral tariffs; and elasticities of composite demand, supply and substitution.

The base year chosen for the analysis was 2007. The wheat market in 2007 was more or less representative of global equilibrium in the market (no major wheat export bans occurred in 2007). In addition, none of the KRU countries had yet joined the WTO in 2007. The bilateral trade flow data at the HS[1] four-digit level that corresponds to wheat comes from the trade statistics for international development (Trademap, 2012) of the International Trade Centre in Geneva (Switzerland).

Bilateral tariff data came from numerous sources. The *ad valorum* equivalents for specific MFN, non-MFN and other preferential tariffs were obtained from the market access map database (MacMap, 2012) of the International Trade Centre and the Centre d'Etude Prospectives et d'Informations Internationales, WTO's Tariff Download Facility Tariff (Tariffdata, 2012), WTO's Tariff Analysis Online (TAO, 2012) and World Integrated Trade Solution by the World Bank (WITS, 2012).

Aggregate import demand elasticity values for most of the countries were adopted from Kee *et al.* (2004). Some of the import demand elasticity values were approximated, due to lack of data. Approximations were based on the rationale provided by Kee *et al.*, (2004). Full details of the elasticities used can be found in Burkitbayeva (2013).

An export supply elasticity value of 1.5 (Francois and Hall, 2003) was adopted for the major exporters of wheat (Australia, the USA, Canada, Argentina, the EU and Brazil); for other countries and regions, the value 0.5 was adopted. This corresponded to the assumption of a small country (Holzner, 2008). Supply elasticity values for the KRU countries vary between zero and 1.5, depending on the three alternative scenarios explained below. An elasticity of the substitution value of 5 (Francois and Hall, 2003) was adopted for all countries and regions in the model. The value of 5 is often used in the literature (Fukita *et al.*, 2000).

Three scenarios were examined in this research. Scenarios vary based on the ability of the wheat sector in the acceding countries to respond to changes in market conditions. The first scenario assumes that the KRU countries are able to respond to changes in demand quickly, meaning that they respond in a similar fashion to the wheat sectors in modern market economies. Therefore, the KRU countries take the export supply elasticity value of 1.5 (as with the rest of the major wheat export suppliers). The second scenario assumes that the KRU wheat sectors are unable to respond fully to demand changes, meaning they do not operate fully as market economies (MacKay and Kerr, 2007). In this case, the KRU countries were each assigned an export supply elasticity value of 0.5. The third scenario is a worst-case scenario that assumes the KRU countries' wheat sectors are non-responsive to demand changes (export supply elasticity equal to zero), meaning the KRU countries are unable to adjust their production output and respond to changes in the prices they receive. Supply response elasticities of less than 1.5 in the KRU countries might occur due to a limited price transmission between world and domestic markets. Countries in transition, such as the KRU countries, often exhibit weak or limited price transmission. This means that firms in transition economies are less price responsive than are firms in modern market economies. Price transmission inadequacies usually occur due to problems in infrastructure and institutional barriers. Small values of supply elasticities assume that the KRU countries will be unable to respond to changes in prices due to constraints on infrastructure and due to the availability of inputs and managerial capacity (Hobbs *et al.*, 1997).

Results

Table 18.1 summarizes the liberalization effects for the first scenario, in which the response of the KRU wheat sectors is expected to be similar to those in modern market economies (i.e. supply elasticity of 1.5). Liberalization effects are reported in US thousand dollars (US$1000) and in percentage of total cereal production value for a better visualization of the magnitude of the changes. Values in the form of percentages of gross cereal production in 2007 for aggregated regions were dropped and separate countries were reported. The three tested scenarios did not lead to significantly different results. Different supply elasticity values for the KRU countries had marginal impact on the outcome. Further discussion of the results is based on the first scenario outcomes. Summaries of the effects for the other two scenarios can be found in Burkitbayeva (2013).

In general, the welfare effects of the wheat trade liberalization as a result of Russia and Ukraine joining the WTO are modest. Total welfare effects in US$1000 were related to individual countries' gross cereal production values in 2007 (Table 18.1). Net welfare effects as a share of gross cereal production value appear to be insignificant for most of the countries. Kazakhstan and Russia experienced positive net welfare effects, while Ukraine experienced negligible net welfare effects. As net exporters, the KRU countries experienced an increase in wheat prices due to higher market access.

Changes in trade flows appear to be consistent with the theory. More trade occurs in freer markets. Table 18.2 reports the changes in trade flows after the liberalization. Trade flow changes are reflected in percentage terms.

The results in Table 18.2 show changes in trade flows for the KRU countries due to the tariff reductions manifest. The increase in trade between Russia and Kazakhstan cannot be attributed to tariff reductions but rather to a big

Table 18.1. Summary of effects for selected countries, scenario I (US$1000). (From results extracted from the GSIM model output and authors' own calculations.)

	Canada	EU (27)	USA	Kazakhstan	Russia	Ukraine
	Welfare (US$1,000)					
Producer surplus	–3,350.3	–11,457.2	–7,452.6	72,003.7	116,083.1	5,492.6
Consumer surplus	402.6	24,056.7	3,587.8	–44,908.1	–84,017.7	–5,919.9
Tariff revenue	0.2	–22,144.3	–141.3	0.0	0.7	0.7
Change in subsidy payments	0.0	0.0	0.0	0.0	0.0	0.0
Net welfare effect	–2,947.6	–9,544.8	–4,006.1	27,095.7	32,076.1	–426.6
	Other (%)					
Change in overall consumer prices	–0.07	–0.09	–0.06	2.27	0.95	0.25
Change in output	–0.10	–0.10	–0.10	3.40	1.40	0.30
Producer price for home good	–0.07	–0.04	–0.05	2.27	0.93	0.22
Market price for home good	–0.07	–0.04	–0.05	2.27	0.93	0.22
	Welfare effects (per cent of 2007 gross cereal production value)					
Producer surplus	–0.042	–0.017	–0.010	2.738	0.826	0.115
Consumer surplus	0.005	0.036	0.005	–1.708	–0.598	–0.124
Tariff revenue	0.000	–0.034	0.000	0.000	0.000	0.000
Change in subsidy payments	0.000	0.000	0.000	0.000	0.000	0.000
Net welfare effect	–0.037	–0.014	–0.005	1.030	0.228	–0.009

Table 18.2. Percentage changes in trade values of world wheat flows from WTO liberalization. (From results extracted from GSIM model, 2013.)

Country of destination	Country of origin: Canada	EU (27)	USA	Kazakhstan	Russia	Ukraine
	Per cent change					
Australia	0.1	–0.1	0.0	0.0	0.0	0.0
Canada	0.2	0.0	0.1	0.0	0.0	0.0
China	0.3	0.2	0.3	0.0	0.0	0.0
EU-27	0.1	0.0	0.1	46.5	–4.9	–1.3
Kazakhstan	0.0	0.0	0.0	–3.3	3.4	0.0
Russia	0.0	3.0	0.0	–8.5	–1.8	1.7
Ukraine	1.2	1.1	1.1	–10.5	–3.8	–0.2
USA	0.2	0.1	0.2	0.0	0.0	31.6

Notes: EU (27) = European Union.

difference in relative changes in prices in Russia and Kazakhstan. Ukraine's value of exports to the USA increased by 31.6%, while Kazakhstan's value of exports to the EU increased by 46.5%. The EU-applied tariffs for high-quality wheat and medium- or low-quality wheat from WTO members were €0/tonne (t) and €95/t, respectively. High-quality wheat is identified as having minimum 14% protein content, while everything else below 14% is considered medium- or low-quality wheat. In the future, as a result of WTO accession, KRU wheat exporters might benefit from the absence of restrictions for high-quality wheat. In general, major wheat exporting countries such as Australia, Canada, the EU and the USA do not appear to be impacted negatively to any important degree by the WTO accession of Russia and Ukraine.

Conclusions

The accession of the KRU countries to the WTO can be expected to result in observable changes in trade flows in the world wheat market. Although net welfare effects appear to be relatively small, there is the potential for growth in wheat trade between KRU countries and their trade partners. Increases in the export opportunities of the KRU countries would come at the expense of major wheat exporters such as Canada, the EU and the USA as their competitiveness erodes in some of the major importing markets. The loss, however, does not appear to be significant in terms of dollar value. Major exporters will be able to compensate for market access erosion by diverting and distributing exports among other importing markets. The market presence of Canada, the EU and the USA is more diverse in terms of geography than is that of the KRU countries of the Black Sea region. Canada, the EU and the USA (traditionally large exporters) have established themselves as reliable wheat exporters in contrast to the KRU countries, which continue to exhibit volatile production practices and use export bans and restrictions.

Accession to the WTO can bring market access opportunities for the KRU countries. The scenarios where the wheat industry supply response in the KRU countries was muted did not alter the results to any important degree. It is, however, questionable whether producers can capitalize on the opportunities provided by WTO membership. This depends on factors such as future productivity growth, domestic policies and adaptability to climate changes and other socio-economic factors. Infrastructure is one of the major constraints for the KRU countries, which are expending considerable resources to improve their infrastructures for storing and transporting grain. If the KRU countries can overcome their present constraints and can use their wheat producing resources efficiently, their presence in the world wheat trade arena will remain strong and gain in prominence.

WTO membership is likely to bring considerable benefits for wheat producers and other participants in the wheat supply chains of the KRU countries. The international wheat market is large and diversified, so no major players in the international wheat market should suffer disproportionally.

The removal of the barriers to market access associated with being outside the multilateral system will, hopefully, assist in the process whereby Kazakhstan, Russia and Ukraine may become major contributors to the international wheat economy. In times of increasing concerns regarding global food security, this should be a welcome outcome.

Note

[1] Harmonized Commodity Description and Coding System (HS) of tariff nomenclature is an internationally standardized system of names in accordance with the World Customs Organization.

References

Antonyuk, O.V. and Kerr, W.A. (2005) Meeting TRIPS commitments in Ukraine: an important challenge in the quest for WTO accession. *Journal of World Intellectual Property* 8, 271–282.

Babkin, K.A., Kuznecov, A.B., Korchevoi, E.A., Pronin, V.V. and Samohvalova, V.A. (2012) *Consequences of Russia's Accession to the WTO.* WTO Inform, Moscow.

Burkitbayeva, S. (2013) Accession of Black Sea region wheat producers to the WTO: implications for world wheat trade. MSc thesis, University of Saskatchewan, Saskatoon, Canada.

Falkus, M.E. (1966) Russia and the international wheat trade, 1866–1914. *Economica* 33, 416–429.

Francois, J.F. and Hall, H.K. (2003) *Global Simulation Analysis of Industry-Level Trade Policy.* Tinbergen Institute/CERP US Department of Commerce, Washington, DC.

Fukita, M., Krugman, P. and Venables, A. (2000) *The Spacial Economy – Cities, Regions and International Trade*. MIT Press, Cambridge, Massachusetts.

Gaisford, J.D. and Iourkova, I. (2007) International trade policy and re-investment in the Russian economy. In: Gaisford, J.D., Mayevsky, V. and Kerr, W.A. (eds) *Revitalizing Russian Industry: The Road Ahead After Fifteen Years of Transition*. Nova Science Publishers, New York pp. 97–109.

Gaisford, J.D. and Kerr, W.A. (2004) The Doha Round: a new agreement on agriculture? In: Burakovski, I., Handrich, L. and Hoffmann, L. (eds) *Ukraine's WTO Accession: Challenges for Domestic Economic Reforms*. Physica-Verlag, Heidelberg, Germany, pp. 197–212.

Gaisford, J.D., Hobbs, J.E. and Kerr, W.A. (2001) Making markets work: modelling agri-food systems in transition. *Canadian Journal of Agricultural Economics* 49, 181–202.

Hobbs, J.E. and Boyd, S.L. (2007) From farm to fork: food supply chains in the Russian Federation. In: Gaisford, J.D., Mayevsky, V. and Kerr, W.A. (eds) *Revitalizing Russian Industry: The Road Ahead After Fifteen Years of Transition*. Nova Science Publishers, New York, pp. 221–236.

Hobbs, J.E., Gaisford, J.D. and Kerr, W.A. (1993) Transforming command economy distribution systems. *Scottish Agricultural Economics Review* 7, 135–140.

Hobbs, J.E., Kerr, W.A. and Gaisford, J.D. (1997) *Transformation of the Agrifood System in Central and Eastern Europe and the New Independent States*. CABI, Wallingford, UK.

Holzner, M. (2008) GSIM measurement of the effects of the EU accession of the Balkans and Turkey on agricultural trade. *South East European Journal of Economics and Business* 3, 7–14 .

Holzner, M. and Peci, F. (2012) *Measuring the Effects of Trade Liberalization in Kosovo*. The Vienna Institute for International Economic Studies, Vienna.

Jammes, O. and Olarreaga, M. (2005) *Explaining SMART and GSIM*. The World Bank, Washington, DC.

Kee, H.L., Nicita, A. and Olarreaga, M. (2004) Import demand elasticities and trade distortions. *The Review of Economics and Statistics* 90, 666–682.

Kerr, W.A. (2010) GATT–1947: a living legacy fostering the liberalization of international trade. *Journal of International Law and Trade Policy* 11, 1–11.

Kerr, W.A. (2012) Taming the bear: the WTO after the accession of Russia. *Journal of International Law and Trade Policy* 13, 150–159.

Leudjou, R. (2012) The Doha Round and food security in the dairy sector in Cameroon: a Global Simulation Model (GSIM) approach. *Journal of International Law and Trade Policy* 13, 115–130.

MacKay, E. and Kerr, W.A. (2007) Transition to wheat?: Institutional change and the evolution of economic systems. In: Gaisford, J.D., Mayevsky, V. and Kerr, W.A. (eds) *Revitalizing Russian Industry*. Nova Science Publishers, New York, pp. 5–18.
MacMap (2012) *MacMap*. International Trade Centre, Geneva, Switzerland. Available at: http://www.macmap.org (accessed 20 February 2014).
Mutambatsere, E. (2006) *Trade Policy Reforms in the Cereals Sector of the SADC Region: Implication on Food Security*. Department of Applied Economics and Management, Cornell University, Ithaca, New York.
Pugachev, N. (2012) *Ukraine's Accession to the WTO: Accession Conditions. Presentation*. Ministry of Agriculture of Russia Training Centre, Moscow.
Schmitz, A. and Bawden, D.L. (1973) *The World Wheat Economy: An Empirical Analysis*. Giannini Foundation Monograph No 32. Giannini Foundation Agricultural Economics, Berkeley, California.
Serletis, G. and Fetzer, J. (2008) *Modeling the Impact of the US Tobacco Quota Buyout*. Working Paper No 2008-06-A. US International Trade Commission (USITC), Washington, DC.
TAO (2012) Tariff Analysis Online. World Trade Organization, Geneva, Switzerland. Available at: https://tao.wto.org/ (accessed 22 December 2014).
Tariffdata (2012) Tariff data. World Trade Organization, Geneva, Switzerland. Available at: http://tariffdata.wto.org/ (accessed 22 December 2014).
Trademap (2012) Trade maps. International Trade Centre, Geneva, Switzerland. Avaiable at: http://www.trademap.org (accessed 20 February 2014).
USDA (2009) Foreign Agricultural Service Production, Supply and Distribution Online. US Department of Agriculture (USDA), Washington, DC. Available at: http://www.fas.usda.gov/psdonline/ (accessed 20 February 2014).
Vanzetti, D., Fernandez, S. and Chau, V. (2005) *Banana Split: How EU Policies Divide Global Producers*. Policy Issues in International Trade and Commodities Study Series Vol 31. United Nations, New York and Geneva.
WITS (2012) *World Trade Integrated Solutions*. World Bank, Washington, DC. Available at: http://wits.worldbank.org/ (accessed 22 December 2014).
Worz, J., Pindyuk, O., Holzner, M. and Astrov, V. (2008) *Russia's WTO Accession in the Meduim and Long Run – A Global Simulation Model (GSIM) Approach*. The Vienna Institute for International Economy, Vienna.
WTO (1996) *Accession of the Republic of Kazakhstan: Memorandum on the Foreign Trade.* World Trade Organization (WTO), Geneva, Switzerland.
WTO (2011) *Understanding the WTO*, Fifth Edn. World Trade Organization (WTO), Geneva, Switzerland.

19 Wheat Export Restrictions in Kazakhstan, Russia and Ukraine: Impact on Prices along the Wheat-to-Bread Supply Chain

Linde Götz,* Ivan Djuric and Thomas Glauben
Leibniz Institute of Agricultural Development in Transition Economies (IAMO), Halle (Saale), Germany

Abstract

Kazakhstan, Russia and Ukraine (KRU) are among the grain-exporting countries that implemented wheat export restrictions between 2007 and 2011 during the global commodity price peaks. This chapter provides an overview on the price effects induced by wheat export controls in the KRU region. It becomes evident that the domestic price effects of export controls were heterogeneous among the KRU countries, and that domestic prices were only partially insulated from international price developments. Also, export controls increased, rather than decreased, domestic wheat price volatility. Furthermore, the effectiveness of export controls as an instrument to protect against high food prices is questionable, particularly in the case of wheat, which is transformed to an end consumer product in a complex supply chain. In all three KRU countries, the intermediate milling industry did not transmit the price increases of the wheat price to the flour price. Instead, the milling industry increased the flour price proportionally, which increased bread production costs and led to higher bread prices throughout the KRU region.

Acknowledgements

The authors gratefully acknowledge financial support by the German Federal Ministry of Food and Agriculture and the Federal Office for Agriculture and Food. The authors are grateful to Martin Petrick for providing helpful comments on an earlier version of this chapter.

Introduction

During the recent price booms on world agricultural markets in 2007/08 and 2010/11, many countries attempted to insulate their domestic markets from price developments on the world market and to stabilize domestic prices through trade policy interventions (Djuric *et al.*, 2014). Exporting countries implemented export controls by decreasing or even banning exports, whereas importing countries reduced or even completely eliminated import restrictions to reduce the influence of high world market prices on domestic price levels (Martin and Anderson, 2012). Governments can reduce domestic prices by decreasing exports and increasing domestic supply. Cancelling or reducing import tariffs creates additional incentives to import goods, which

*Corresponding author; e-mail: goetz@iamo.de

can decrease export prices and increase domestic supply. One way to respond to high world market prices is through trade-oriented policy measures aimed at curbing domestic food price inflation. During the 2007/08 food crisis, roughly 37 countries implemented export barriers and 59 countries removed import restrictions (FAO, 2008). Among these countries were the three large grain-exporting countries of Kazakhstan, Russia and Ukraine (KRU), all of which were members of the former Soviet Union.

Export controls are generally implemented by three means: export taxes, quotas and bans. Export taxes increase the costs of exporting, depending on the size of the tax, thereby reducing competitiveness on the world market (they can also make exports unprofitable if the tax is set too high). Export quotas limit exports to the size set by the quota. Export bans forbid exports, such that no quantity of the respective product can be exported (Mitra and Josling, 2009; Sharma, 2011). For example, Kazakhstan implemented a wheat export ban in 2008, Russia implemented wheat export taxes in 2007/08 and a wheat export ban in 2010/11, and Ukraine established a wheat export quota system in 2006–2008 and 2010/11.

This chapter provides an overview on the price effects of export restrictions implemented by the KRU countries. It explains governmental interventions on wheat export markets in each of the KRU countries and describes price and export developments in general. The *intended* price effects of export controls on the domestic market and the *accidental* price effects on the world market are addressed by explaining their theoretical background and providing empirical evidence. This chapter focuses on the development of prices along the wheat-to-bread supply chain and simulates the profits achieved during export controls. Finally, it points out the dramatic consequences of grain export controls for the KRU countries and for global food security.

Export Restrictions in the KRU Countries

This section provides an overview on the chronology of the export controls implemented in the KRU countries and describes the price and export developments on their domestic wheat markets. Export restrictions were introduced in Kazakhstan during the food crisis of 2007/08, whereas trade remained open during the 2010/11 international commodity price peak. In light of the very high world market prices and strongly increasing wheat exports, the Kazakhstani government introduced a grain export licensing system in September 2007 that was controlled by the Ministry of Agriculture. Under this system, grain was allowed to be exported only by companies that were able to satisfy certain conditions, such as having their own production and storage facilities, and by companies exporting at least 5000 tonnes (t) of grain. This regulation was valid until January 2012.

In response to significant grain and bread price increases in September 2007, the Kazakhstani government formed a stabilization fund for wheat intervention. The state-controlled Food Procurement Corporation secured about 180,000 t of wheat for intervention purposes (USDA, 2008). In addition, the government signed a Memorandum of Understanding (MOU) with grain traders, proclaiming that domestic wheat prices should not be increased until the new harvest. According to the MOU Agreement, wheat traders were to deliver a minimum of 1.2 million tonnes (Mt) of wheat to the domestic market before the next harvest (AgraFood East Europe, 2008a). However, when the 1.2 Mt of wheat were already sold on the domestic market by April 2008, the domestic wheat price started to skyrocket (AgraFood East Europe, 2008b). In response, the government immediately introduced an export ban to protect domestic consumers from high food prices, which remained in force until September 2008.

From Fig. 19.1, it is evident that the export ban was unsuccessful when reducing the domestic wheat price. On the contrary, the domestic wheat price was temporarily higher than the world wheat market price. According to official statistics, however, no wheat exports occurred during the export ban, although substantial wheat flour exports could be observed. Additionally, Fig. 19.2 shows that the export ban was introduced when the Russian wheat producer price peaked in 2008, suggesting that the export ban insulated domestic prices in Kazakhstan from domestic prices in Russia, rather than from world market price developments.

Wheat exports in Russia were limited by an export tax of 10% that was established in November 2007 and was increased to a prohibitive level of

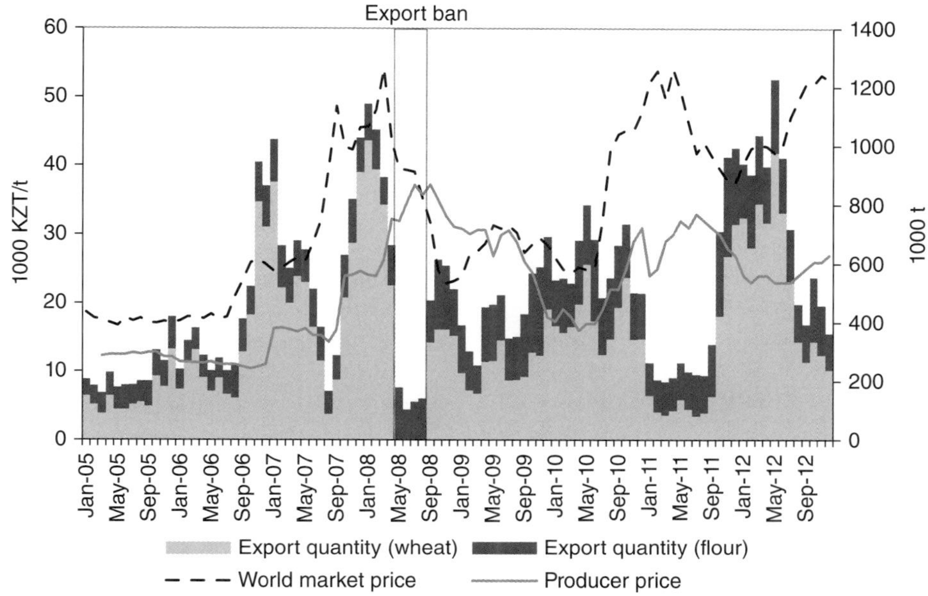

Fig. 19.1. Development of world market price, producer price and export quantities in Kazakhstan, 2005–2012. (From APK-Inform, 2013; GTIS, 2013; HGCA, 2013; authors' illustration.)
Note: Wheat grower prices for Kazakhstan are regional prices for the Almaty region.

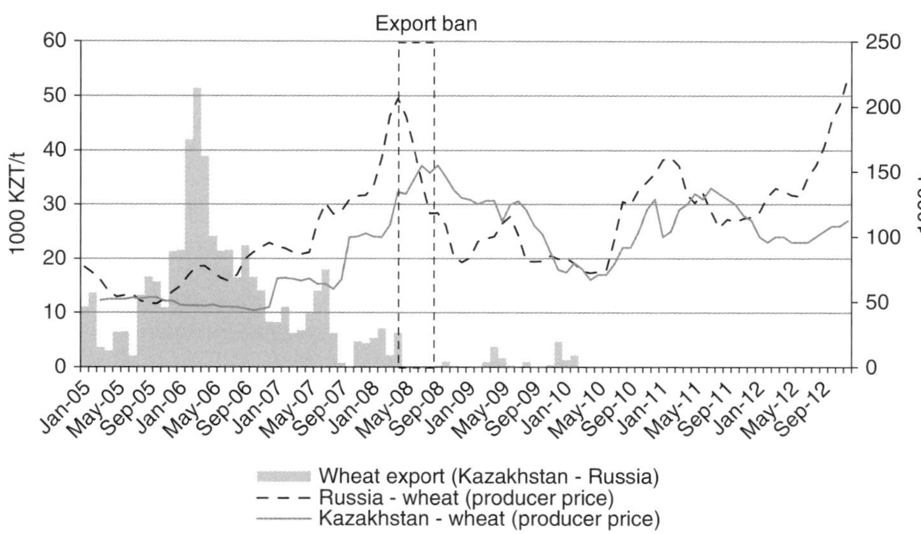

Fig. 19.2. Development of producer price for Russia and Kazakhstan, and Kazakhstani exports to Russia, 2005–2012. (From APK-Inform, 2013; GTIS, 2013; HGCA, 2013; authors' illustration.)
Note: Wheat grower prices for Kazakhstan are regional prices for the Almaty region.

40% in December 2007. This export tax remained in effect until July 2008. Later, Russia restricted wheat exports again during the 2010/11 commodity price peak.

Figure 19.3 shows that the export tax in Russia was only partially successful when reducing domestic wheat prices. In particular, early 2008 domestic prices increased beyond world

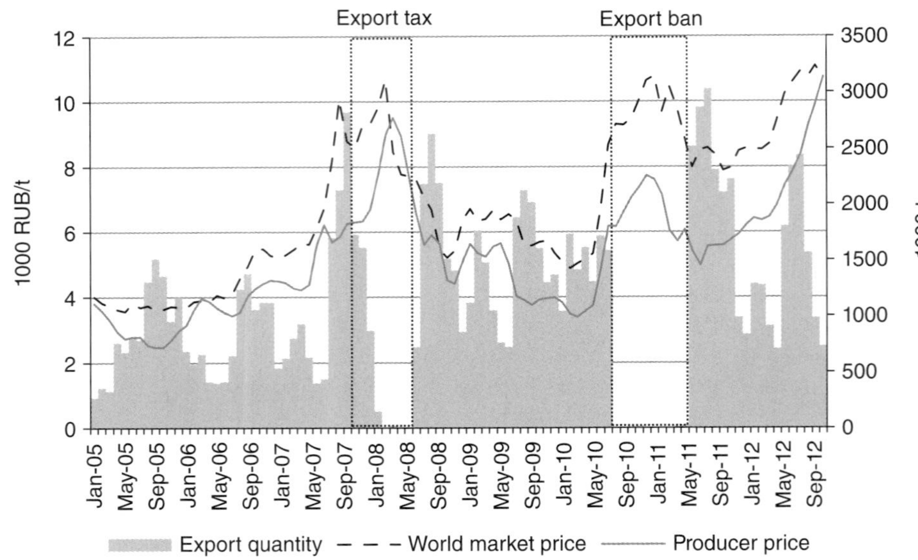

Fig. 19.3. Development of world market price, producer price and export quantities in Russia, 2005–2012. (From APK-Inform, 2013; GTIS, 2013; HGCA, 2013; authors' illustration.)
Note: Prices for Russia are average prices for the Central Federal District.

market price levels. The strong increase in domestic wheat prices might have been caused by increased wheat storage in expectation of even higher world market prices. During the export ban, domestic prices were continuously lower than was the world market price.

An export quota system was implemented in Ukraine during both world market price peaks within a governmental licensing system. Export quotas varying between 3000 t and 1.2 Mt were in force from October 2006 until May 2008, and again from October 2010 until May 2011. In addition, Ukraine implemented wheat export taxes of 9% in July 2011, which were removed in October 2011. The size of the quota was changed repeatedly and the quota system was extended multiple times, both of which created high market uncertainty. For example, the wheat export quota in 2010 came into effect so rapidly that ships already loaded with wheat could not leave the harbour, thereby causing high additional costs to the exporters (APK-Inform, 2013). As a consequence, contracts could not be fulfilled, which eroded the reputation of traders exporting from Ukraine. There were also questions about the transparency in the distribution of export quotas in 2010. Figure 19.4 shows that the difference between the world and the domestic wheat price increased substantially when the export quota system was in effect (Götz *et al.*, 2012; Goychuk and Meyers, 2013).

Theoretical and Empirical Price Impacts of Export Controls

Export restrictions influence the level of the price prevailing on the domestic market. In addition, export controls may have accidental effects on the world market price level and may induce additional price volatility.

Domestic price effects

As indicated above, by decreasing the export quantity, export controls aim to increase domestic market supplies to reduce domestic wheat market prices. This price development can be theoretically explained within a market diagram. As an example, Fig. 19.5 illustrates the domestic price and quantity effects of an export quota within a partial equilibrium framework.

Suppose we observe a situation on the world market similar to 2006, when adverse

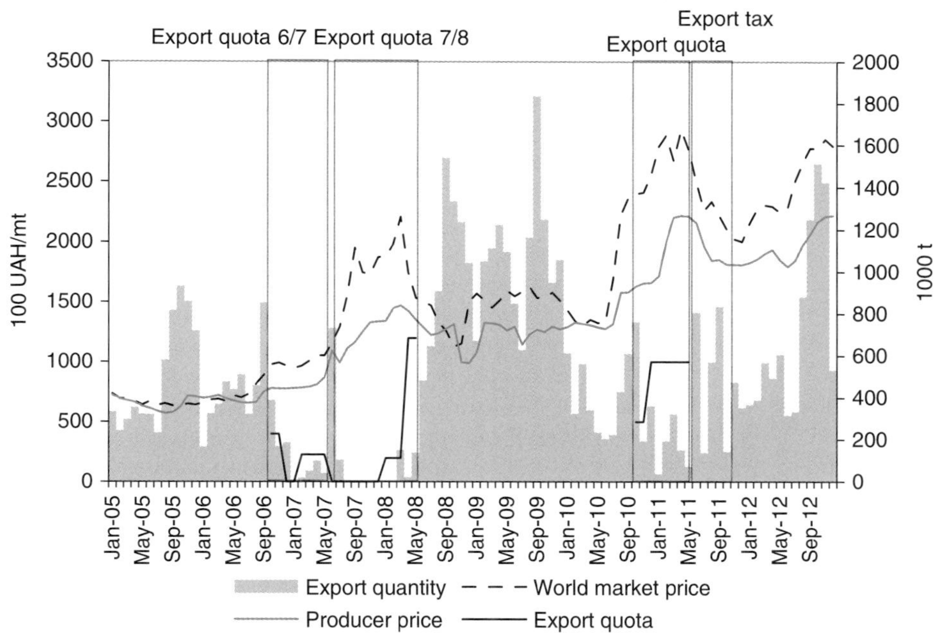

Fig. 19.4. Development of world market price, producer price and export quantities in Ukraine, 2005–2012. (From APK-Inform, 2013; GTIS, 2013; HGCA, 2013; authors' illustration.)
Note: Wheat grower prices for Ukraine are national averages.

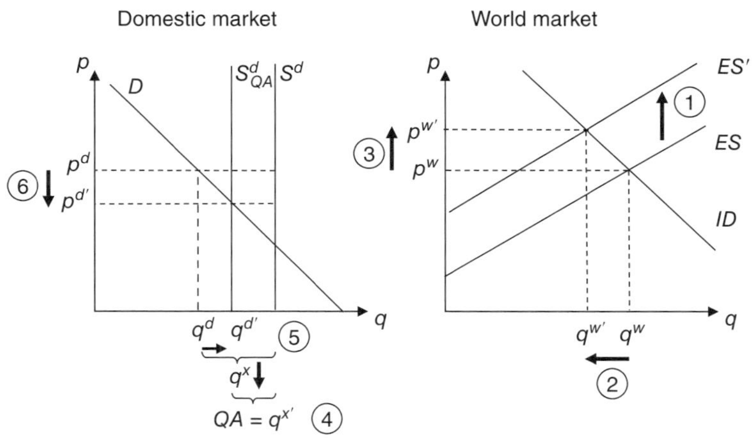

Fig. 19.5. Price and quantity effects of an export quota on the domestic market. (From authors' illustration adapted from Götz *et al*., 2012.)

weather conditions led to a relatively low wheat harvest in Ukraine and Australia (Grueninger and von Cramon-Taubadel, 2008), thus reducing the world wheat export supply. This can be depicted graphically by the movement of the export supply (*ES*) of wheat on the world market, from *ES* to *ES'* (Fig. 19.5, 1). Thus, the amount of wheat traded on the world market decreases from q^w to $q^{w'}$ (Fig. 19.5, 2) and the world market price increases from p^w to $p^{w'}$ (Fig. 19.5, 3). If Ukraine implements an export quota the size of *QA*, Ukrainian wheat exports decrease from q^x to $q^{x'} = QA$ (Fig. 19.5, 4). Suppose that export controls are implemented at the beginning of the harvest; thus, domestic supply S^d is fully inelastic and the domestic supply of wheat increases from q^d to $q^{d'}$ (Fig. 19.5, 5). As a result, the domestic price level decreases from p^d to $p^{d'}$ (Fig. 19.5, 6). The more the exports are reduced, compared to the open trade regime, the larger is the increase of supply on the domestic market and the more the domestic price should decrease.

However, as Figs 19.1, 19.3 and 19.4 show, the domestic wheat prices in the KRU countries generally continue to increase in times of export controls, provided that the world market price is increasing. Due to the temporary nature of these restrictions, domestic wheat prices remain related to the world wheat market price, even in times of export controls. Thus, the price-reducing effect of export restrictions might be counteracted by the increasing price effects induced by rising world market prices. Empirically, we observe domestic wheat prices in the KRU countries increasing in parallel to elevating world market prices during export controls, but rising to a lesser extent than when trade is freely possible.

The domestic price effects of export restrictions can be identified within a price transmission model that captures the transmission of price changes from the world market to the domestic market. The insulation of domestic prices from world market price developments decreases long-run price transmission and weakens the integration between these prices. Temporary changes in price transmission caused by export restrictions can be identified by non-linear, regime-switching price transmission models. This approach is followed by Götz *et al.* (2013a) for Russia and Ukraine in a Markov-switching error correction model, and by Götz *et al.* (2012) for Ukraine in a smooth transition cointegration model. The degree of price insulation resulting from export restrictions in the KRU countries is measured as the percentage decrease of long-run price transmission. As Fig. 19.1 makes evident, even the domestic price level increases beyond the world market price level. The results of the price transmission models suggested that wheat export controls reduced the transmission of price increases from the world wheat market to the respective domestic market by 15%–20% for Russia, 30% for Ukraine and 0% for Kazakhstan compared to when wheat exports were possible.

World market price effects

Export restrictions are criticized for their additional price-increasing effects on the world market (Dollive, 2008; Dawe and Slayton, 2011; Martin and Anderson, 2012). In general, an export quota and an export ban are not permitted by the World Trade Organization (WTO), and only an export tax is in line with WTO regulations. However, temporary export restrictions *applied to prevent or relieve critical shortages of foodstuffs or other products essential to the exporting contracting party* are exempted from this rule (Sharma, 2011). Even small countries have a price-increasing effect on the world market if many small exporting countries implement export controls simultaneously (Martin and Anderson, 2012).

Suppose a large exporting country limits its exports to insulate its domestic price from the increased world market price level, $p^{w'}$. This further decreases the wheat export supply on the world market to ES'(Fig. 19.6, 1) and further increases the world market price to $p^{w''}$(Fig. 19.6, 2). Even small exporting countries have a price-increasing effect on the world market if many of them implement export controls simultaneously (Martin and Anderson, 2012).

Anderson and Nelgen (2012) and Martin and Anderson (2012) investigate the feedback effects of price-insulating behaviour on the world market price. According to their calculations, the world wheat market price increased by about 70% between 2005/06 and 2008. These authors estimated that 19%–29% of this price increase was caused by the feedback effects of increasing export barriers by exporters, as well as by removing import barriers by importing countries worldwide.

Effects on price volatility

Export restrictions influence both price levels and price volatility. As indicated above, implementing export restrictions, particularly in the case of export quotas in Ukraine, brought a dramatic increase in market uncertainty, for several reasons: the Ukrainian export quotas were implemented at short notice, the size of the quota was changed and prolonged several times, and the quota licences were distributed in a non-transparent way. Thus, export restrictions make market conditions unreliable and difficult to foresee, thereby increasing market risk and ultimately leading to additional price volatility.

The effects on domestic market uncertainty were investigated by Götz *et al.* (2013c), using the generalized autoregressive conditional heteroscedasticity (GARCH) approach. Figure 19.7 depicts the development of volatility in the Ukrainian wheat market, measured as the conditional variance of the return price series for each price observation, given the variance in the previous period. It is evident that the Ukrainian wheat price is highly susceptible to external shocks, as reflected in the many pronounced spikes of the conditional variance. Also, high volatility coincides with political market interventions. For example, as shown in Fig. 19.7, volatility spike A coincided with the announcement

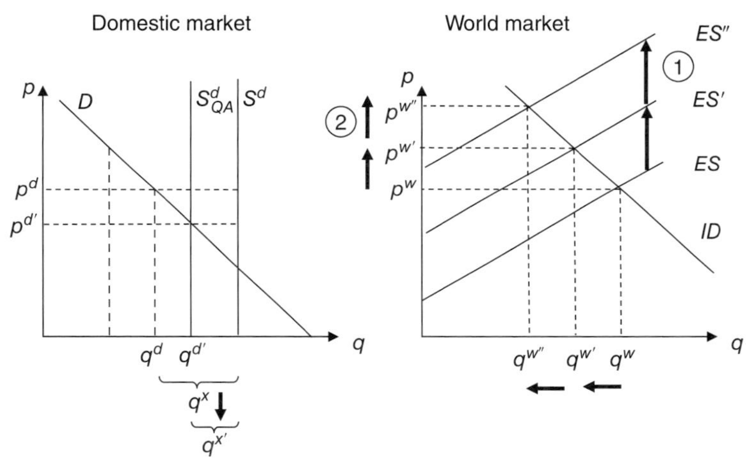

Fig. 19.6. Feedback price effect of an export quota on the world market. (From authors' illustration.)

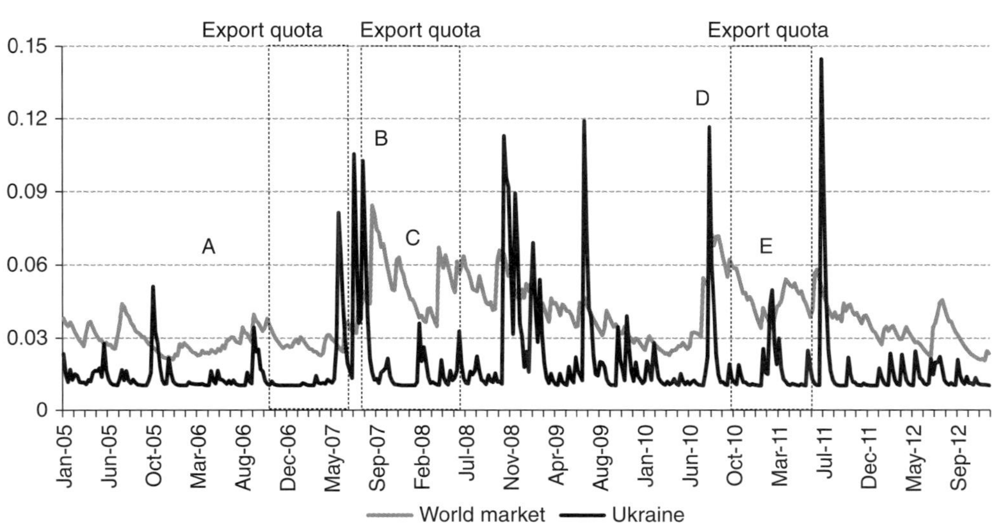

Fig. 19.7. Development of price volatility in the Ukrainian wheat market. (From Götz *et al.*, 2013c.)

of an export quota system by the Ukrainian government. The time of spike B matched with the temporary lift of export quotas and their unexpected reintroduction. High volatility phase C coincided with the quota increase that was announced but not realized. It also coincided with the commission meetings concerning the distribution of the export licences. Spike D occurred when an export ban was introduced in Russia, which induced extensive discussions in the Ukrainian media about whether export controls would also be introduced in Ukraine. Spike E occurred when an extension of the export quota in Ukraine was announced. These results confirm that export restrictions temporarily increase price volatility that is caused by increased market uncertainty.

Price Developments along the Wheat-to-Bread Supply Chain

Export restrictions are generally motivated by the aim to secure sufficient supply on domestic markets and to curb domestic food price inflation. In the case of wheat export controls, domestic wheat price reductions must be transmitted by intermediaries such as mills, bakeries and retailers along the wheat-to-bread supply chain (Fig. 19.8) to have a decreasing effect on the end consumer bread price and to reduce domestic food price inflation. In the following analysis, we provide an overview on wheat, flour and bread price developments for each of the KRU countries. Figures 19.9, 19.10 and 19.11 illustrate the development of prices for the amount of wheat and flour required to produce one loaf of bread (0.5 kg).

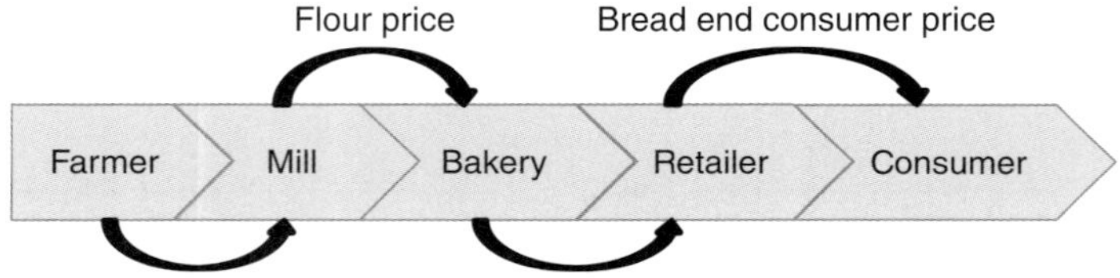

Fig. 19.8. Wheat-to-bread supply chain. (From Götz *et al.*, 2013b.)

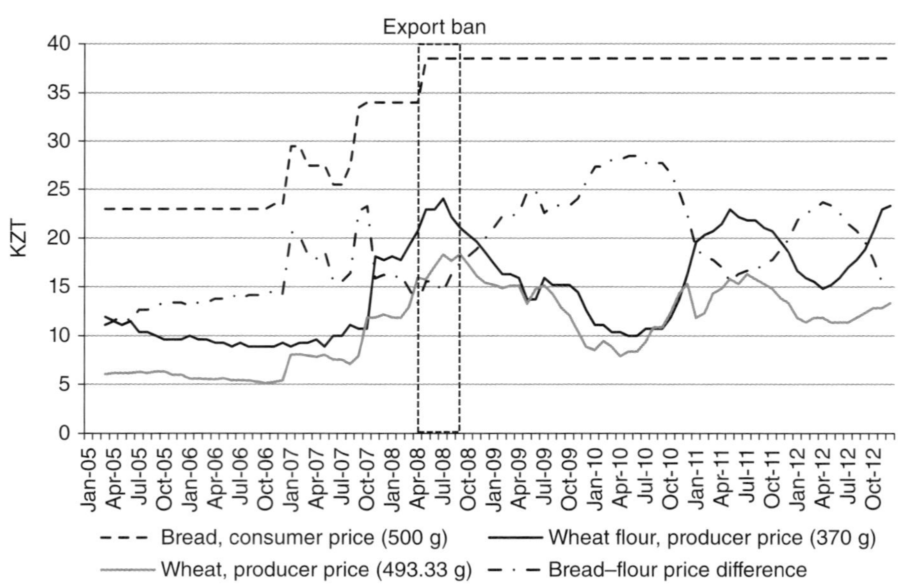

Fig. 19.9. Wheat, flour and bread price developments in Kazakhstan. (From APK-Inform, 2013; authors' illustration.)
Note: Kazakhstani tenge (KZT) is the national currency of Kazakhstan.

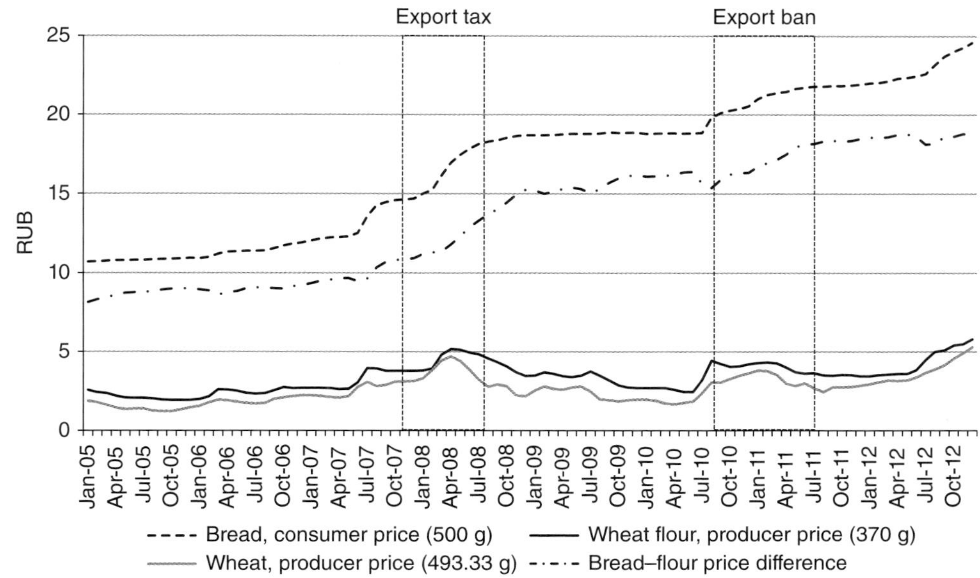

Fig. 19.10. Wheat, flour and bread price developments in Russia. (From APK-Inform, 2013; authors' illustration.) *Note*: Russian rouble (RUB) is the national currency of Russia.

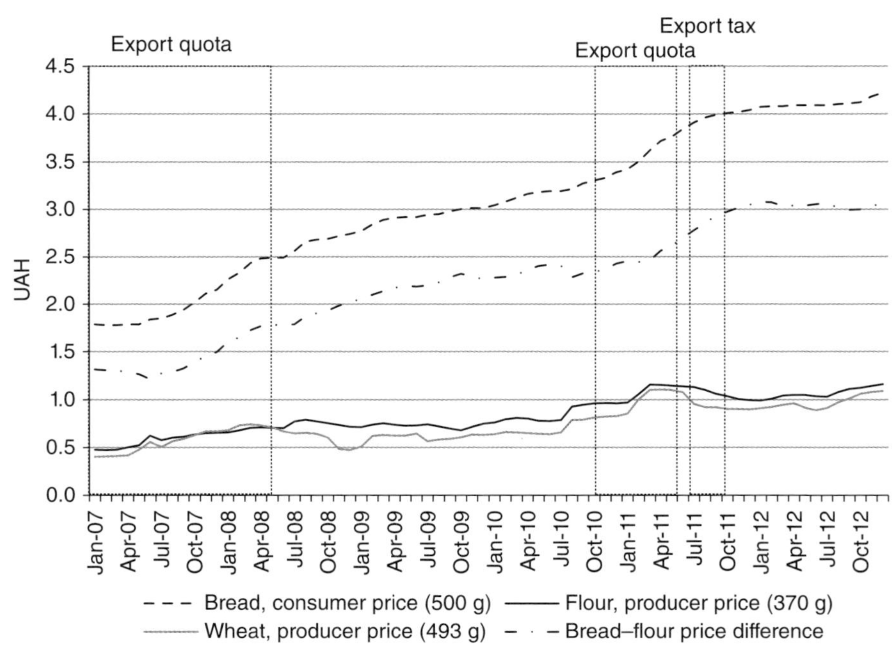

Fig. 19.11. Wheat, flour and bread price developments in Ukraine. (From APK-Inform, 2013; authors' illustration.) *Note*: Ukrainian hryvnia (UAH) is the national currency of Ukraine.

During the export ban in Kazakhstan, the wheat producer price and the bread end consumer price both increased substantially, whereas the flour price decreased after the initial price increases were observed. This resulted in a strong increase of the bread–flour price difference toward the end of the export ban. Under the export tax system in Russia, wheat and flour prices increased, which contributed to the rise of the bread end consumer price. In contrast, wheat and flour prices did not increase during the export ban, whereas the bread end consumer price rose. The increase in the bread flour price margin observed during the export ban resulted from the increase of other bread production costs, especially labour and energy costs. Relatively moderate increases of wheat, flour and bread prices are observed in Ukraine during both export quota systems. The increase in the bread flour price margin during both phases of export quotas (similar to Russia and Kazakhstan) can be traced back to the input cost increase of flour, labour and energy.

The flour wheat price margin for the KRU countries is presented in Fig. 19.12. It becomes evident that the flour wheat price difference increased during the export controls for all three countries. In Russia and Ukraine, the increase is observed particularly toward the end of the period with controlled exports. The flour wheat price difference increased right after the implementation of the export ban in Kazakhstan, while later it decreased, particularly toward the end of the export ban. To investigate whether the increase in the flour price, relative to the wheat producer price, is justified by increased flour production costs, we calculated the mills' profits and simulated the flour production costs (includes other costs such as labour, electricity, fuel, water and additional materials, with the total costs equally about 81% for wheat flour and 19% for other costs) (Djuric *et al.*, 2012). The simulation was conducted for large industrial mills that had their own silos and bought wheat during harvest at rather low prices. Therefore, we conducted the simulation based on the wheat price during the previous harvest plus respective storage costs.

Our first results indicated that the profits of the large Russian mills amounted to an average of about 177 RUB/t prior to the governmental interventions (January 2006 to November 2007). During the export tax system, the mills' profits increased by about 20%. In the aftermath of the interventions, the average mill's profit dropped considerably by about 65% compared to the profits obtained during the interventions. In summary, the large mills in Russia could increase their profits and thus could benefit from export controls. After the export controls were removed, the profits of the large mills decreased substantially. Qualitatively, similar results were obtained for Kazakhstan and Ukraine; however, mills' profits seem to be significantly higher in Kazakhstan compared to Russia and Ukraine.

Similarly, we calculated bread production costs and simulated the bread margin, which was the difference between the bread end consumer price and the bread production costs. The first results suggested that the bread price

(a) Kazakhstan

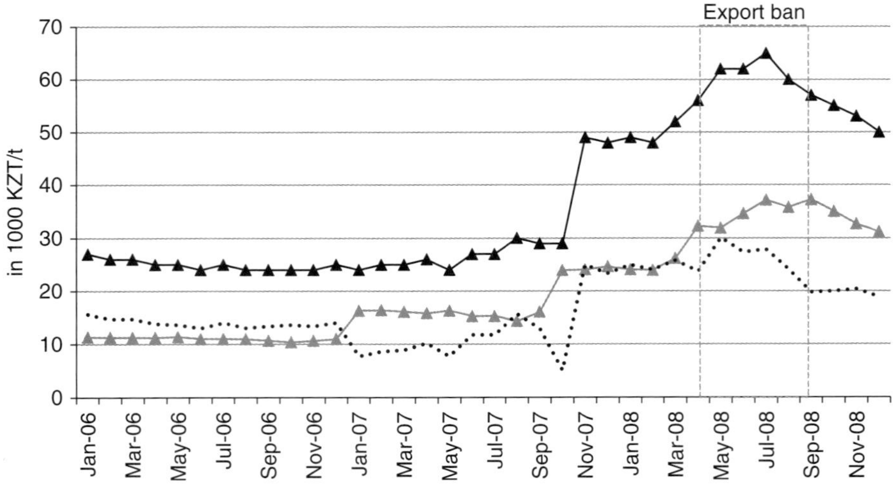

Fig. 19.12. Flour wheat price margin in the KRU countries. (From APK-Inform, 2013; authors' illustration.)

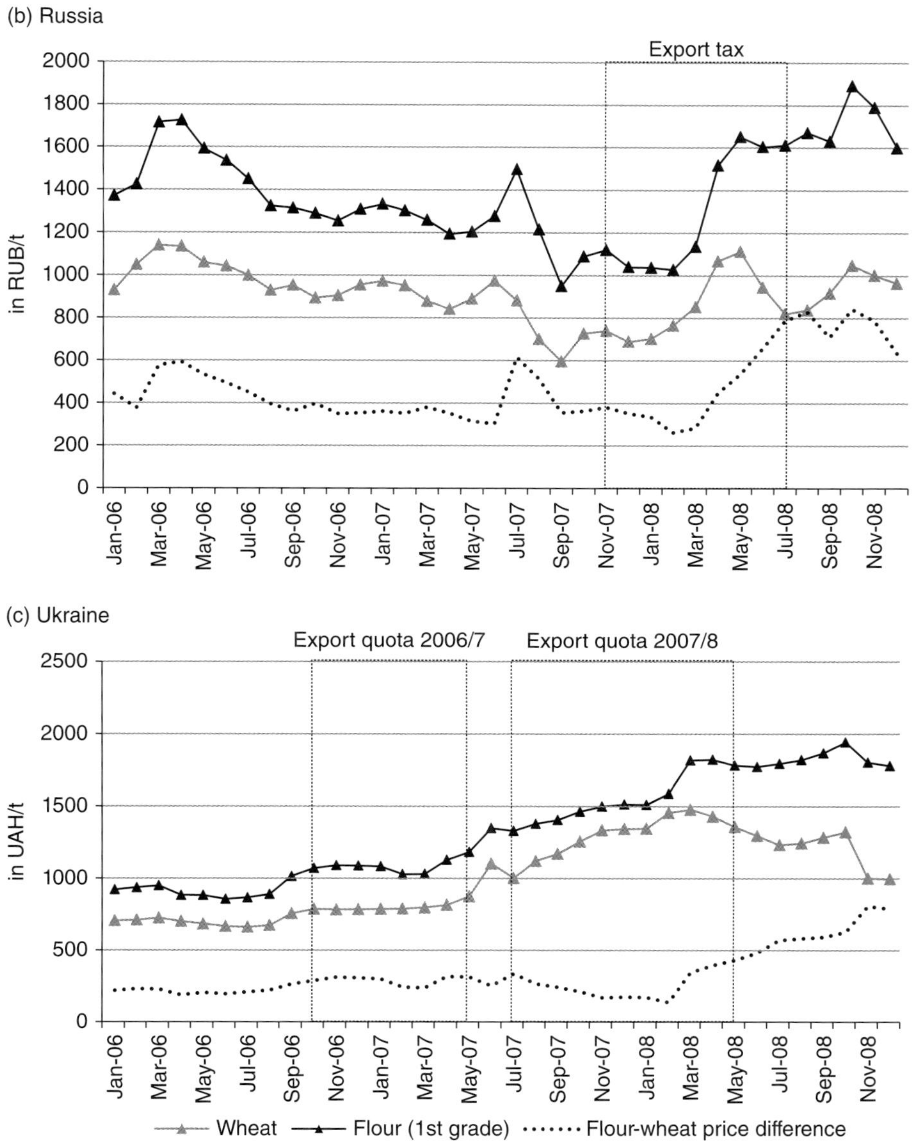

Fig. 19.12. Continued.

increases observed in times of export controls in the KRU countries were caused by rising production costs. Thus, small bakeries faced both increased flour costs and production costs, which decreased the bread profit margin. Therefore, bakeries seemed not to have achieved additional profits during the export controls in the KRU countries. In contrast, Djuric *et al.* (2012) found that bakeries were able to achieve considerably higher profits during the wheat export ban in Serbia in 2007/08. Thus, not only millers but also bakeries benefited from the governmental interventions.

Conclusions

This chapter has provided an overview of the price effects of export restrictions for wheat implemented by the KRU countries during the 2007/08 and 2010/11 commodity price peaks. It was shown that the domestic prices for the KRU countries were relative to the world market prices, even with export controls. Thus, the export restrictions had only limited success in insulating domestic prices. Also, heterogeneity was observed in the domestic price effects of the export controls. In particular, results of the price transmission analyses showed that the export ban in Kazakhstan could not decrease the transmission of world wheat market price changes to the domestic wheat producer price, whereas the transmission of price increases on the world market was reduced 15%–20% by the export tax in Russia and 30% by the export quota in Ukraine. However, considering the feedback price effects of price-insulating behaviour worldwide during the 2007/08 commodity price peak that increased the world wheat market price by about 20%–30%, the wheat export controls in the KRU countries did not curb domestic wheat price increases successfully (Anderson and Nelgen, 2012; Martin and Anderson, 2012). From a global point of view, the domestic wheat price in the KRU countries would have increased to about the same degree without any price-insulating behaviour from 2006 to 2008 (Götz *et al.*, 2012). Export controls also seem to be ineffective when stabilizing domestic wheat prices. Rather, analysing the development of price volatility showed that the export quota system in Ukraine increased domestic price risk substantially and created additional price variations.

Decreased domestic wheat prices, foregone export revenues and increased domestic price risks all create economic losses and additional costs to farmers and traders, and thus reduce incentives for investments in grain production. This is particularly problematic since the KRU countries bear high additional grain production potential and could play a significant role in heightened global grain production and trade, assuming they make substantial investments in grain production (Petrick and Oshakbaev, Chapter 2, this volume). The European Bank for Reconstruction and Development (EBRD) estimates that investments of US$1000–2000/ha are required to mobilize fully the grain production potential in Ukraine (Harmgart, 2011).

Furthermore, wheat export restrictions do not seem to be an appropriate means to achieve a reduction in domestic food price inflation. In all three KRU countries, the intermediate milling industry did not transmit the price-reducing effects from the wheat price to the flour price. Rather, large industrial mills increased flour prices during the export control regime against the background of higher wheat spot market prices. Also relevant for the large milling industry was the generally low wheat prices that prevailed during the previous harvest. Ultimately, the increased flour costs forced bakeries in the KRU countries to increase bread prices.

Thus, the effectiveness of export controls as an instrument of protection against high food prices is highly questionable, particularly in the case of wheat, which is transformed to an end consumer product in a complex supply chain with several intermediaries. Considering their far-reaching global consequences, the WTO should mandate that export restrictions be regulated at a multilateral level. Instead of aiming to insulate domestic agricultural prices, governments should allow domestic prices to increase and help lower-income consumers to cope with high food prices. Consumer-oriented crisis measures, food subsidies, food vouchers and direct income transfers can be better targeted and are more effective.

References

AgraFood East Europe (2008a) Kazakhstan in U-turn over grain quotas. Weekly issue No 313, 13 March 2008.

AgraFood East Europe (2008b) Kazakhstan reconsiders export duties. Weekly issue No 317, 10 April 2008.

Anderson, K. and Nelgen, S. (2012) Trade barrier volatility and agricultural price stabilization. *World Development* 40, 36–48.

APK-Inform (2013) *AgriNews*, various issues.

Dawe, D. and Slayton, T. (2011) The world rice market in 2007–2008. In: Prakash, A. (ed.) *Safeguarding Food Security in Volatile Global Markets*. Food and Agriculture Organization of the United Nations, Rome, pp. 164–174.

Djuric, I., Götz, L. and Glauben, T. (2012) Export restrictions on wheat markets in Serbia during global commodity price peaks – did consumers really benefit? Invited Paper, *IATRC Annual Meeting 'New Rules of Trade?'* 9–11 December, San Diego, California.

Djuric, I., Götz, L. and Glauben, T. (2014) Are export restrictions an effective instrument to insulate domestic prices against skyrocketing world market prices? The wheat export ban in Serbia. *Agribusiness*, doi:10.1002/agr.21398. Available at: http://onlinelibrary.wiley.com/doi/10.1002/agr.21398/abstract;jsessionid=A1071121B1ACA9684317AF8262A9A8AC.f01t02 (accessed 18 December 2014).

Dollive, K. (2008) *The Impact of Export Restraints on Rising Grain Prices*. Office of Economics Working Paper No 2008-09-A. US International Trade Commission, Washington, DC.

FAO (2008) *Crop Prospects and Food Situation*. FAO No 5. Food and Agricultural Organization (FAO), Rome.

Götz, L., Qiu, F., Gervais, J.P. and Glauben, T. (2012) Export restrictions and multiple spatial price equilibria when international prices spike: the wheat export quota in Ukraine. Invited Paper, *IATRC Annual Meeting*, 9–11 December, San Diego, California. Available at: http://ageconsearch.umn.edu/handle/143179 (accessed 18 December 2014).

Götz, L., Djuric, I. and Glauben, T. (2013a) *The Wheat Export Ban in Serbia. Are Export Restrictions an Effective Instrument to Dampen Food Price Inflation?* IAMO Policy Brief No 10. Leibniz Institute of Agricultural Development in Transition Economies (IAMO), Halle (Saale), Germany.

Götz, L., Glauben, T. and Brümmer, B. (2013b) Wheat export restrictions and domestic market effects in Russia and Ukraine during the food crisis. *Food Policy* 38(1), 214–226.

Götz, L., Goychuk, K., Glauben, T. and Meyers, W.H. (2013c) Export restrictions and market uncertainty: Evidence from the analysis of price volatility in the Ukrainian wheat market. Selected Paper, *AAEA & CAES Joint Annual Meeting*, 4–6 August, Washington, DC. Available at: http://ageconsearch.umn.edu/handle/150308 (accessed 18 December 2014).

Goychuk, K. and Meyers, W.H. (2013) Black Sea and world wheat market price integration analysis. *Canadian Journal of Agricultural Economics* 62(2), 245–261.

Grueninger, M. and von Cramon-Taubadel, S. (2008, May) *Competitive Agriculture or State Control: Ukraine's Response to the Global Food Crisis.* Sustainable Development Department of the Europe and Central Asia (ECA) Region of the World Bank, World Bank, Washington, DC.

GTIS (2013) *Global Trade Atlas*. Global Trade Information Services, Inc, Columbia, South Carolina. Available at: http://www.gtis.com/english/GTIS_WTA.html (accessed 1 September 2013).

Harmgart, H. (2011) Financing food – challenges and opportunities for the transition region. In: *Proceedings of the IAMO-Forum*, 23–24 June, Halle (Saale), Germany. Available at: http://www.iamo.de/fileadmin/uploads/forum2011/Keynote_presentations/Harmgart_IAMO_Forum_2011.pdf(accessed1September 2013).

HGCA (2013) Internet database. Available at: http://www.hgca.com (accessed 19 July 2013).

Martin, W. and Anderson, K. (2012) Export restrictions and price insulation during commodity price booms. *American Journal of Agricultural Economics* 94, 422–427.

Mitra, S. and Josling, T. (2009) *Agricultural Export Restrictions: Welfare Implications and Trade Disciplines*. IPC Position Paper, Agricultural and Rural Development Policy Series. International Food and Agricultural Trade Policy Council (IFATPC), Washington, DC.

Sharma, R. (2011) *Food Export Restrictions: Review of the 2007–2010 Experience and Considerations for Disciplining Restrictive Measures*. FAO Commodity and Trade Policy Research Working Paper No 32. Food and Agriculture Organization (FAO), Rome.

USDA (2008) *Republic of Kazakhstan – Grain and Feed, Grain Update*. USDA Foreign Agricultural Service, Global Agriculture Information Network, GAIN Report No KZ8001. UStates Department of Agriculture, Washington, DC. Available at: http://www.fas.usda.gov/gainfiles/200802/146293592.pdf (accessed 19 July 2013).

20 Agricultural Land Policy of Ukraine: State Legislation and Efficiency Analysis

Olga Murova*
Texas Tech University, Lubbock,Texas, USA

Abstract

This chapter focuses on the land market in Ukraine by providing a detailed description of land laws. The main objective is twofold: (i) to provide a descriptive analysis of land legislation and policies in Ukraine; and (ii) to investigate how land policies impact the technical efficiency of crop production in Ukraine. The two approaches used to investigate the impact of land policy on the technical efficiency of the Ukrainian crop sector are stochastic frontier analysis and data envelopment analysis. Regions with high technical efficiency are examined in terms of land policies. Leasing is a common practice in Ukraine, and the extent to which agricultural enterprises and state enterprises rent farmland impacts efficiency significantly. The amount of land leased by all types of agricultural enterprises increases technical efficiency, while farming enterprises working rented land decrease technical efficiency.

Introduction

Significant changes have occurred since Ukraine gained its independence from the Soviet Union in 1991. The 1990s saw the implementation of economic reforms, including policy and institutional reforms. The policy reforms were aimed at macroeconomic stabilization and price and trade liberalization. Institutional reforms targeted enterprise privatization. Since 2000, Ukraine has experienced increased production, higher living standards, positive growth rates and lower inflation (for the interested reader, the privatization of agriculture in Bulgaria (Meekhof *et al.*, 1994) provides an interesting comparison to this chapter).

From 2000 through 2008, Ukraine had positive growth in gross domestic product (GDP). In 2009, GDP fell 14.8% from the previous year (State Statistics Service of Ukraine, 2011). Since then, the Ukraine economy has shown positive growth.

Agriculture is an important part of the Ukraine economy. Ukraine is very proud of its fertile black soil, which accounts for around 27% of the world's black soil and holds great potential for agricultural production (Gmyrya and Vlasyuk, 2013). About 57% of the total land area is arable, and an additional 11% is utilized as permanent pastureland. Unfortunately, this valuable resource has been abused, due to long and improper land and crop management.

Since 2000, general growth in the Ukraine economy has been supported by positive growth in agricultural production. Figure 20.1 shows the overall increasing trend in total agricultural production, subject to small yearly fluctuations. In terms of agricultural production, 2011 turned out to be a record year, with an estimated

*E-mail: olga.murova@ttu.edu

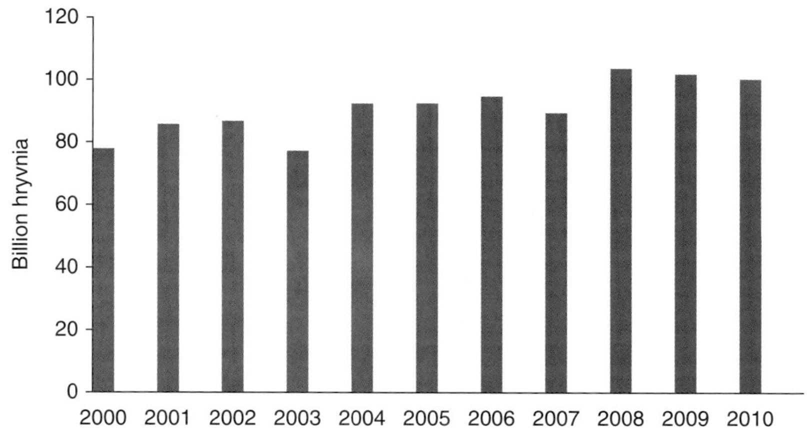

Fig. 20.1. Gross agricultural output, billion hryvnia (UAH) in 2005 UAH prices. (From State Statistics Service of Ukraine, 2011.)

17.5% growth in volume (Worldwide News Ukraine, 2012). In the same year, Ukraine harvested the largest amount of grain in the 20 years since its independence. This put Ukraine among the world's top three grain exporters (Worldwide News Ukraine, 2012).

In spite of the growth in Ukrainian agriculture, there are still many policy areas such as agricultural labour, rural development, environmental policy and food safety that need improvement. This chapter provides a descriptive analysis of land legislation and policies in Ukraine and analyses the impact of land policies on the technical efficiency (TE) of agricultural production in Ukraine.

Agricultural Land Market

Land market can be defined as *the circulation of parcels of land when ownership rights are changing hands based on the existing social legal rules* (Lazareva, 2013). By this definition, the land market is represented not only through seller–buyer relations but also through inheritance, exchanges, leases, settlements or other types of relationships. A land market includes all forms of ownership rights for land parcels. In Ukraine, however, land market means the sale or purchase of land.

Ordinarily, the term *agricultural land* implies arable land used in the production of commodities. In Ukraine, this definition also includes parcels of household holdings, subsidiary enterprises, agricultural enterprises and farm enterprises. However, legislative documents of the existing moratorium on the sale of agricultural land do not include all existing types of agricultural land ownership.

At the time of the reforms in 1991, owners of agricultural land in Ukraine received land share certificates as an official document of ownership; thus, a more appropriate wording in legislative documents would be sales of land shares rather than sales of agricultural land (Lazareva, 2013).

In developed economies, legislative land law defined zoning areas for agricultural activity, private and public construction, industrial usage and transportation, environmental zones and refuges and historic and cultural areas. This zoning practice united the national economy so that questions related to land use were solved in the common interest. However, in Ukraine, valuable land was not distributed in the best interest of the country. Future laws should exist that would not allow changes in appropriated land use. The national market in agricultural land shares needs to be protected from unfair competition through the enactment of antidumping policies. Also, the price of agricultural land in Ukraine should correspond to world prices. In 2013, the price of agricultural land in Ukraine was around 17,000–20,000 hryvnia per hectare (UAH/ha) (Lazareva, 2013). In US dollars (US$), this would be around US$2100–2400/ha, whereas a fair market price for agricultural land in Ukraine is around US$3000–9000/ha (Kozachenko, 2001).

An important issue related to the agricultural land market in Ukraine is the inaccessibility to long-term credit on the agricultural market (Gmyrya and Vlasyuk, 2013). This situation was caused by insufficient resources being used as collateral on the part of the borrower, since land could not be used as collateral in Ukraine banks. This problem led to short-term credit at very high interest rates.

When reforming the credit system, a single source of investment, such as a government land bank (e.g. similar to the Farm Credit Corporation of Canada), needs to be established that would offer credit at low interest rates and would accept land as collateral. In addition, local and foreign creditors need to compete for clients by offering credit at the same or better rates than a government land bank. In Ukraine, the State Land Bank was established in 2013.

Unless legislative changes are made, the leasing practices of small shareholders accepting small in-kind payments as of 2014 will not help land prices to increase. There is the fear that this practice might lead to agroholdings dominating the agricultural sector through purchases of large blocks of cheap land. On the positive side, agroholdings will help to modernize the technology of the agricultural sector.

The necessary condition for the development of a national agricultural land market is the creation of land cadastres and land registries. Ukraine is still in the initial phase of creating a land cadastre. The problem is that as long as parcels of agricultural land remain land shares where physical borders cannot be shown on a map, these parcels cannot be entered into land cadastres or land registries. In addition, the cadastre price of these land parcels cannot be determined economically without knowledge of the parcels' legal definitions and technical characteristics.

In summary, the following actions may hasten the creation of an efficient land market in Ukraine: (i) create a national cadastre and registry of property rights of ownership; (ii) adopt a legislative system to protect landownership rights; (iii) reform the credit system by organizing a fully functioning government land bank; and (iv) develop economic and financial mechanisms to control and regulate the land market.

Land Policy

Prior to the 1991 dissolution of the USSR, 98% of all agricultural land in Ukraine was managed by large-scale collective and state farms and only 2% was cultivated by small-scale individual sector farmers. Ukraine legalized private ownership of land in 1991. However, despite the new laws, landowners cannot sell or buy land, due to the moratorium imposed on the sale of land.

Between 1998 and 2000, Ukraine implemented several legislative decrees allowing new forms of legal land ownership, such as partnerships and cooperatives. By 2002, about 30% of Ukraine's agricultural land was privately owned, with members of the former collective agricultural enterprises owning about 65% of it (Liefert and Liefert, 2008). In 2010, Ukraine citizens owned approximately 30 million hectares (Mha) of Ukraine's agricultural land and about 12–13 Mha were owned by the state (Fedorchenko and Lapa, 2011).

In 2001, the President of Ukraine signed the *Land Code Law* (Verkhovna Rada of Ukraine, 2001). While this legislation gave citizens the right to own and lease agricultural land, it did not stipulate conditions for the sale of land. In 2002, the law *On Land Protection* and law *On the State Control over Use and Protection of Land* established the legal, economic and social bases for carrying out state control over land tenure and the protection of land (Verkhovna Rada of Ukraine, 2002a,b). However, no structure was created and no agency was put in place to ensure proper implementation of these laws. Another law regarding subsidiary farms came out in 2003. A subsidiary farm is defined as a business activity, conducted without creating a legal entity, by a physical person individually or by persons who are family or relatives, or living together with the purpose of producing, processing, consuming, or selling agricultural products (Verkhovna Rada of Ukraine, 2003a). Since the term *subsidiary farm* is defined very broadly in the legislation, it includes the majority of farms in Ukraine. This law determines the legal, organizational, economic and social principles for farm operations.

The law *On Lease of Land* came into effect in 2003 (Verkhovna Rada of Ukraine, 2003b). Land plots under state and communal ownership are leased out on the basis of a decision by the state executive body or local government by signing a lease agreement for a land plot. Individual citizens lease land on the basis of a lease agreement between the landowner (lessor) and lessee. Lease agreements are required: (i) to follow a specific format; (ii) to be notarized; and (iii) to be registered with the local district department of land resources. As a result of land privatization, 6.9 million citizens became owners of agricultural land shares, each with an average area of 4 ha (Kotykova and Vlasenko, 2013). In 2013, the majority of these land parcels were rented by large agricultural enterprises, mainly farming enterprises, utilizing around 18.8 Mha of agricultural land, 90% of which was leased from small landowners (Kotykova and Vlasenko, 2013). Renting land from small landowners has many negative consequences for large agricultural producers. Long-term credit is unavailable on the leased properties, which limits the implementation of new equipment and technology. Often, leased land is subleased, divided or attached to another agricultural enterprise. Renting land can have long-term negative consequences, since renters often disregard natural landscape boundaries between soil types and violate established crop patterns.

Collective agricultural enterprises (CAEs) dominated farm structure in Ukraine until 1999. Since then, three main forms of farm ownership have developed: (i) agricultural enterprises (individual corporate) farms; (ii) private/smallholder farms; and (iii) rural household plots. According to a 2006 survey of 1400 corporate farms, private/smallholder farms and rural household plots, the average size of a corporate farm is 1700 ha, a private/smallholder farm is 140 ha and an average rural household plot is 1.7 ha (Lerman *et al.*, 2006). There is a great deal of variability of farm income in Ukraine. Farm incomes of private/smallholder farmers and rural household plot owners depend on farm size. Households with 1–2 ha of land make around 5000 UAH (US$625) annually per household; farms larger than 50 ha make from 20,000 UAH (US$2500) or more annually per farm (Lerman *et al.*, 2006). Families of private/smallholder farmers enjoy much larger incomes than do other rural worker households: 54,500 UAH (US$6813) for farming households and 9750 UAH (US$1218) for rural worker (employee) households (Lerman *et al.*, 2006). The same survey shows that private/smallholder farmers rent land from other landowners, while most rural worker households lease land they cannot cultivate. Corporate farms rely primarily on land rentals. According to the State Agency of Land Resources, owners of land shares signed over 4.61 million lease contracts in 2011; the aggregate rent on these leases was US$0.71 billion. Tenants use agricultural output to pay for utilization of the land; in 2011, approximately 65% of land rentals were paid in kind (State Agency of Land Resources, 2012). The average annual payment for leasing land is around 100 UAH/ha (US$12/ha) (Lerman *et al.*, 2006).

The moratorium still stands as a major impediment to Ukrainian economic development. Land ownership is highly fragmented, and the moratorium keeps the land market in this fragmented state. The absence of a transparent and easily accessible national system of land location and ownership precludes capital and foreign investment. There have been numerous propositions to lift the moratorium; however, the land cadastre and land registry will have to be functional prior to lifting the moratorium. In 2012, a new law *On the Introduction of Amendments to the Land Code of Ukraine* came into effect that extended the moratorium on land sales until 2016 (Verkhovna Rada of Ukraine, 2012).

The law *On State Registration of Ownership Rights to Immovable Property and Their Encumbrances* came into effect in 2004 (Verkhovna Rada of Ukraine, 2004). It stipulates the creation of a single cadastre of land parcels and immovable property. Disputes, however, arose between the Ministry of Justice and the State Committee for Land Resources as to which agency would oversee the registration of landownership rights. Some argued for the creation of a single body in charge of the cadastre, while others proposed dividing the two functions between the two agencies. Finally, it was decided to create two systems: one for registration of title to property and the other for registration of the land cadastre. The Ministry of Justice and the State Agency for Land Resources, respectively, maintain the two systems. In 2010, the law *On State Registration of Ownership Rights to Immovable Property and Their Encumbrances* (Verkhovna Rada of Ukraine, 2004) specified the registration of titles under established procedure to take effect in 2012. The law *On State Land Cadastre* was passed by the Ukrainian Parliament in 2011 (Verkhovna Rada of Ukraine, 2011).

By the middle of 2012, over 93% of all citizens in Ukraine with certificates of the right to a land share received state acts (deeds) to the right of private ownership of land parcels. State acts specify the parcel area, list the names of the adjacent parcel owners and include a plan of the individual parcel of land.

Analysis of Technical Efficiency in Ukraine

Policies of Ukraine are directed toward the creation of a transparent and organized system of land management in preparation for lifting the moratorium on sales of agricultural land. This section of the chapter analyses the technical efficiency of the Ukraine crop sector to predict the impact of policy changes on agricultural technical efficiency.

Methods and data

Agricultural technical efficiency is used in this study as an economic indicator of how well the agricultural crop sector is performing in Ukraine based on the latest agricultural policy changes. The performance of Ukraine crop production through technical efficiency is first estimated and then analysed. Two approaches are used to estimate technical efficiency: data envelopment analysis (DEA) and stochastic frontier analysis (SFA). These approaches allow the estimation of a static measure of technical efficiency. Two methods are used to compensate for the advantages and disadvantages each has over the other.

DEA is estimated to determine the amount by which output can be increased without requiring additional amounts of inputs. It is measured as a ratio of observed output to the maximum level of output to reach the frontier. It is calculated as

$$max_{\varphi\lambda}\,\phi\, s.t. -\phi y_i + Y\lambda \geq 0,\ x_i - X\lambda \geq 0, \lambda \geq 0, \quad (20.1)$$

where $1 \leq \phi \leq \infty$ and $\phi - 1$ is the proportional increase in outputs that could be achieved by the *i*th firm (Coelli *et al.*, 1998). The value of $1/\phi$ defines a technical efficiency (TE) score.

The goal of SFA is to explain TE performance using economic and policy variables. To estimate this model, the following Cobb–Douglas production function is used

$$\ln y_i = \sum \beta_i \ln x_i + \upsilon_i - u_i, \quad (20.2)$$

where $i = 1, 2, ..., n$, and where $\ln y_i$ is the logarithm of the total value of output for the crop sector for farms in region *i*, $\ln x_i$ is the logarithm of input quantities in region *i*, υ_i is a random error, $\upsilon_i \sim N(0, \sigma_\upsilon^2)$, and u_i is the measure of technical inefficiency, $u_i \sim N(\mu, \sigma^2)$, truncated from the left at zero. This model was independently proposed by Aigner *et al.* (1977) and Meeusen and Van Den Broeck (1977). SFA allows simultaneous estimation of causal factors that explain technical inefficiencies u_i

$$u_i = z_{it}\delta + w_{it}, \quad (20.3)$$

where z_{it} are the exogenous variables which explain inefficiencies that can be farm or policy characteristics, δ is the vector of the parameters to be estimated and w_{it} is a random variable. Substituting u_i back in Eqn 20.2 yields

$$\ln y_i = \sum \beta_i \ln x_i + z_{it}\delta + w_{it} + \upsilon_i \quad (20.4)$$

where $i = 1, 2, \ldots, n$.

The data in Table 20.1 for this analysis were collected from the State Statistics Service of Ukraine website (State Statistics Service of Ukraine, 2011). Cross-sectional data for 25 Ukrainian regions were collected for 2010.

The crop sector in Ukraine was selected for this analysis because it represented a larger share of total agricultural production than did the livestock sector. For example, in 2010, the crop sector share was 58.6%, while the livestock sector share was 41.4% (Prokopenko, 2011). In addition, the 2010 performance of the crop sector can be compared to its 1994 performance (Murova *et al.*, 2004).

The data on variables used to estimate technical efficiency by region (Table 20.2) are collected for all types of agricultural enterprises, including household holdings. The term *all agricultural enterprises* includes government enterprises, non-government enterprises, individual enterprises, cooperatives, agroholdings, private/smallholder farms and household plots. Variables collected for all agricultural enterprises used in this analysis include total crop output, labour, fertilizer

Table 20.1. Summary statistics of Ukraine data, 2010. (From State Statistics Service of Ukraine, 2011.)

	Mean	Standard deviation	Minimum	Maximum
Scalar output	58,677.6	856.2	1,012.2	4,224.2
Land	24,343.9	507.4	179.3	1,945.6
Fertilizer	10,606.4	240.3	14.3	314.0
Machinery/equipment	41,582.0	853.0	142.0	3,303.0

Table 20.2. Data for total value of output (UAH million), land, fertilizer (1000 ha) and machinery and equipment (thousand kw/h), by region (oblast) in Ukraine, 2010. (From State Statistics Service of Ukraine website, 2011.)

Region	Total value of output, 2005 prices (UAH million)	Land (1,000 ha)	Fertilizer (1,000 t)	Machinery and equipment (thousand kw/h)
Republic of Krym	2,064.8	757.6	2,529.0	1,553.0
Vinnyts'ka	4,224.2	1,410.6	9,140.0	2,823.0
Volyns'ka	1,481.0	348.4	1,336.0	880.0
Dnipropetrovs'ka	3,491.3	1,945.6	6,161.0	3,033.0
Donets'ka	2,633.3	1,330.9	3,505.0	2,106.0
Zhytomirs'ka	1,859.6	554.2	2,560.0	1,169.0
Zakarpats'ka	1,012.2	179.3	143.0	142.0
Zaporiz'ka	2,319.8	1,505.7	4,201.0	2,505.0
Ivano-Frankivs'ka	1,012.2	234.8	740.0	357.0
Kyyivs'ka	3,405.4	1,023.4	5,745.0	2,584.0
Kirovohrads'ka	2,688.5	1,504.1	5,067.0	2,339.0
Luhans'ka	1,365.9	879.3	3,000.0	1,380.0
L'vivs'ka	2,102.4	412.6	2,479.0	757.0
Mykolayivs'ka	2,414.2	1,369.8	3,954.0	1,620.0
Odes'ka	3,329.4	1,550.0	5,252.0	2,714.0
Poltavs'ka	3,548.6	1,705.5	8,620.0	2,664.0
Rivnens'ka	1,738.5	375.0	2,162.0	704.0
Sums'ka	1,809.1	912.8	5,019.0	1,485.0
Ternopils'ka	1,867.1	673.9	4,988.0	907.0
Kharkivs'ka	2,495.2	1,411.9	6,187.0	2,485.0
Khersons'ka	3,107.8	1,076.6	3,093.0	1,667.0
Khmel'nyts'ka	2,728.9	832.5	6,114.0	1,618.0
Cherkas'ka	2,805.1	1,131.6	7,724.0	2,040.0
Chernivets'ka	1,156.6	259.5	861.0	327.0
Chernihivs'ka	2,016.5	958.3	5,484.0	1,723.0
Average	58,677.6	24,343.9	10,606.4	41,582.0

and land. Total crop output is measured by the value of total crop production in hryvnia (UAH) for the year 2010 in 2005 prices. Land is measured as the total agricultural land used by all agricultural enterprises in crop production and is measured in thousand hectares. Fertilizer is measured in thousand tonnes (t) as the amount of mineral fertilizers used. The machinery/equipment variable represents the total engine capacity of tractors, trucks and combines, and all other engines used in all agricultural enterprises in the crop sector measured in thousand kilowatts per hour (kw/h).

Unfortunately, some of the data needed for this study are unavailable for all regions in Ukraine. Additional variables used in this study to represent land policy variables include the total number of all agricultural enterprises z_1, the total number of farming enterprises z_2, the amount of land leased by all agricultural enterprises z_3 and the amount of land rented by farming enterprises z_4. Data for these variables were collected for 25 regions in Ukraine.

Results

A likelihood ratio (LR) test is used to determine a functional form of the production function that would fit the Ukraine data appropriately. Two functional forms of the production function considered in this study are the trans-log and the Cobb–Douglas. The values of two log likelihood functions are used to calculate the LR value: LR = –2 [15.6 – 14.9] = 1.4. The LR value is compared with the upper 5% for the χ^2 of 14.07. The null hypothesis that the Cobb–Douglas frontier is an appropriate representation of data is not rejected in favour of trans-log specification.

Model 1 in Table 20.3 shows the final test statistics and the magnitudes of Cobb–Douglas coefficients. Initially, labour was included in the model; however, this led to a negative coefficient for one of the variables. This outcome suggested the presence of multicollinearity; thus, correlations between variables were examined. There is a high correlation between land and labour, and between machinery and labour. Due to the unavailability of additional data, only the labour variable was dropped from the model. This improved the results for both models. In the first model, the elasticity of crop output with respect to machinery/equipment in Ukraine is found to be significant and equal to 0.271.

The land and fertilizer variables are not statistically significant. For land, this is a surprising result. However, land used in agricultural production decreased in Ukraine after its independence in 1991. Of Ukraine's total 60 Mha land area, only 30.4 Mha were used in crop production in 2000 (Lindeman, 2004). By 2010, land used in crop production decreased another 20%.

The amount of fertilizer used in Ukraine for crop production between 1990 and 2010 decreased significantly. For example, the fertilizer application rate for wheat plunged from 149 kg/ha in 1990 to 24 kg/ha in 2000 (Lindeman, 2004). Because of high fertilizer prices, farmers in Ukraine still rely to a considerable degree on the application of manure as an organic fertilizer.

A second stochastic frontier model was estimated with four additional variables that represented land policy variables. These exogenous variables showed how current land policies impacted technical efficiency in the regions (Model 2, Table 20.3). The explanatory power of this model increased after the four variables were added. As in the first stochastic frontier model, machinery shows statistical significance. Two exogenous variables, land leased by all agricultural enterprises and land rented by farming enterprises, were also significant.

It was found that an increase in the amount of land leased by all agricultural enterprises increased efficiency. The category of all agricultural enterprises includes a variety of private, collective and state enterprises. In terms of equipment and labour, many do not have the sufficient expertise and economic means to organize efficient crop production, so they lease their land to other enterprises.

Land rented by farming enterprises decreases the efficiency of crop production in Ukraine.

Table 20.3. Stochastic frontier Cobb–Douglas production function estimates. (From author's calculations.)

Variables	Model 1	Model 2
X variables		
Intercept	4.656[a] (0.421)	4.654[a] (0.940)
Land	0.155 (0.179)	0.125 (0.282)
Fertilizer	0.061 (0.104)	0.120 (0.122)
Machinery	0.271[b] (0.203)	0.224[b] (0.172)
Z variables		
z_0 – Constant	–	4.293 (3.639)
z_1 – Total number of ag. enterprises	–	0.823 (1.686)
z_2 – Total number of farming enterprises	–	–1.485 (1.296)
z_3 – Land leased by ag. enterprises	–	–0.746[a] (0.33)
z_4 – Land rented by farming enterprises	–	1.051[a] (0.327)

Notes: The standard errors are shown in parentheses.[a]Significant at the 1% significance level; [b]significant at the 10% significance level.

Farming enterprises rent land from small landowners and from other farming enterprises. This involves high transaction costs when signing a large number of short-term rental agreements. As mentioned earlier, other negative consequences include the unavailability of long-term credit for leased property, and leased land can be subleased, divided or attached to another farm or enterprise. Farming enterprises do not always rent land of a good quality, since rental land often is abused in terms of improper land and crop management. Overall, farming enterprises rent 70%–80% of the land they cultivate.

The DEA model was estimated for the 2010 production year with the same set of variables as those in the SFA Model 2 (Table 20.4). The average TE scores for the Ukrainian crop sector estimated by the SFA (0.902) and the DEA (0.937) approaches were somewhat similar. DEA scores were a bit higher than SFA scores. This outcome is consistent with the results of previous research (Hjalmarsson *et al.*, 1996; Hamrouni and Ratsimbanier, 2012).

Table 20.4. Technical efficiency scores by region with SFA and DEA methods, 2010. (From State Statistics Service of Ukraine website, 2011.)

	SFA	DEA
Region	Technical efficiency (TE) scores	
Republic of Krym	0.956	0.920
Vinnyts'ka	0.957	0.948
Volyns'ka	0.943	0.979
Dnipropetrovs'ka	0.963	0.874
Donets'ka	0.968	0.920
Zhytomirs'ka	0.819	0.973
Zakarpats'ka	0.992	1.000
Zaporiz'ka	0.885	0.864
Ivano-Frankivs'ka	0.875	1.000
Kyyivs'ka	0.983	0.942
Kirovohrads'ka	0.883	0.872
Luhans'ka	0.564	0.869
L'vivs'ka	0.962	0.993
Mykolayivs'ka	0.980	0.863
Odes'ka	0.991	0.862
Poltavs'ka	0.985	0.926
Rivnens'ka	0.867	1.000
Sums'ka	0.747	0.946
Ternopil's'ka	0.907	0.964
Kharkivs'ka	0.744	0.919
Khersons'ka	0.983	0.917
Khmel'nyts'ka	0.919	0.949
Cherkas'ka	0.965	0.934
Chernivets'ka	0.977	0.986
Chernihivs'ka	0.729	1.000
Average	0.902	0.937

The two methods do not identify extreme performers consistently, with the SFA parametric method being regarded as somewhat more reliable for policy analysis (Fiorentino *et al.*, 2006). The stochastic frontier model shows that the Zakarpats'ka, Odes'ka, Poltavs'ka, Kyyivs'ka and Khersons'ka regions had the highest TE scores during 2010. Close evaluation of these regions revealed that in terms of total production of crops, the Kyyivs'ka, Odes'ka and Khersons'ka regions increased production relative to the previous year, while other leading regions showed a slight decrease in total crop production. The increase in crop production in the Kyyivs'ka, Odes'ka and Khersons'ka regions reflected an increase in production by household holdings in these regions of 106.5%, 128% and 108.9%, respectively (Prokopenko, 2011). Data envelopment analysis shows that the Zakarpats'ka, Ivano-Frankivs'ka, Rivnens'ka and Chernihivs'ka regions determine an efficient frontier (the efficiency of the other regions was measured as a distance to this efficiency frontier).

To understand the results of both approaches, an examination of production practices in all leading regions is in order. A comparison of the proportion of land used in crop production held by the populace and land in all other agricultural enterprises revealed that in all of the most efficient regions determined by the DEA approach, the amount of land held by the populace was greater than the amount of land in all other agricultural enterprises. For example, comparing populace holdings in the different regions shows that the Zakarpats'ka region constitutes 6.7 times more land in crop production than all other agricultural enterprises taken together, followed by the Ivano-Frankivs'ka region with 3.5 times more and the Rivnens'ka region with 2.1 times more. This is not the case with the leading regions ranked by SFA analysis; the majority of land in the Kyyivs'ka, Odes'ka, Poltavs'ka and Khersons'ka regions is concentrated in all other agricultural enterprises and not in the hands of the general population.

All four regions selected by DEA show a large share of crop production by household holdings. This form of farming is still prevalent in almost all Ukrainian regions. Household plots are less susceptible to changes in input prices because they employ more manual labour and use mainly organic fertilizer. They also allow a high degree of adaptation in landscape conditions and type of crops. Individual household plots allow the application of environmentally friendly agricultural production. Many individual owners of household plots describe their production practices as organic production.

Household plot owners produce 95.9% of all crop output in the Zakarpats'ka region and 86.7% of all crop output in the Ivano-Frankivs'ka region; for the Rivnens'ka region, this percentage is 77.3%, and for the Chernihivs'ka region, it is 55.2% (Prokopenko, 2011). Agricultural enterprises own a smaller amount of land in these regions relative to other regions, but farming enterprises use the majority of this land.

SFA's best performers, the Kyyivs'ka, Odes'ka, Poltavs'ka and Khersons'ka regions, produce around one-half of their total crop output in agricultural enterprises. In the Kyyivs'ka region, 49% of crop output is produced by agricultural enterprises; in the Odes'ka region, the percentage is 52.5%; in the Poltavs'ka region, around 53%; and in the Khersons'ka region, around 47%. Thus, for these regions, changes in current land policy are equally important for private/smallholder farms and for agricultural enterprises. In the Kyyivs'ka region, 89% of the land is owned by agricultural enterprises; in the Odes'ka region, ownership is 81%; in the Poltavs'ka region, 88%; and in the Khersons'ka region, 80%.

A comparison was made of the technical efficiency of the Ukrainian crop sector for 1994 and 2010 (Murova *et al.*, 2004). In 1994, Ukrainian average TE scores were 0.731 (SFA) and 0.836 (DEA). In 2010, the technical efficiency in Ukraine had increased to 0.902 (SFA) and 0.937 (DEA). This increase in technical efficiency shows that there has been progress in more efficient utilization of resources used in crop production. Thus, it can be concluded that production practices are improving and that the economic policies in Ukraine have had a positive impact on technical efficiency overall.

Two approaches show that while technical efficiency varies considerably across Ukrainian regions, the majority of the Ukrainian regions are efficient in crop production. Nine regions in the country operate at levels below the average TE level of crop production. The underperforming regions include the northern regions with low soil fertility and where crops require a longer growing period. Both approaches show an increase in technical efficiency between 1994 and 2010, due to reforms and the establishment of an operational market in land that is a free market, to a certain degree.

SFA reveals the leading Ukrainian regions that increased crop production in 2010. It shows that agricultural enterprises in these regions produce slightly more than 50% of the entire crop output. Individual citizens on their household plots produce the rest of the crop output. All agricultural enterprises in the leading regions of Ukraine account for more than 80% of all agricultural land used in crop production. These agricultural enterprises lease more than 90% of their land. Farming enterprises in these regions represent on average 77%–96% of total agricultural enterprises. They rent around 70%–80% of the land used for crop production.

DEA analysis reveals that in the most efficient regions, land is accumulated in private hands. The amount of land in private hands in these regions is two to six times greater than the amount of land in all other agricultural enterprises. The majority of crops (55%–95%) are grown on individual household plots 1–4 ha in size. Even though the majority of crops are produced on the household plots, agricultural enterprises use a large proportion of land in these regions, the majority of which they rent.

Both the unavailability of a land market and an undeveloped credit system have encouraged the current practice of leasing agricultural land. Leasing agricultural land is cheap, and 65% of all lease payments are made in kind. Leasing of agricultural land by all agricultural enterprises leads to increased technical efficiency in crop production.

Conclusions

Land reform in Ukraine occurred in three stages in our analysis. The first stage occurred between 1980 and 1991, when some agricultural enterprises began changing their management practices by becoming more independent and using self-financing. During the second stage, which lasted from 1992 until 1999, collective ownership of land was transformed to common ownership. In the third stage, 2000 onward, land shares have changed into private ownership through state acts/deeds. By June 2012, around 93% of all land share certificates in Ukraine were converted into the state legislative acts. A new stage of land reform will begin when limitations on the sale of land are lifted. So far, Ukraine has not defined fully the institutions and procedures that will determine its future land market. Designing an efficient system of land registration will influence the future direction of land policy in Ukraine.

While there is an abundance of agricultural land for lease in Ukraine, high transaction costs make renting inefficient for farming enterprises. The leasing of land by all types of agricultural enterprises increases the technical efficiency of crop production. The unavailability of liquid currency and credit from Ukrainian banks makes it impossible for common citizens to start new enterprises. Foreign investors cannot buy land in Ukraine. The only way for foreigners to gain a foothold in agricultural land is to rent it directly from farmers. It is a common practice now to sign a medium-term (up to 25 years) or long-term (up to 99 years) rental agreement. Foreign investors register their companies as Ukrainian-owned companies and sign rental agreements with private farmers. In some cases, large investors have to sign hundreds or even thousands of such agreements.

All structures of the farming sector – agricultural enterprises, private/smallholder farms and household plot owners – still remain problematic in terms of technical efficiency in Ukraine, and are constrained by the lack of appropriate policies and inadequate markets (Gumeniuk *et al.*, 2009). There are relatively few medium-sized commercial farms in Ukraine, and subsistence farms are too small to be commercially viable in the long run.

The absence of a land market reduces the growth potential of Ukrainian agriculture and leads to an overall loss to the national welfare. The government sets land rents at a fixed percentage on the nominal value of the land that does not reflect its real market value. In addition, there are concerns that oligarchic groups and the government may manipulate land legislation to fulfil their own interests. The development of the land market in Ukraine will be strongly influenced by future government legislative policies.

References

Aigner, D.J., Lovell, C.A.K. and Schmidt, P. (1977) Formulation and estimation of stochastic frontier production function models. *Journal of Econometrics* 6(2), 21–37.

Coelli, T., Rao, D.S.P. and Battese, G.E. (1998) *An Introduction to Efficiency and Productivity Analysis*. Kluwer Academic Publishers, New York.

Fedorchenko, M. and Lapa, V. (2011) Agricultural land market in Ukraine: prospects and expected outcomes. Available at: http://www.myland.org.ua (accessed 28 September 2012).

Fiorentino, E., Karmann, A. and Koetter, M. (2006) *The Cost Efficiency of German Banks: A Comparison of SFA and DEA*. Discussion paper, Series 2: Banking and Finance, No 10. Deutsche Bundesbank, Eurosystem.

Gmyrya, V. and Vlasyuk, S. (2013) Land market of agricultural purpose in Ukraine: current status and problems of development. *Economist* (Ukrainian Journal) 3, 34–35.

Gumeniuk, K., Mishchenko, N., Fischer, G. and Van Velthuizen, H. (2009) *Agro-ecological Assessment for the Transition of the Agricultural Sector in Ukraine. Part II. Methodology and Results for Base Line Climate*. International Institute for Applied Systems Analysis (IIASA), Laxenburg, Austria. Available at: http://webarchive.iiasa.ac.at/Admin/PUB/Documents/IR-09-050.pdf (accessed 9 November 2012).

Hamrouni, A. and Ratsimbanier, H. (2012) The extent of voluntary information in the annual report and the share price volatility: application of DEA and SFA on the French market. *Journal of Modern Accounting and Auditing* 8(7), 951–966.

Hjalmarsson, L., Kumbhakar, S.C. and Heshmati, A. (1996) DEA, DFA, and SFA: a comparison. *Journal of Productivity Analysis* 7, 303–327.

Kotykova, O. and Vlasenko, I. (2013) Future development of agricultural land market in Ukraine. *Economist* (Ukrainian Journal) 3, 36–37.

Kozachenko, L. (2001) Today is not the best time for marketing land. *Zemlevporyadnyi Visnuk* 4, 6–7.

Lazareva, O. (2013) Conditions of land market's competitiveness ensuring. *Economist* (Ukrainian Journal) 3, 42–43.

Lerman, Z., Sedik, D., Pugachev, N. and Goncharuk, A. (2006) *Ukraine after 2000: A Fundamental Change in Land and Farm Policy?* Discussion Paper No 7.06. Available at: http://ageconsearch.umn.edu/bitstream/7140/2/dp060007.pdf (accessed 24 October 2012).

Liefert, B. and Liefert, O. (2008) *The Russian Federation, Part 3. The International Bank for Reconstruction and Development*. The World Bank, Washington, DC, pp. 135–174.

Lindeman, M. (2004) *Ukraine: Agricultural Overview*. Production Estimates and Crop Assessment Division, Foreign Agricultural Service, US Department of Agriculture (FAS/USDA), Washington, DC.

Meekhof, R., Schmitz, A. and Penov, I. (1994) The grain sector: wheat is king. In: Schmitz, A., Moulton, K., Buckwell, A. and Davidova, S. (eds) *Privatization of Agriculture in New Market Economics: Lessons from Bulgaria*. Kluwer Publishing, Norwell, Massachusetts, pp. 235–262.

Meeusen, W. and Van Den Broeck, J. (1977) Efficiency estimation from Cobb-Douglas production functions with composed error. *International Economic Review* 18(125), 435–444.

Murova, O., Trueblood, M.A. and Coble, K.H. (2004) Measurement and explanation of technical efficiency performance in Ukrainian agriculture, 1991–1996. *Journal of Agricultural and Applied Economics* 36(1), 185–198.

Prokopenko, O.M. (2011) Total agricultural output of Ukraine for the year 2010. Available at: http://www.ukrstat.gov.ua (accessed 26 October 2012).

State Agency of Land Resources (2012) Lease of land in Ukraine. Available at: http://www.dazru.gov.ua/terra/control/uk/index (accessed 17 October 2012).

State Statistics Service of Ukraine (2011) State Statistics Service of Ukraine website. Available at: http://urkstat.gov.ua/ (accessed 16 November 2011).

Verkhovna Rada of Ukraine (2001). *Land Code Law*. Verkhovna Rada, Kiev.

Verkhovna Rada of Ukraine (2002a) *On Land Protection*. Verkhovna Rada, Kiev.

Verkhovna Rada of Ukraine (2002b) *On the State Control over Use and Protection of Land*. Verkhovna Rada, Kiev.

Verkhovna Rada of Ukraine (2003a) *Law of Ukraine on Subsidiary Farms*. Verkhovna Rada, Kiev. Available at: http://www.lexadin.nl/wlg/legis/nofr/oeur/lxweukr.htm (accessed 23 October 2012).

Verkhovna Rada of Ukraine (2003b) *On Lease of Land*. Verkhovna Rada, Kiev.

Verkhovna Rada of Ukraine (2004) *On State Registration of Ownership Rights to Immovable Property and Their Encumbrances*. Verkhovna Rada, Kiev.

Verkhovna Rada of Ukraine (2011) *On State Land Cadastre*. Verkhovna Rada, Kiev.

Verkhovna Rada of Ukraine (2012) *On the Introduction of Amendments to the Land Code of Ukraine*. Verkhovna Rada, Kiev.

Worldwide News Ukraine (WNU) (2012) In 2011 Ukraine boosts agricultural production by 17.5 percent. Available at: http://wnu-ukraine.com/news/economy-business/?id=1271 (accessed 21October 2012).

21 Dynamics of Agricultural Production and Land Use in Post-Soviet Ukraine

Denys Nizalov,[1]* Suzanne Thornsbury,[2] Scott Loveridge,[3] Mollie Woods[3] and Olha Zadorozhna[4]

[1]Kiev School of Economics, Kiev, Ukraine; [2]Economic Research Service, US Department of Agriculture, Washington, DC, USA; [3]Michigan State University, East Lansing, Michigan, USA; [4]Lazarski University, Warsaw, Poland

Abstract

Multiple economic and institutional factors drive change in the mix of agricultural commodities produced over time in any country. We examine grain, fruit and vegetable markets in Ukraine since 2004, with a focus on the evolution of property rights and their impact on planting decisions. Institutional uncertainty about property rights impacts both the use of land in agriculture and crop mix, and leads to underinvestment in crops with a longer investment cycle and higher value added. As an alternative to land sales, a well-developed rental market has emerged. The rental market requires minimal upfront investment in land acquisition but limits incentives for long-term investments. Farm-level data from the State Statistics Committee of Ukraine on 21,000 farms between 2004 and 2011 show that despite higher revenue per hectare, the share of fruit crops has remained virtually unchanged since 2004 (State Statistics, 2013). In contrast, the share of oilseeds produced has doubled over the same period.

Acknowledgements

This chapter was partially supported by NSF Award CNH 0909378. The views expressed in this chapter are those of the authors and may not represent the position of their corresponding institutions.

Introduction

Multiple economic and policy factors have influenced post-liberalization changes in Ukrainian agriculture. Productivity and output adjustments vary substantially between sectors, and the national enterprise mix has evolved over time. Despite higher per hectare revenue and relatively high profit, the share of fruit crops in the national agricultural output mix has declined since 2004, and about two-thirds of the producers have left the industry. The share of vegetables has remained unchanged, despite a lower number of producers and a lack of reported profit in these markets. In contrast, grain crop production has expanded rapidly and the sector is often mentioned as having made a successful transition to the market economy. The share of oilseeds has doubled, influenced heavily by high international prices for both primary and processed products.

*Corresponding author; e-mail: nizalov@kse.org.ua

The evolution of property rights and the development of support industries (i.e. transportation, handling, input supplies and financial services) have had important impacts on changes in the national agricultural output mix beyond a simple comparison of returns between alternative crop enterprises. This chapter provides a descriptive assessment of the transition in the fruit and vegetable industries in comparison with the grain and oilseed industries; identifies land reform policies that have influenced relative adjustments; and describes factors that will impact the future development of the fruit and vegetable industries in Ukraine.

The Agricultural Sector

Ukraine is often characterized as the breadbasket of Europe, and has some of the richest arable land in the world. Approximately one-third of all high-quality European arable land is found in Ukraine (twice as much as in France and three times more than in Germany). However, in 2012, agriculture contributed only 9.2% of the Ukrainian gross domestic product (GDP), a figure that has declined steadily since the dissolution of the Soviet Union in 1991 (State Statistics, 2013). While some of the decline may be attributed to market transition and post-Soviet recovery of non-agricultural sectors, the 2012 national agricultural output was only 79% of the 1990 level (Fig. 21.1).

The agricultural sector in Ukraine includes two distinct groups of producers: commercial (corporate farms and private/smallholder farms) and subsistence (household) farms. The first group is composed of commercial producers who sell to domestic and international markets. The second group is made up of individual households who produce primarily for their own household consumption or limited supplemental income. Almost all rural households are involved with subsistence or supplementary farming. However, for most rural households, a primary source of revenue is off-farm employment, retirement pensions or remittances. Some output from households is sold in local markets, and a small portion of the rural household population is employed by commercial farms. About 38% of commercial farms (particularly the larger corporate farms) are successors to the former state-owned collective farms (Deininger et al., 2013). A new type of commercial farm emerged following the dissolution of the Soviet Union in 1991 – private/smallholder farms that rely on sales of agricultural products as their major source of income. In 2012, the commercial sector was composed of 8121 large corporate farms (average size about 2000 hectares (ha)) and 47,745 private/smallholder farms (average size about 84 ha), jointly cultivating about 21.4 million ha (Mha) (State Statistics, 2013). Another 4.6 Mha were cultivated by approximately 9.3 million individual households (State Statistics, 2013). The average garden plot size for individual households producing

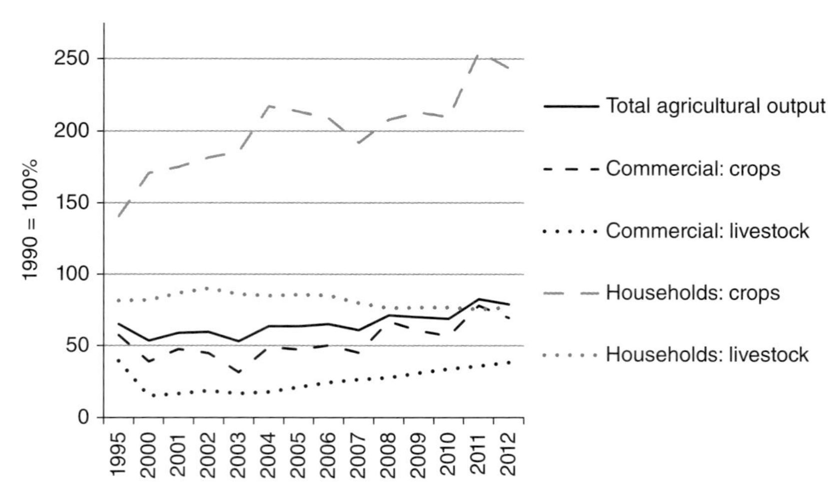

Fig. 21.1. Index of agricultural output, Ukraine, 1995–2012. (From State Statistics, 2013.)

for their own consumption is 0.21 ha, and the average garden plot size for individual households selling some of their output is 1.1 ha. Based on the Household Budget Survey, individual households account for about 16.4 Mha of land (State Statistics, 2013), with over 70% of this land being rented to commercial farms or other households (State Statistics, 2013).

While large commercial (state-owned collective) farms supplied the majority of output before the dissolution of the Soviet Union, household production has played a major role since the economic transition began in 1991. When comparing the household sector with the commercial sector (corporate farms and private/smallholder farms), only the crop output from the household sector has notably increased since 1995. Crop and livestock output from the commercial sector has increased slightly since 2007, but growth remains modest. The output index for livestock from the commercial sector has not been higher than 50% since 1995 (1990 = 100). Likewise, output of livestock from the household sector and crops from the commercial sector remain below 100% (i.e. levels are below those of 1990).

The household sector expanded rapidly between 1991 and 2000, as many rural and some urban households depended on subsistence farming as a strategy for coping with the economic changes due to the transition reform. While the garden plots of individual households were small, they were also numerous. Over the same period, output from commercial farms decreased, resulting in the national agriculture output share from individual households increasing from 29.6% in 1990 to 61.6% in 2000 (Table 21.1). The household sector share remained at 49.3% in 2012, even as the commercial sector began to recover.

The contribution by individual households to total production varies significantly between commodities. In 2012, the household sector contributed only about 22% of the grains and 15% of the sunflowers produced in Ukraine. In contrast, the household sector in 2012 contributed 97%, 87% and 82% of the potatoes, vegetables and fruit, respectively. Market-oriented commercial farms (corporate farms and private/smallholder farms) contributed about 51% of all national agricultural output in 2012. Almost all of the commercial farms were involved in grain production. The major commercial farm crops included grains (50%), oilseeds (above 20%), animal feed (8%) and root crops (8%).

In commercial agriculture (corporate farms and private/smallholder farms) that excludes individual households, private/smallholder farms dominate the number of Ukrainian farms. In 2012, the State Statistics Committee of Ukraine recorded 55,866 commercial farms (8121 corporate and 47,745 private/smallholder farms), including approximately 15% reporting no land

Table 21.1. Per cent share of agricultural output from individual households, Ukraine, selected years, 1990–2012. (From State Statistics, 2013.)

Indicators	1990	1995	2000	2005	2010	2011	2012
				Per cent			
Agricultural output	29.6	45.9	61.6	59.5	51.7	48.2	49.3
All crops	18.9	36.5	50.7	51.4	46.4	43.3	45.0
Grains	2.8	8.1	18.4	24.3	24.2	22.1	21.9
Sugarbeet	0	2.6	12.2	21.5	7.9	8.5	8.7
Sunflower	2.4	4.4	12.5	21.2	17.5	15.9	15.0
Potato	71.4	95.8	98.6	98.8	97.4	96.9	96.7
Vegetables	26.9	72.7	83.1	89.3	88.1	84.3	85.7
Fruit and berries	53.6	83.6	81.8	88.2	83.6	84.2	81.6
Grapes	20.5	26.5	30.0	41.8	36.3	35.4	35.8
All livestock	40.8	58.9	79.0	73.8	61.2	59.4	58.2
Meat	28.9	51.7	73.7	63.2	44.9	43.3	42.5
Milk	24.0	45.3	71.0	81.2	80.3	79.7	77.7
Eggs	37.8	55.6	66.2	50.5	39.9	37.2	37.3
Wool	11.2	30.1	61.4	78.3	83.1	83.3	85.1

in production. Over 32,000 of the private/smallholder farms reported operating less than 100 ha (Fig. 21.2). These private/smallholder farms represent over two-thirds of the commercial operations but less than 5% of the land area.

State Statistical Data

To describe changes in crop production patterns, we rely on detailed micro-level data collected regularly by the State Statistics Committee of Ukraine. The Committee maintains a well-developed system of state statistics linked to the registry of business entities. All agricultural producers are required by law to file several statistical forms integrated into reporting systems used for farm accounting. Electronic submission was established in 2011 to simplify the reporting process. The reporting forms, instructions and methodology available online make the self-reporting process relatively straightforward. Local State Statistical Offices track the reports. In addition to data reporting, the same statistical forms are required as support documents when obtaining credit or state subsidies. These forms are also requested by the private sector in many other cases as part of the background documentation on a farm. As stipulated by law, the state statistics system maintains strict confidentiality of identifiable individual data.

The discussion in this chapter draws on data from two forms: Form 29SG *Total Yield of Agricultural Crops* and Form 50SG *Economic Results of Agricultural Producers* (State Statistics, 2012a,b). While both forms are collected annually, data from Form 29SG were available only for 2008 and 2011. Data on land area and yield by crop are reported through Form 29SG for all individual production units, in contrast to the legal entity that owns the corporate farm/agricultural enterprise, and can operate multiple production units or farms. There were approximately 54,000 and 51,000 production unit level observations from Form 29SG in 2008 and 2011, respectively (Table 21.2). The sample captures over 90% of all Ukraine commercial farms (both corporate farms and private/smallholder farms) based on data from census results reported in the *Statistical Year Book* (State Statistics, 2013).

Annual data from 50SG were available for each year between 2004 and 2012. Form 50SG was submitted by large commercial farms that contained more than 200 ha for land holdings; had either 50 animals or at least 20 full-time employees at the place of registration, which is the legal entity; and provided data that may have included multiple production locations. Crop and livestock output and sales data as well as cost of production were reported through Form 50SG. Compiled data from Form 50SG represented about 21,000 unique farm identifications (IDs) between 2004 and 2012. There were observations for an average of over 8200 commercial farms in each year. Farms in the sample operated approximately 18 Mha annually between 2004 and 2012.

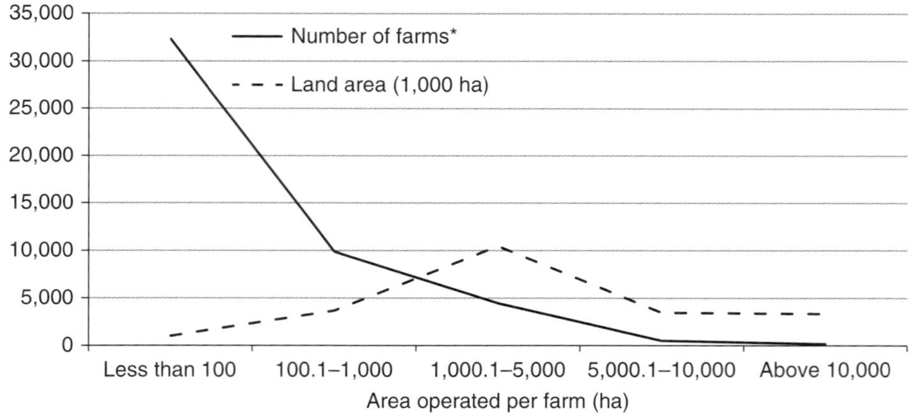

Fig. 21.2. Number of commercial farms and total land area, by farm size, 2012. (From State Statistics, 2013.) *Note*: *Does not include farms reporting zero land area.

Small commercial farms dominate the sector in terms of farm numbers, but not in terms of land area. Farms in the Form 50SG sample operated over 80% of the agricultural land cultivated commercially in 2012. The share of farms planting specific crops and the share of respective land area varied substantially (Table 21.3). For example, in 2012, approximately 93% of the commercial operations that filed Form 50SG produced grains, and those farms had almost 50% of their acreage planted to grains. In contrast, less than 4.5% of the same farms produced fruit, and those that did had only 0.2% of their acreage planted to fruit. While shares were relatively stable for grains over time, the share of sunflower hectares reported by the larger producers (i.e. those who completed Form 50SG) increased both in terms of the share of farms (from 52.6% in 2004 to 70.1% in 2012) and the land area used for the crop (from 9.7% in 2004 to 18.1% in 2012). During the same period, the share of farms growing fruit declined from 11% to 4.3%. Likewise, the share of farms planting vegetables decreased from 15.4% to 6.6%. Since land area planted to vegetables and output remained steady, this suggests that large-scale commercial vegetable production has become more concentrated.

Table 21.2. Number of commercial operations reported, Ukraine, 2004–2012. (From Form 29SG and Form 50SG, State Statistics, 2012a,b.)

Year	Statistical Yearbook	Form 29SG[a]	Form 50SG[b]
2004	58,575	–	8,892
2005	57,877	–	7,961
2006	57,858	–	7,220
2007	58,387	–	8,098
2008	59,059	54,016	8,485
2009	57,152	–	8,396
2010	56,493		8,515
2011	56,133	50,887	8,482
2012	55,866	–	8,335

Notes: [a]Form 29SG reports commercial operations involved in any land cultivation; [b]Form 50SG reports commercial operations larger than 200 ha.

Transition in the Grain Sector

The grain sector, which includes staple and legume crops (such as maize, wheat, barley, rye, buckwheat, pea, oat, millet, sorghum and rice) and oilseeds (primarily sunflower; also rape and soy), has been judged largely a success story for Ukrainian agriculture (e.g. Bezlepkina *et al.*, 2013). Ukraine was the 11th largest producer of wheat globally between 2010 and 2012, and the eighth largest exporter of wheat from 2010 to 2011 (FAOSTAT, 2012). In the same periods, Ukraine was the largest world producer and eighth largest exporter of sunflower seeds. All are annual crops, and the share of grains and oilseeds in the national enterprise mix has grown steadily.

While the household sector has had an increased role in national grain production, the

Table 21.3. Share of farms and share of area planted by sector from farms operating more than 200 ha, 2004–2012. (From Form 50SG, State Statistics, 2004–2012.)

	Grains		Sunflower		Fruit		Vegetables	
Year	Land	Farms	Land	Farms	Land	Farms	Land	Farms
				Per cent				
2004	46.0	96.9	9.7	52.6	0.4	11.0	0.2	15.4
2005	47.3	97.3	10.9	53.3	0.4	10.8	0.2	11.6
2006	46.6	93.9	11.8	52.4	0.3	8.7	0.2	12.2
2007	45.2	91.5	10.8	52.3	0.3	7.1	0.1	7.8
2008	52.3	92.7	14.0	59.8	0.2	5.9	0.2	8.7
2009	51.8	92.8	14.1	59.8	0.2	5.2	0.2	6.7
2010	48.3	91.3	15.9	63.7	0.2	4.9	0.2	7.0
2011	51.3	91.6	16.5	67.4	0.2	4.7	0.2	8.7
2012	48.9	92.7	18.1	70.1	0.2	4.3	0.2	6.6

majority of output remains in the hands of large commercial farms. Prior to 1991, less than 3% of Ukrainian grain output came from individual households, as large state-owned collective farms dominated the sector. The household share increased to about 25% in 2006 before decreasing to just under 22% in 2012 (State Statistics, 2013).

Commercial farms across all size categories allocated approximately 50% of their acreage to grains in 2011, with the very largest operations (over 3000 ha) accounting for 55% of the national grain output. Among the oilseed crops, expansion in sunflower production has been particularly notable (Fig. 21.3). A larger share of land among all size farms was planted to sunflowers in 2011 compared with 2008 (increasing from approximately 16% of the crop mix in 2008 to approximately 19% in 2011).

Grains have been an important, and a growing, source of foreign exchange for Ukraine since 2000, as exports from the cereals sector have trended sharply upward (Fig. 21.4). The value of cereal exports in US dollars (US$) has exceeded US$1 billion in every year since 2005, with the exception of 2007, when export restrictions were imposed

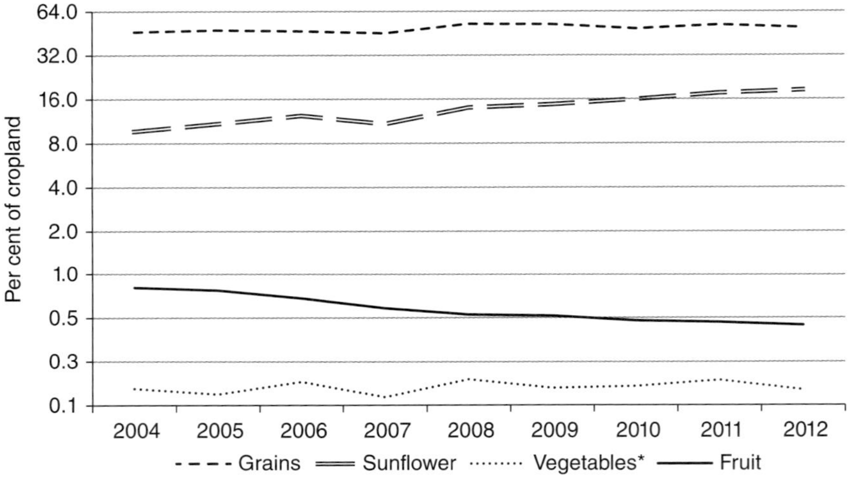

Fig. 21.3. Per cent share of cropland, by sector, farms of more than 200 ha, 2004–2012. (From Form 50GS, State Statistics, 2004–2012.)
Note: *Does not include potatoes or other root crops.

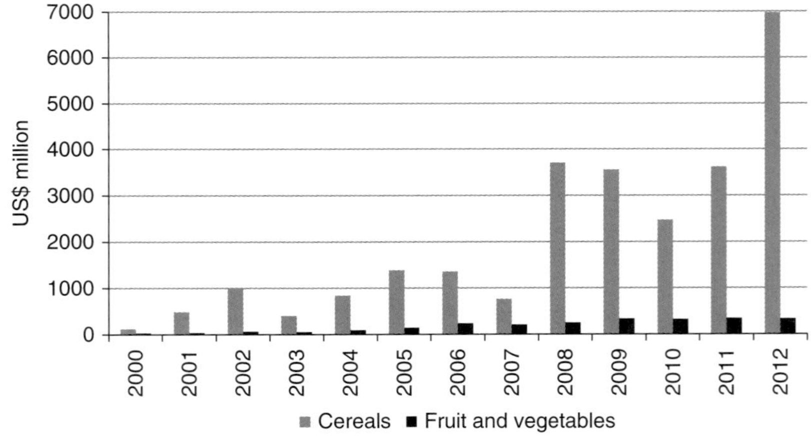

Fig. 21.4. Ukraine cereal, fruit and vegetable exports to the world, 2000–2012. (From World Integrated Trade Solution, 2011, HS chapters 7, 8, 10 only.)

(for a recent review of Ukrainian grain exports, see Kobuta *et al.*, 2012). In 2012, cereal exports reached a record of almost US$7 billion.

Transition in the Produce Sector

Transition in the produce sector (vegetables and fruit) is slower in developing and is far less encouraging. Potatoes are the most important vegetable crop in Ukraine. Other important vegetables include tomatoes, cabbage and cucumbers. Household producers contribute approximately 85% of the national vegetable production and approximately 97% of the potato production in Ukraine. More than one-half of the household output is targeted toward home consumption; the rest is traded at local farmers' markets. Per capita consumption of potatoes was 139 kg/year in 2011 compared to 163 kg for all other vegetables (Bezlepkina *et al.*, 2013). Area planted to potatoes and other root crops increased across most farm sizes (an exception was farms operating less than 5 ha) and the share of root crops in the crop mix increased from approximately 6.5% to 9%. As expected, vegetables are widely grown across Ukraine, with pockets of commercial concentration in the Kherson, Mykolaiv and Odessa oblasts, where irrigation canals have remained in place from former Soviet systems. Commercial vegetable acreage tends to cluster around these irrigation canals.

An emerging growth trend is developing in vegetable production (Fig. 21.5). Larger commercial farms are investing some profits from grain sales into irrigation upgrades and into the construction of greenhouses, targeting produce supplies to supermarket chains. To help increase the area of greenhouse production, the government has instituted a policy to compensate 50% of the construction costs if energy-saving technologies are included (Bezlepkina *et al.*, 2013). As a result of this policy, household producers and private/smallholder farms are incorporating greenhouse production of vegetables (primarily cucumbers and tomatoes) as a profitable and low-risk line of business where financial resources from microcredit institutions and banks are readily available.

Development in the fruit sector differs notably from the vegetable and grain sectors. A majority of fruit crops are perennials, with long production and investment cycles. Modern orchard systems require substantial investment and there are multi-year lags before trees come into marketable production.

Similar to the production of vegetables, the share of individual households in total output remains high. Households produced 54% of the fruit and berries in 1990 and 82% in 2012. Most of this fruit was sold to the processed fruit market through traders or brokers acting as the aggregators in supplying fruit and berries to local processors (e.g. juice concentrate, frozen) and food retailers. Again, like vegetables, the increase

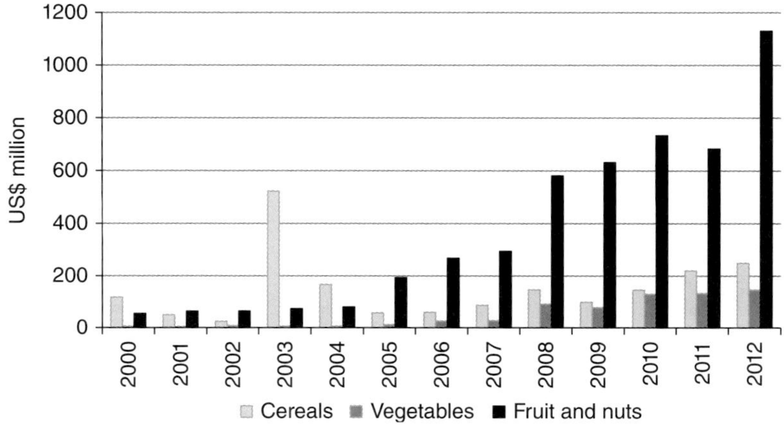

Fig. 21.5. Ukraine cereal, fruit and vegetable imports from the world, 2000–2012. (From World Integrated Trade Solution, 2011, HS chapters 7, 8, 10 only.)

in food standardization in both domestic and international markets is putting pressure on traders, processors and retailers to link with large-volume suppliers. High transaction costs of consolidation and variable quality limit household sector participation in the nascent commercial supply chains. So while both the share of total fruit output from the commercial sector and fruit production as a share of the commercial crop mix are small, the majority of fruit targeted to emerging modern supply chains is concentrated among the commercial farms.

The 2011 share of fruit in the on-farm crop mix ranged from 4.3% of area for the smallest farms to 0.2% of area for farms of more than 3000 ha (Table 21.4). However, in terms of total area, farms operating between 1000 and 2000 ha continue to have the greatest amount of land planted in orchards (Table 21.5). In 2011, farms in this size category produced about 30% of the fruit output from commercial farms, with 95% of fruit output produced on commercial farms operating more than 50 ha. Average yields were highest for commercial farms between 15 ha and 50 ha, and fruit production area expanded over 50% on these farms between 2008 and 2011. The largest farms (more than 3000 ha) had the smallest growth rate in orchard area.

Table 21.4. Average on-farm enterprise shares for commercial farms, Ukraine, 2011. (From Form 29SG, Statistical Year Book of Ukraine, 2013.)

Hectares	Grains	Sunflower	Non-root vegetables	Fruit	Root crops
			Per cent		
<5	52.2	6.0	2.4	4.3	3.7
5–15	49.9	14.9	1.4	2.6	1.9
15–50	43.9	23.1	0.7	1.4	0.8
50–200	48.7	21.4	0.8	1.7	1.0
200–500	49.3	21.2	0.6	1.5	0.9
500–1000	51.5	20.9	0.3	1.0	1.1
1000–2000	52.0	20.6	0.1	1.0	1.2
2000–3000	49.8	20.9	0.2	0.6	1.4
>3000	51.5	19.2	0.1	0.2	2.3
Total	48.2	19.4	0.9	1.8	1.4

Table 21.5. Crop mix and land use for commercial farms, Ukraine, 2011. (From Form 29SG, Statistical Year Book of Ukraine, 2013.)

	Total area per crop		Output (t)		Yield (t/ha)		Change in crop output 2008–2011	
Farm size	Non-root vegetables	Fruit	Non-root vegetables	Fruit	Non-root vegetables	Fruit	Non-root vegetables	Fruit
Hectares	Hectares		Tonnes		(t/ha)		Per cent	
<5	256	532	2,961	1,484	10.34	2.32	–41	19
5–15	754	1,522	11,561	6,017	13.75	3.92	–6	344
15–50	3,223	5,670	60,505	20,986	15.2	4.57	25	183
50–200	6,409	13,809	161,723	60,735	16.8	3.25	26	107
200–500	8,480	20,036	235,501	95,798	17.9	3.55	81	80
500–1,000	5,485	18,508	150,900	95,419	19.26	3.42	–10	7
1,000–2,000	3,948	33,876	118,981	192,032	16.83	2.59	–21	7
2,000–3,000	5,768	17,748	179,428	118,852	14.63	2.29	27	41
>3,000	10,167	15,288	358,900	70,590	16.58	2.13	92	2
Total	44,489	126,990	1,280,459	661,912	15.25	3.25	–40.86	19

Between 2008 and 2011, the area planted to orchards decreased on the smallest commercial farms of less than 5 ha and on operations of 500 ha or more. At the same time, output increased for the larger commercial farms as production was becoming more intensive. Between 2008 and 2011, the area of newly planted orchards increased by 6.5%. Annual data for farms operating over 200 ha confirmed these trends, with a steady increase in fruit output per hectare from 1865 kg in 2004 to almost 4230 kg in 2012 (a 127% increase), while area fell by an average of 10% annually over the same time period.

In 2011, the fruit crop farming area was geographically widespread. Fruit crops were particularly heavy in western and southern Ukraine and throughout the central and eastern oblasts. In contrast to increasing exports in the grain sector, exports in the fruit and vegetable sectors have been relatively flat since 2000. The total value of exports was only US$141 million in 2005, reaching a high of US$350 million in 2011 (World Integrated Trade Solution, 2011). Over the same period, imports of fruit to Ukraine have expanded dramatically, particularly since 2008. Fruit imports were greater than US$1 billion in 2012 (Fig. 21.5).

Despite decreasing acreage and increasing imports, the average annual rate of revenue growth has been greater for fruit and vegetables than it has been for grains and slightly less than it has been for sunflowers. Once costs are considered, increasing profitability for sunflowers, particularly since 2008, is even more evident. This higher profitability is consistent with the observed shift of farming activity moving into the oilseed sector (Fig. 21.6). Profitability in the fruit sector is higher consistently than that of grains and much less variable than that of vegetables (Fig. 21.6). Profit, output and firm industry entrance and exit are not independent of each other. The general trends are illustrative of the patterns in the development of agricultural sectors without attempting to assign causality, which suggests that lack of profitability has not been the driving factor behind the lack of development in the fruit sector.

Land Reform

Expected returns that impact the enterprise mix selected by producers include consideration of risks and risk preference, in addition to cost and revenue. In addition to price and yield uncertainty, producers in transition economies recognize greater uncertainty over changes in the institutional protection of property rights. Given a higher investment and a multi-year time horizon to recoup costs, long-term property rights are a critical source of uncertainty that may have hampered development in the fruit sector. The level of protection and definition of property rights themselves continued to change throughout the transition period.

In 1990, all land belonged to the state and was operated by collective enterprises (Article 11, Constitution of the Ukrainian Soviet Social Republic, 1978). With the parliamentary resolution *On Land Reform* (Parliament of Ukraine, 15 March 1991) and the law *On Forms of Land*

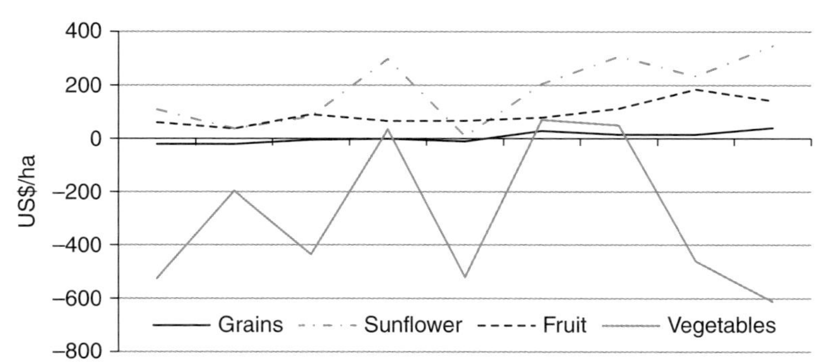

Fig. 21.6. Profit by crop sector, 2004–2012. (From State Statistics, 2013.)

Tenure (Parliament of Ukraine, 30 January 1992), farms and other industrial and trade entities were given a right to manage their business and to finance activities independently from the state, and non-state ownership of land was formally established. Former members of Soviet state-owned and collective farms were granted joint ownership rights to the land and other farm assets. By 1995, about 40% of the agricultural land had been transferred to corporate farming enterprises (Fig. 21.7). Citizens who were not members of the collective farms were given the right to lifelong possession and inheritance of parcels of housing, garden and recreational land that were eventually privatized. As a result, every citizen of Ukraine became entitled to a share of land; however, specific parcels of agricultural land were not identified and other property rights (use and disposal) were not realized at that point.

Several changes in direction for the establishment of individual property rights for agricultural land were implemented between 1991 and 2000. Initially, even though the land of state-owned collective farms was parcelled among former members, the managers of these former state farms exercised power over land distribution and management. The land market moved closer to its current shape in 1999, when the Presidential Decree *On Urgent Measures to Accelerate Reform in Agriculture* gave owners of land parcels the right to leave a collective enterprise without the consent of a general assembly and gave owners the right to keep their land parcel (Parliament of Ukraine, 1999). This allowed for the transformation of collective enterprises into other forms of ownership (e.g. private enterprises, limited liability companies), but a lack of support and clear procedures limited the number of applications. The 1999 decree has been characterized as the beginning of establishing a second round of reforms, in contrast to the more ad hoc policies of 1990 (Lerman *et al.*, 2007).

A functional sales market for land did not exist until 2002, when a new *Land Code* came into effect that entitled legal entities, territorial units and the state to hold land property rights (Parliament of Ukraine, 2001). A procedure was established to identify the size and borders of land parcels, and individuals were released from the need for third-party consent of the collective enterprises. Together with adoption of the new *Land Code*, a moratorium on land sales was established in 2002. It was meant as a temporary measure, due to the absence of a clear land inventory and absence of clear legal procedures for exchange, sales, purchase or gifting of land. The general population strongly opposed the establishment of free markets for trading agricultural land in fear of the potential for large-scale land grabbing. The moratorium was initially set to expire on 1 January 2005, but has been extended multiple times and currently is set to expire 1 January 2016.

As an alternative, a well-developed rental market for land has emerged (in 2014, it was the only legal way to exchange rights for land cultivation). By 2011, more than 92% of the land cultivated by commercial farms was rented from private landowners and from local governments. The average size of an individually owned land plot rented to commercial farms or to other households was about 5.7 ha (State Statistics, 2013). The share of land rented by commercial farms varies with respect to farm size. Farms below 5 ha operate primarily on their own land, while more than 90% of land is rented by farms that operate on more than 200 ha (Table 21.6).

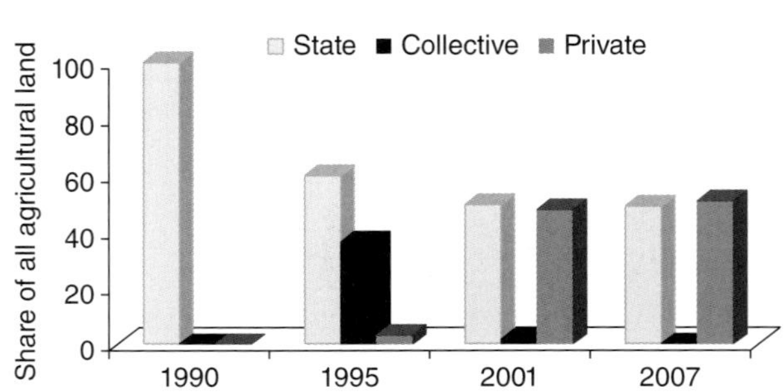

Fig. 21.7. Changes in landownership structure, selected years, 1990–2007. (From State Agency for Land Resources of Ukraine, 2014.)

Despite the large number of rental transactions, there is much diversity among the geographic regions of Ukraine. Of the 490 rural rayons (standardized administrative land units) in Ukraine, there were 21 rayons where less than 5% of farm operators owned acreage in 2011, and 38 rayons where more than 40% of the farm operators owned acreage (Fig. 21.8). The average across all areas was 24.5% of operators who owned some acreage in 2011.

The rental market for agricultural land has important benefits and a number of limitations that continue to impact the development of the agriculture and crop mix in Ukraine.

Table 21.6. Per cent share of owned land by farm size, Ukraine, 2011. (From State Statistics of Ukraine, 2013.)

Farm size	Total land area	Total area owned by farms	Share of owned land
Hectares	1,000 ha	1,000 ha	Per cent
<5	86.9	49.5	57.0
5–15	70.2	52.8	75.2
15–50	509.2	397.9	78.1
50–200	662.3	167.9	25.4
200–500	1,260.5	101.7	8.1
500–1,000	1,804.1	79.6	4.4
1,000–2,000	3,589.2	134.1	3.7
2,000–3,000	3,075.9	158.1	5.1
>3,000	10,301.8	520.5	5.1
Total	21,360.0	1,662.3	7.8

An important benefit is that a rental market reduces entry and exit costs and enhances overall productivity growth in the sector (Deininger *et al.*, 2013). Farms that exit tend to be underperforming and are replaced by new farms with higher productivity. The rental market requires minimal upfront investment in land acquisition.

A critical limitation is that rental land cannot be collateralized, which limits access to financial resources (Ciaian *et al.*, 2012). This constraint is often more binding on medium-size farmers who have higher per cent shares of fruit and vegetable crops. Individual household producers who have moveable assets or some real estate can be served by the informal credit market or by the microfinance institutions, with housing or future crops used as collateral. The largest farms have greater access to financing through formal domestic and international markets. A 2005 FAO survey of Ukraine's farms found that 63% of corporate farm managers reported borrowing funds compared to only 34% of medium-size farmers who borrowed funds (Lerman *et al.*, 2007). Modern commercial fruit and vegetable production is very resource-intensive during start-up, with requirements for irrigation, climate control, cost of trees and site preparation. Thus, restrictions on access to capital may have a greater negative impact on growth in the fruit and vegetable sectors among mid-sized operations.

Another limitation of rental markets is related to specifics of the institutional setting in

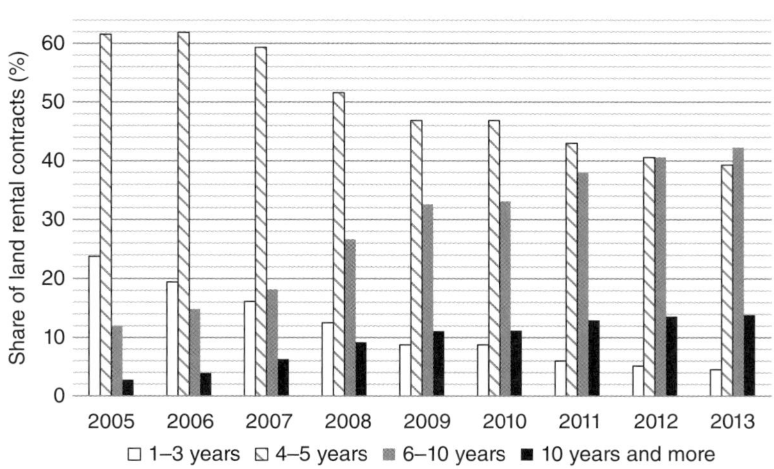

Fig. 21.8. Share of operators who owned some farmland, Ukraine, 2011. (From Form 50SG and Form 29SG, State Statistics, 2013.)

Ukraine and of unfinished land reforms. The ban on land sales was established as a temporary measure; uncertainty will remain over property rights and parameters of the land market until the ban is lifted, which is slated for 2016. The government has proposed several scenarios for regulating the land market that stipulate alternative limitations to maximum land purchase and holding size, a different set of pre-emptive rights for buying land (e.g. state versus neighbours versus tenants) and limitations on the rights of legal entities and foreign citizens to buy land.

This institutional uncertainty increases the risk of land taking at some future point, when an operating land sales market may be established. Both owners and tenants rely on strategies that reduce risk and, as a consequence, limit investment for land improvements. In anticipation of further market adjustments, landowners have been reluctant to lock themselves into long-term rental agreements. Over time, as the ban on land sales has continued and the rental market has grown, the average contract length has slowly increased. The share of shorter contracts has decreased since 2005 because other regulations have changed, so that by 2013, about 44% of all rental agreements in Ukraine were for 5 years or less.

When considering investment decisions in a hectare of land that is leased for approximately 5 years, such lease is well below the time needed to recover investments in orchards, irrigation or many other land improvement projects. While the average length of land rental contract has increased, it remains a limitation. Between 2005 and 2013, the share of contracts for longer than 10 years increased from 2.7% to 13.9%. Without a long lease, tenants prefer crops where returns can be captured easily. In the commercial farms, grains dominate vegetable and fruit crops that require irrigation or greenhouses for large-scale production. As a result, Ukraine's agriculture is facing underinvestment in high value-added crops and in land improvements.

Potential Future Developments

If better access to financial resources and expected improvements in the protection of property rights were to develop, fruit production could become a more attractive industry for investment. Besides changes related to functioning land markets, other shifts in the institutional environment have influenced the crop mix with Ukraine agriculture. Ukrainian productivity in many agricultural sectors remains below levels of other European countries, and a large share of agricultural land has not yet been converted to commercial production (World Bank, 2013). Continued transition will provide an opportunity for productivity and output growth in many crops. Like other developing economies, improved off-farm employment opportunities will eventually reduce available labour inputs, and perhaps will raise the wage rate. Individual household and other small landowners will have additional incentives to transfer some of their land plots to larger operations.

Liberalization of the banking, insurance and financial services sectors and the telecommunication and business services sectors took place between 2001 and 2008 as a precondition for the accession of Ukraine to the World Trade Organization (WTO). During this period, there was significant productivity growth in the manufacturing sector as transaction costs were reduced (Shepotylo and Vakhitov, 2012). In the agricultural sector, this liberalization was associated with the entrance of international operators and investors, who rented land. This entry brought new technologies and access to international commodity and financial markets for larger operations, allowing them to increase investments in modern production systems for all crops.

Another factor that impacted the crop mix in 2011 was the regulation of grain exports. Over time, most of the large producers and trading companies made investments in grain storage and in logistics infrastructure, which increased the benefit of vertical integration, decreased dependence on local monopolies for these services and added to sunk costs that limited incentives for exiting the industry. At the same time, the government made several attempts to restrict exports of grain. This intervention introduced additional risk to grain production and provided incentives for the diversification of crop mix.

Further opening of the European markets to Ukrainian products and to the expansion of supermarket retail chains in Ukraine will continue to push quality management and control as a priority for fruit and vegetable growers who would like to access these relatively lucrative markets. The introduction of hazard analysis and critical control points (HACCP), GlobalGAP or similar standards are a precondition for entry into the European markets, and some quality requirements must be introduced by the government as part of free trade agreements with the European Union. A potential effect of introducing higher standards could be additional pressure on small commercial producers, unless they can access resources needed for investment or unless they can consolidate their operations via cooperatives or unions.

Finally, recent conflicts with the Russian Federation, including the annexation of Crimea and war in the Donetsk and Lugansk regions, have added another source of uncertainty, which will restructure import and export flows as well as access to some seaports and international markets. A significant portion of Russian grain exports are shipped via Ukrainian ports, so conflict may impact the opportunity or price for Russia to access international agricultural commodity markets, and will, in turn, affect the prices of and changes in the crop mix for the entire Kazakhstan, Russia and Ukraine (KRU) region.

References

Bezlepkina, I., van Berkum, S. and Rau, M. (2013) Prospects for Ukraine's agrifood sector: implications for Dutch trade relations. LEI report 2013-045. Wageningen, Netherlands. ISBN/EAN: 978-90-8615-638-2.

Ciaian, P., Kancs, A., Swinnen, J., Van Herck, K. and Vranken, L. (2012) Institutional factors affecting agricultural land markets. Factor Markets Working Paper No 16. Centre for European Policy Studies (CEPS), Brussels.

Constitution of the Ukrainian Soviet Social Republic (1978) Article 11, Constitution of the Ukrainian Soviet Social Republic. Government of Ukraine, Kiev.

Deininger, K., Nizalov, D. and Singh, S. (2013) Are mega-farms the future of global agriculture? Exploring the farm size–productivity relationship for large commercial farms in Ukraine. Policy Research Working Paper No 6544. World Bank, Washington, DC.

FAOSTAT (2012) FAOSTAT. Food and Agriculture Organization of the United Nations, Rome. Available at: http://faostat.fao.org/ (accessed 5 November 2013).

Lerman, Z., Sedik, D., Pugachev, N. and Goncharuk, A. (2007) *Rethinking Agricultural Reform in Ukraine*. Studies on the agricultural and food sector in Central and Eastern Europe, No 38, ISBN 3-938584-18-1. Available at: http://nbn-resolving.de/urn:nbn:de:gbv:3:2-2454 (accessed 5 November 2013).

Kobuta, I., Sikachyna, O. and Zhygadlo, V. (2012) *Wheat Export Economy in Ukraine*. Policy Studies on Rural Transition No 2012-4. FAO, Rome.

Parliament of Ukraine (1991) *On Land Reform*. Parliament of Ukraine, Kiev.

Parliament of Ukraine (1992) *On Forms of Land Tenure*. Parliament of Ukraine, Kiev.

Parliament of Ukraine (1999) *On Urgent Measures to Accelerate Reform in Agriculture*. Parliament of Ukraine, Kiev.

Parliament of Ukraine (2001) *Land Code of Ukraine, Law No 2768-III* (25 October 2001). Parliament of Ukraine, Kiev.

Shepotylo, O. and Vakhitov, V. (2012) Services liberalization and productivity of manufacturing firms: evidence from Ukraine. World Bank Policy Research Working Paper Series No 5944. World Bank, Washington, DC.

State Agency for Land Resources of Ukraine (2014) Available at: http://land.gov.ua/en. (accessed 27 February 2014).

State Statistics (State Statistics Committee of Ukraine) (2012a) *Total Yield of Agricultural Crops*. State Statistics Form 29SG.

State Statistics (State Statistics Committee of Ukraine) (2012b) *Economic Results of Agricultural Producers*. State Statistics Form 50SG.

State Statistics (State Statistics Committee of Ukraine) (2013) *Statistical Year Book of Ukraine for 2012 (Ed. Osaulenko)* (published in Ukrainian). ISBN 978-966-2224-52-8.

World Bank (2013) World development indicators. World Bank, Washington, DC. Available at: http://data.worldbank.org/data-catalog/world-development-indicators (accessed 5 November 2013).

World Integrated Trade Solution (2011) World Bank, Washington DC. Available at: http://wits.worldbank.org/wits/Presentation.html (accessed 5 November 2013).

22 Competitive Analysis of Pulse Production in Russia, Ukraine and Kazakhstan

Marlene Boersch[1]* and Sandra Kirby[2]
[1]Mercantile Consulting Venture Incorporated, Winnipeg, Manitoba, Canada; [2]University of Winnipeg, Winnipeg, Manitoba, Canada

Abstract

This chapter explores the state of pulse production in Russia, Ukraine and Kazakhstan (RUK). There is need for background and context to increasing pulse production in some emerging markets, and in this chapter we look specifically at what impact RUK might have on the Canadian/North American pulse industry. The chapter covers fairly basic questions: where is there potential for pulse markets and where are the bottlenecks? It explores production costs, marketing costs and structures, policy, trade and market access and future capacity projections. Much of this is exploratory; both publically and privately available documents were used to fill in the information gaps. The end result is a competitive analysis summary.

Acknowledgements

Information obtained from both public and private documents is used with permission from Saskatchewan Pulse Growers (2012) and from Pulse Canada (2012). The contact for the original study is Greg Cherewyk, COO, Pulse Canada.

Agriculture Overview

The agricultural sectors of Russia, Ukraine and Kazakhstan (RUK) have been struggling to rebuild following the dissolution of the Soviet Union in 1991 and the sudden loss of state agricultural subsidies. Between 2000 and 2010, the agricultural planted area decreased 25% in Russia and 5% in Ukraine. However, the transition to a more market-oriented system has resulted in increased efficiencies, and productivity is improving slowly despite persistent resource constraints (FAOSTAT, 2012). For example, Kazakhstan reintroduced 2 million acres into crop production in 2011 (Table 22.1).

Russia's main agricultural region expands from the central district in European Russia bordering Ukraine and Belarus to western Siberia. Over 90% of the national grain output is grown in the central, southern, Volga and Siberian districts. Approximately 48.5 million acres is planted to row crops or temporary fallow; the remainder is permanent meadow or pasture. Russia's prime crop producing area is comparable to the state of Nebraska in the USA, and receives roughly 30 centimetres (cm) of precipitation

*Corresponding author; e-mail: mboersch@mercantileventure.com

Table 22.1. Descriptive statistics of the RUK region, 2011. (From Mercantile Venture Inc, 2012.)

	2011		
Descriptive statistics	Russia	Ukraine	Kazakhstan
GDP (US$)	1.86 trillion	165 billion	186 billion
GDP per capita (US$)	13,000	3,600	11,000
GDP growth (annual %)	4	5	8
Export goods and services (per cent GDP)	28	53	42
Population total (urban and rural)	142 million	46 million	16.5 million
Rural population (per cent total population)	26	31	46
Employment agriculture (per cent total employment)	10	16	29
Land area (km^2)	16,376,870	579,320	2,699,700
Land cereal production (km^2)	323,310	141,845	150,684
Agricultural land (per cent of land area)	13	71	77
Arable land (per cent of land area)	7	56	9
Agricultural land irrigated (per cent total agricultural land)	2	5	–
Agricultural machines, tractors/100 km^2 arable land	27	103	–
Fertilizer use (kg/ha arable land)	16	30	2
Agriculture, value added (per cent GDP)	4	8	5
Cereal yield (kg/ha in 2009)	2,282	3,000	1,200

during the growing season. Wheat accounts for over one-half of Russia's grain production and covers up to 10 million acres. Other major grains include barley and maize. Minor grains include rye, oats, buckwheat and millet. Russia's major oilseed is sunflower seed.

In Ukraine, approximately 17 million acres are classified as agricultural land. With a climate comparable to the US state of Kansas, it receives roughly 35 cm of precipitation during the growing season. Its arable land is suitable for both winter and spring crops. Ukraine is the third largest grain exporter in the world, with winter wheat, spring barley and maize being its main grain crops. In addition, the production of sunflower seeds and sugarbeet is very high. Ukraine is the world's largest producer of sunflower oil. The primary constraint on agricultural competitiveness is the policy framework that lacks transparency, predictability and market orientation. Others constraints include a lack of private investment and inefficient public and private resource allocation.

In Kazakhstan, 31 million acres of agricultural land are permanent pasture and about 10 million acres are classified as arable land. Kazakhstan's soils are highly variable, with only 5% of all cropland regularly fertilized. Annual precipitation is about 50 cm. Although the region is subject to occasional drought, grain fields are not irrigated. Reduced tillage allows for a decrease in clean fallow to preserve moisture, and reasonably high yields can be achieved during years of adequate rainfall. As a result, yield, production and quality are marked by frequent and sharp year-to-year fluctuations. Wheat is Kazakhstan's major crop and its largest export crop. Minor crops include barley, cotton, sugarbeet, sunflower, flax and rice.

RUK Pulses

Since the Soviet dissolution in 1991, there has been a robust increase in both planted area and production of pulses, although the share of pulses in Russia's total crop production remains small. The pulse seeded area nearly doubled from 1.1 million hectares (Mha) in 2007 to about 2.02 Mha in 2013. Russia's major pulse production regions are Volgograd, Altay, Saeatov, Samara and Orenburg. Despite the steady growth of seeded area, pulse production and yield continue to fluctuate from year to year according to the weather (USDA, 2013).

An increasing percentage of pulses produced in Russia are sold for export. In 2012, this reached 32% of production. Russia increased its exports of pulses from 73,680 tonnes (t) in 2008 to 770,846 t in 2012. However, in 2013, pulse exports totalled 364,210 t, 34% lower than in 2012 (USDA, 2013).

Peas comprise the bulk of Russia's pulse exports. In 2012, Russia exported 590,643 t of peas, or 77% of all pulses exported from Russia that year, along with 161,321 t of chickpeas and 18,786 t of lentils. India and Turkey are the main importers of pulses from Russia.

In Ukraine, the main production regions for pulses are central and south Ukraine and the Crimea. From 2000 to 2012, Ukraine produced between 300,000 t and 650,000 t of pulses annually. Peas account for 90% of the pulse crops produced and beans and chickpeas account for 10%.

Peas also account for the bulk of Ukraine's pulse exports. From 2009 to 2012, Ukraine exported between 155,000 t and 270,000 t of peas annually. The top importers of Ukraine's pulse crops are India, the European Union (EU), Pakistan and Turkey.

Pulse production in Kazakhstan reached about 100,000 t in the 1990s, but decreased to almost zero in the decade that followed. In 2012, pulse production in Kazakhstan had rebounded to about 80,000 t. The main pulse production region is the Kostonay Province in northern Kazakhstan. Figure 22.1 illustrates the growth in pulse production in the RUK region.

Production Costs

Establishing cost of production (COP) on incomplete data is difficult, but not impossible. To get the most competitive results relative to Canadian pulse production, data from very large corporate farm enterprises coupled with data from Agribenchmark (2013) and personal contacts were used. We analysed the export data based on port access and export capacity.

Russia

We analysed the 2012 production costs for several of Russia's crops, with the averages presented in Table 22.2. Both spring and winter wheat are included to reflect the presence of both types of wheat in the Russian production pattern. It is estimated that spring wheat/spring barley areas are switched to alternate crops more easily because they have a closer yield comparison than do higher-yielding winter crops.

The COP data generated for Russia's pulse crops show mixed results in terms of profitability, especially compared to maize, sunflower, barley and winter wheat. The lower return per acre performance of pulses is due to the relatively low yields combined with relatively high variable

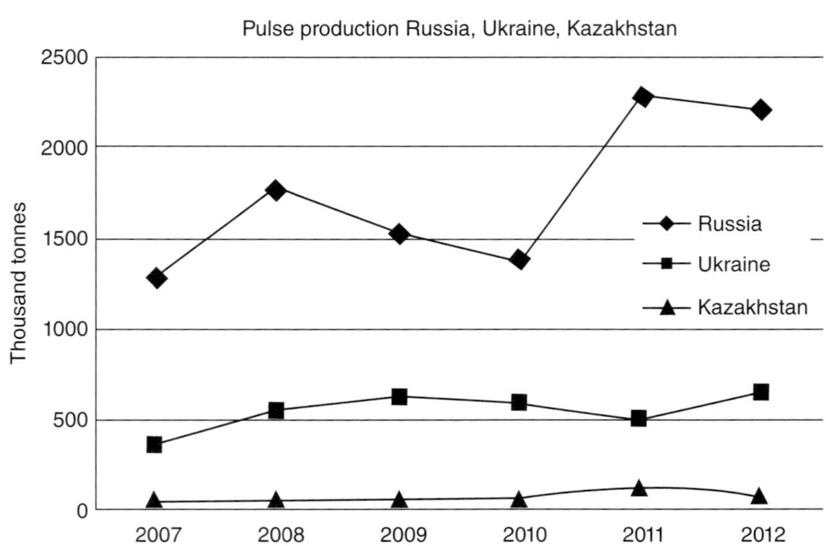

Fig. 22.1. Pulse production (per 1000 t) in Russia, Ukraine and Kazakhstan, 2007–2012. (From USDA, State Statistical Committees, 2012.)

Table 22.2. Cost of production of major crops and pulse crops (in CAN$) in Russia, 2012. (From Mercantile Venture Inc, 2012.)

	Spring wheat	Winter wheat	Spring barley	Maize	Sunflower	Peas	Lentils	Chickpea
					CAN$			
Variable cost/acre	118.48	76.78	65.71	177.74	106.56	160.02	115.00	156.50
Fixed costs	148.81	105.46	139.64	205.89	137.90	131.81	96.39	107.00
Total costs	267.29	182.24	205.35	383.63	244.46	291.83	211.39	263.50
Yield bu/acre	38.66	44.61	59.48	105.15	65.44	25.28	13.40	16.40
Price CAN$/t	252.00	231.90	396.62	304.00	399.46	320.00	773.06	736.16
Crop revenue/acre	265.16	281.54	513.63	811.98	468.81	220.15	281.92	328.57
Margin per acre	–2.13	99.30	308.28	428.35	224.35	–71.67	70.53	65.07

costs of production. Also, profitability is affected by the historically high prices for maize, barley, wheat, sunflower and rapeseed. Of the pulse crops, lentils and chickpeas performed better than peas, but still ranked only fifth and sixth of the top eight crops. Peas ranked last.

The COP data for peas are especially high in the categories for seed, chemicals and fertilizer. The same is true for lentils and chickpeas. For pea seed COP, there seems to be several reasons for that. Commercial seed is said to be very difficult to procure, and the seeding rate for peas is about double that of wheat. Chemical costs also are high for both fungicides and herbicides, indicating disease problems. The background to the relatively high fertilizer costs was field applications of nitrogen, phosphorus and potash (NPK), as the soils had deficiencies in all three. Two large agroholdings also indicated the need for a programme of costly and high applications of lime (CaO) due to significant problems with pH. This makes sense, as peas are sensitive to high pH numbers.

Another important observation is that pea yield is at 65% of the spring wheat yield and at 57% of the winter wheat yield for Russia. In addition, pea yield is at 43% of the spring barley yield and at 24% of the maize yield for Russia.

In areas where winter wheat can be produced, peas do not tend to make sense as a crop choice, because of the larger yield difference between the crops. We therefore are concerned primarily with spring crop areas as potential production competition to Canadian pulse crops.

Within the spring crop production areas, maize offers significant competition to pulses. Maize production areas in Russia and Ukraine are growing because of the high values for maize sold to the US ethanol programme. Maize is also a good fit from an export marketing point of view, as the RUK region has ready access to the EU maize market geographically, and also has a product trait advantage as the EU does not allow into its market genetically modified (GM) products from North America and South America.

Chickpea yield is at only 42% of the spring wheat yield and 37% of the winter wheat yield for Russia. Relative to spring barley yields, chickpea yield calculates to 28%.

Lentil yield calculates to 35% of the spring wheat yield and 30% of the winter wheat yield for Russia. Relative to spring barley yields, lentil yield calculates to 23%. Pulse yields trail those of cereal and oilseed, and therefore require higher per tonne prices to be included in the crop choices purely from a return per acre decision model.

Similarly, sunflower seeds also offer good competition to pulses. Domestic use has been increasing, and the geographic and market access to the EU biodiesel programme is good. Barley competes well with pulses due to good coarse grain prices and the proximity and accessibility of the world's largest feed barley market in Saudi Arabia.

The important additional observations on the RUK grain markets include the following:

- The percentage of RUK participation in the world wheat and coarse grain trade is still increasing, especially to the Middle East markets, where the RUK region holds a significant freight advantage over Canada and other exporters.

- In 2009/10, Russia's wheat prices were below the world wheat price (wheat embargo), which led to the substitution of peas to wheat exports.
- In an environment of high world wheat prices, the substitution of peas to wheat exports will be more difficult.
- World demand of wheat grew by 6.3% between 2007 and 2013, and is still growing, so RUK exports are increasing in the global wheat market.
- World demand of maize grew by 8.9% between 2007 and 2013, and is still growing, so RUK exports are increasing in the global coarse grain market.
- World utilization of oils and fats grew by 19.4% between 2007 and 2013, and is still growing, so RUK exports are increasing in the global oilseed market.
- Pea exports are to some extent a consequence of the export wheat price.

Ukraine

Relative to Russia, the yield differential of pulses to spring wheat is larger, and therefore from a purely market value viewpoint, pea production makes less sense in Ukraine than in Russia. Furthermore, it is possible to grow significantly more winter crops in Ukraine than in Russia, and winter crop areas tend to present higher-yielding options to pulses. In addition, where spring crops are grown, maize, sunflower seeds and barley offer very profitable alternatives to pulses. These crops have yield advantages and have readily accessible and growing markets in the Middle East, North Africa and the EU. As a result, there is more concern about summer, rather than winter, crop areas as potential production competition to Canadian pulses. Basis given market prices against the COP data (Table 22.3), maize exhibited the highest return per acre, distantly followed by winter wheat and winter rapeseed, in 2012. Peas ranked sixth and lentils ranked last of the eight crops shown in Table 22.3.

Relative to Russia in 2012, the yield differential to spring wheat is larger, and from a purely market value viewpoint, pea production makes less sense in Ukraine than in Russia. Furthermore, it is possible to grow significantly more winter crop in Ukraine than in Russia, and as noted above, winter crop areas tend to present higher-yielding options to pulses. In addition, where spring crops are grown, maize, sunflower seeds and barley offer very profitable alternatives to pulses. These crops not only have yield advantages but also have readily accessible and growing markets in the Middle East, North Africa and the EU.

Kazakhstan

The COP data (Table 22.4) for Kazakhstan in 2012 are generally much lower than for Russia and Ukraine, and total costs for peas are very close to those for the spring grains grown in Kazakhstan. Basis given market prices against the COP data, chickpeas exhibited the highest return per acre, followed by spring barley and spring wheat, in 2012. Peas rank last for the year of observation out of the five crops shown. This is due mostly to modest yields.

Table 22.3. Cost of production for major crops and pulse crops (in CAN$) in Ukraine, 2012. (From Mercantile Venture Inc, 2012.)

	Spring wheat	Winter wheat	Winter barley	Maize	Sunflower	Winter rapeseed	Peas	Lentils
					CAN$			
Variable cost/acre	136.18	173.51	220.76	289.70	171.53	277.37	178.74	162.52
Fixed costs	132.51	173.52	168.45	469.08	204.23	319.67	260.13	285.84
Total costs	268.69	347.03	389.21	758.79	375.76	597.04	438.87	448.36
Yield bu/acre	48.10	74.35	78.07	159.32	74.35	41.00	33.00	17.10
Price CAN$/t	295.00	282.43	220.00	311.69	384.34	768.41	320.00	507.00
Crop revenue/acre	386.17	571.49	373.94	1261.39	388.85	800.43	287.39	235.95
Margin per acre	117.48	224.46	–15.27	502.61	13.09	203.39	–151.48	–212.41

Marketing Costs and Structure

On-farm storage remains a problem in the RUK region. Official government reports tend to downplay the problem, but industry players maintain it is still very important to grain flow and seasonalities. Rail capacity remains a problem, but Black Sea port capacity has improved significantly. Inland logistic costs for Russia and Ukraine are fairly similar to those in Canada. However, Russia and Ukraine have very significant ocean freight advantages over Canada into the EU, India and the Middle East (Table 22.5). Sailing times from Canada to China are about 10 days *shorter* than from the Black Sea (Table 22.6).

Kazakhstan does not possess the same Black Sea freight advantages to India and the EU as do Russia and Ukraine. In fact, in addition to significantly higher inland costs, Kazakhstan competes with Russia for rail space and with Russia and Ukraine for port terminal space. However, rail and rail/port export capacity continue to grow into neighbouring countries as well as into China. While not an immediate threat, over a longer time horizon, pulse exports from Kazakhstan may well be routed via rail into both Turkey and China. Once these trade routes have been established, growth in pulse production in Kazakhstan likely will accelerate and specifically threaten Canadian lentil and pea exports into Turkey and China.

Policy, Trade and Market Access

Russia – policy

State support to agriculture in Russia is modest compared to that of other developed countries. The majority of state support for agriculture (up to 80%) is provided by federal entities. A Russian agrobusiness survey found that 72% of respondents believed that many companies would not be adequately profitable in the short term without subsidies (FAOSTAT, 2009).

Ukraine – policy

In Ukraine, the state implements market interventions for key agricultural commodities with a variety of domestic measures. Official minimum prices are set for grains and oilseeds, while livestock

Table 22.4. Cost of production for major crops and pulse crops (in CAN$) in Kazakhstan, 2012. (From Mercantile Venture Inc, 2012.)

	Spring wheat	Spring barley	Rapeseed	Peas	Chickpea
			CAN$		
Variable cost/acre	34.53	53.57	51.79	44.01	66.87
Fixed costs	47.26	40.43	52.31	28.25	35.65
Total costs	81.79	94.00	104.10	72.26	102.52
Yield bu/acre	30.63	52.05	11.24	12.94	11.90
Price CAN$/t	173.52	173.04	499.00	261.52	720.00
Crop revenue/acre	144.66	196.08	127.21	92.10	233.25
Margin per acre	62.86	102.08	23.10	19.84	130.73

Table 22.5. Summary of logistics costs (in US$) for the RUK region, 2012. (From Mercantile Venture Inc, 2102.)

Country	Country storage and handling	Inland freight: rail	Inland freight: truck	Government impacts on costs	Port costs	Inland logistics costs
			US$/t			
Russia	15.96	26–48	26	18%VAT:7.47–10.69	30	64–79 (60)
Ukraine	15	15	15–25	Export tax: 23	20	65–75 (60)
Kazakhstan	15	140–160	–	Transportation subsidy: 40	30	160–180 (150–160)

Table 22.6 Comparison of sailing times from Canada and Black Sea/Baltic to India, China and EU (Spain), 2012. (From GSAN (Global Shipping Network, Sea Rates, Internet-sourced), 2012.)

Country of origin	Loading port	Discharge port	Distance	Sailing time
	City	City/country	Nautical miles	Days
Canada	Vancouver	Mumbai, India	9,507	28 days
	Vancouver	Chennai, India	8,660	26 days
	Vancouver	Kolkata, India	8,664	26 days
	Vancouver	Dalian, China	5,170	15 days
	Vancouver	Shanghai, China	5,103	15 days
	Montreal	Shanghai, China	11,492	34 days
	Vancouver	Valencia, Spain	8,803	26 days
	Vancouver	Gijon, Spain	8,466	25 days
	Montreal	Valencia, Spain	3,549	11 days
Black Sea (Russia)	Novorossiysk	Mumbai, India	4,291	13 days
	Novorossiysk	Chennai, India	4,956	15 days
	Novorossiysk	Kolkata, India	4,627	17 days
	Novorossiysk	Dalian, China	8,844	26 days
	Novorossiysk	Shanghai, China	8,395	25 days
	Novorossiysk	Valencia, Spain	1,997	6 days
Baltic (Russia)	St Petersburg	Gijon, Spain	1,743	5 days

products are subsidized. Interventions to the export market are implemented to restrain domestic food prices during periods of world food price spikes.

Kazakhstan – policy

In Kazakhstan, the State Food Contract Corporation (SFCC), created in 1995, purchases and sells grain for state and commercial purposes. The SFCC is the largest holder of domestic grain stocks and strongly influences the domestic market through its price policy. Kazakhstan's government believes it can become a major agricultural exporter and plans to focus on agricultural production (Pomfret, 2008).

Russia – trade

Russia joined the World Trade Organization (WTO) in 2012. State support provided to Russia's agricultural sector will not be limited due to membership in the short term (WTO, 2013). On average, border measures are rather modest in comparison with the Organisation for Economic Co-operation and Development (OECD) countries. The most important border measures are the export taxes on cereals and sunflower seed, a changeable import duty for raw sugar and tariff rate quotas (TRQs) for meat. Large private and public investments have been made in the grain export infrastructure, particularly seaports.

Ukraine – trade

Ukraine became a WTO member in 2008. WTO commitments limit their ability to control and impose restrictions on exports (WTO, 2013).

Kazakhstan – trade

As of 2014, Kazakhstan is not a WTO member, although it has applied for membership. Kazakhstan began negotiations in 1996 and negotiations are ongoing (WTO, 2013). In 2011, China and Kazakhstan entered into a strategic partnership agreement to double their trade by 2015, thereby increasing their trade turnover to US$40 billion.

RUK – market access

As part of the new Commonwealth of Independent States (CIS) market, Russia, Ukraine and

Kazakhstan can export to and import from all CIS member countries tariff free. Russia, Ukraine and Kazakhstan also belong to the Single Economic Space. Dispute resolutions of all preferential trading areas (PTAs) with CIS countries are limited.

European Union – market access

Tariffs for pulses entering the EU from Russia, Ukraine and Kazakhstan fall under the third country duty category (WTO, 2013). Peas, chickpeas and lentils are not charged a tariff, while beans and other dried leguminous vegetables are charged 3.2%.

India – market access

As WTO members, Russia and Ukraine receive preferential area duty rates for imports into India, as opposed to the standard rate of duty charged to WTO non-members such as Kazakhstan. All imports to India are subject to an import permit, a phytosanitary certificate and a no-objection certificate from the plant quarantine authority.

China – market access

As WTO members, Russia and Ukraine receive most-favoured-nation duty rates for imports into China, as opposed to the general rate of duty charged to WTO non-members such as Kazakhstan. China's tax rates for imports apply to cost, insurance and freight value (CIF); duties; and any tax or charge on the import, such as value added tax (VAT). China is currently reforming its VAT; therefore, import tax rates may be subject to change. Peas, beans and chickpeas are subject to quarantine examination and approving procedures.

Future Capacity Projections

Russia

Once pulses fit into the crop rotations in Russia due to improved yields, improved COP profiles or improved relative prices, Russia's competition with Canada in the pulse export market will intensify, especially in India and the EU (Table 22.7). In addition, because of the excellent transportation links into the EU, Russia may well displace much of the EU pulse production going forward (Boersch *et al.*, 2012).

Ukraine

A major expansion of pea acreage in Ukraine in the near future is not expected, primarily because peas (pulses) show much lower (unprofitable) returns than do maize, sunflower and winter crops in Ukraine (Table 22.8). The main reason for this is persistent low yields that have not exceeded an average of 2 t/ha since 2008/09. Specifically, pulses have not gained favour with Ukraine's agroholdings, which grow only the most commercially attractive crops in an environment of limited storage capacity. It is therefore unlikely that the Ukraine area seeded to pulse crops will grow significantly on land controlled by agroholdings that provide about 80% of total pulse production in Ukraine. A maximum of 500,000 t is projected for Ukraine's pea production in 2015 (Boersch *et al.*, 2012).

We note that the research work for this article was done in 2012, well before major changes in commodity markets and the political upheavals in Russia and the Ukraine in 2014. The basic premises based on market information still hold, but it is difficult to tell what impact the political realities will have on these markets.

Table 22.7. Pulse production forecast (in tonnes) for Russia, 2022 versus 2012. (From Mercantile Venture Inc, 2012.)

Russia pulses	2012 production	2022 production forecast	Change
	Tonnes		Per cent
Peas	1,850,000	2,500,000	35
Chickpeas	300,000	500,000	67
Lentils	50,000	80,000	60
Total for all pulses	2,200,000	3,080,000	40

Table 22.8. Pulse production and export forecast (1000 t) for Ukraine, 2015 versus 2011. (From Ukrainian State Statistics Service, 2012.)

Ukraine pulses	Pulse crop production			Pulse exports		
	2011	2015	Change	2011	2015	Change
	1000 t		Per cent	1000 t		Per cent
Peas	364	390	7%	140	150	7
Lentils	1	2	122%	0.005	0.1	1900
Beans	7	10	79%	4	7	91
Chickpeas	13	15	14%	2	4	69

Kazakhstan

The major competitive advantages of Kazakhstan are its low production costs for wheat and pulse crops and its research and development of suitable grain and pulse varieties for the RUK region (Table 22.9). Effort is being made to speed up diversification. Of particular note is the 2012 Memorandum of Understanding (MOU) between the Indian Council of Agricultural Research (ICAR) and KazAgroInnovation (KAI) (ICAR/KAI, 2012). The two parties have agreed to cooperate in a number of areas, including the exchange of germplasm and breeding materials and the development and implementation of shared research projects.

Table 22.9. Pulse production forecast (in tonnes) for Kazakhstan, 2022 versus 2012. (From Mercantile Venture Inc, 2012.)

Kazakhstan pulses	2012 production	2022 production forecast	Change
	Tonnes		Per cent
Peas	65,000	125,000	92
Chickpeas	10,000	50,000	400
Total	75,000	175,000	133

Further developments in Kazakhstan include expanding its rail and rail/port export capacity into neighbouring countries as well as into China (Boersch *et al.*, 2012). In the short term, pulse exports will be routed via the established Black Sea routes. In the long term, it is projected that pulse exports from Kazakhstan will be routed via rail to Turkey and to China. Additional rail truck routes into Pakistan may develop for pulse crops. Once these trade routes have been established, the growth in pulse production in Kazakhstan very likely will accelerate and specifically threaten Canadian exports into Turkey, Pakistan and China.

References

Agri-benchmark (2013) Agri-benchmark website. Available at: http://www.agribenchmark.org/home.html (accessed 30 November 2013).

Boersch, M., Temple, A.P. and Kirby, S. (2012) Competitive assessment of pulse production in RUK. *Pulse Canada* (30 November 2013).

FAOSTAT (2009) Analysis of the agribusiness sector in southern Russia. Available at http://faostat.fao.org (accessed 10 December 2013).

FAOSTAT (2012) http://faostat.fao.org/ (accessed 10 December 2013).

GSAN (2012) Sea rates. Global Ships Agencies Network. Available at: http://www.go2gsan.com/ (accessed 10 December 2013).

ICAR/KAI (2012) Memorandum of Understanding. Indian Council of Agricultural Research (ICAR) and KazAgroInnovation (KAI). Available at: http://kai.gov.kz/company (accessed 20 December 2014).

Mercantile Venture, Inc (2012) *Graphs and Charts for RUK: Competitive Assessment of Pulse Productions*. Mercantile Venture, Inc, Winnipeg, Manitoba, Canada.

Pomfret, R. (2008) Using energy resources to diversify the economy: agricultural price distortions in Kazakhstan. *Comparative Economic Studies* 51, 181–212.

Pulse Canada (2012) RFP commissioning Mercantile Venture, Inc for competitive assessment of pulse production in RUK (Permission for using study granted from Greg Cherewyk, COO, Pulse Canada). Available at: http://www.pulsecanada.com/ (accessed 10 December 2013).

Ukrainian State Statistics Service (2012) Ukraine-processing volumes of legumes by large and medium industrial enterprises.

USDA (2012) State Statistical Committees. US Department of Agriculture, Washington, DC.

USDA (2013) GAIN Report RS1381. US Department of Agriculture, Washington, DC (November).

WTO (2013) World Trade Organization (WTO) website. Available at: http://www.wto.org/ (accessed 10 December 2013).

23 Evaluating the Economic Efficiency of Subsidies Based on the Basic Output Equations for Agricultural Enterprises in the North-western Regions of Russia

David Epstein*
North-West Research Institute of Agricultural Economics, St Petersburg, Russia

Abstract
The existing approaches to estimating the efficiency of subsidies on agricultural production are considered. The basic output regression equations have been enhanced by including more factors (subsidies, production profitability and quality of enterprise management) in addition to the primary ones (labour, land, fixed assets and material inputs). This enhancement allows the derivation of statistically authentic estimates of subsidy efficiency. Calculations based on data from 371 agricultural enterprises for 2007 and 2008 show a statistically significant positive effect of subsidies on production in the Northwestern Federal District of Russia.

Problem Definition

There exists a broad scope of literature on the impact of budget support of agriculture in Russia that can be divided into two types: (i) a general theoretical approach to the issue from a certain point of view that potentially makes concrete estimations possible; and (ii) the estimation of the efficiency of budget support based on some particular parameters in some concrete cases. Very often, it is performed with the help of econometric models. Both types of literature are of significant interest. However, one should bear in mind the principal difficulty of determining the economic effect of budget subsidies. The difficulty lies in singling out the impact of budget subsidies from the entire range of factors that include climate conditions, differences in resources and management inequality. Using econometric models is the only method that yields statistically significant results, which explains their popularity.

When subsidies are used, one approach that relies on microeconomic theory states that the supply function of the subsidized goods moves downwardly by the corresponding value. At the same time, the intersection point with the demand curve moves to the right, which leads to increased supply at a slightly decreased market price (Voelken, 1999; Shaykin *et al.*, 2001). The economic effect is defined in this scenario as the sum of customer benefit (from the decreased price and the increased demand), producer

*E-mail: epsteindb@gmail.com

benefit (from revenue and subsidies for a greater amount of goods) and budget losses (taxpayer losses) required to finance these subsidies. In this case, the financial result depends on the elasticities of the demand and supply curves (Pindyck and Rubinfeld, 1992; Just *et al.*, 2004). An impressive example of estimating the community benefit from the shift in the supply curve can be seen in the study of the effect of introducing tomato harvesters in California (Schmitz and Seckler, 1970). An example of an in-depth analysis of the influence of price subsidies and water consumption subsidies on cotton production, based on demand and supply estimations, is given in Rossi *et al.* (2007). Unfortunately, as far as this author is aware, the impact of subsidies on supply has not been previously considered in the Russian agroeconomics literature.

Uzun and Gataulina (2009) used grouping method and regression equations and position profitability as the main factors defining the effect of subsidies. The marginal output achieved because of various factors, including subsidies, is credited exclusively to profitability, while the effect of subsidies is considered proportional to the share in profit or revenue. However, production growth apparently is not a result of profitability only, but is related to other factors (e.g. increasing resources). The larger farms are mainly responsible for the higher profitability. However, the above work has triggered an attempt to formulate an approach involving a wider set of factors that influence agricultural output and attempt to use this basis to specify the basic output equations and to determine more accurately the effect of subsidies.

Using another approach, computations by Bezlepkina (2004) are based on data from particular regions (Russian agricultural enterprises as a whole, and Moscow oblast dairy farms in particular) and periods (from 1995 through 2000). Using a special type of production function, Bezlepkina's computations for all of Russia showed that the effect of subsidies, expressed as an unknown coefficient in the output production function, was negative but small in absolute terms. At the same time, these computations identified a positive influence of debt on revenue. The latter is not surprising since the debt in the late 1990s was highly correlated with input and output volumes.

Petrick and Zier (2011) used an econometric (regression) model to analyse the impact of subsidies on the number of jobs in agriculture in eastern Germany. This work shows that regression analysis may be used to evaluate the social effect of subsidies that manifests itself when increasing the number of jobs for agricultural workers, as well as when increasing their income. The results of the analysis are influenced by a great number of factors: (i) unknown factors; (ii) exogenous factors, including the subsidies; and (iii) variable and invariable endogenous factors. The model included only the following independent variables: different forms of subsidies, density of population in the region, average salary per year and dummy variables for measuring fixed effects per year (the annual differences of these variables). The resulting R^2 were small (0.4–0.5), but an evaluation of the effects of different forms of subsidies allowed a reliable conclusion on whether the dependent variable – the number of jobs – was affected negatively or positively by each form of subsidy.

Method Development

Beginning with experimental computations and calculating the correlation between the main factors that allegedly impact output and its growth, agricultural enterprises from nine regions (oblasts and republics) of the Northwestern Federal District (NFD) were studied for 2007 and 2008. These years preceded the worldwide economic downturn and witnessed agricultural growth in Russia.

Computations were performed on the data of all the enterprises available in the database of the Committee on Statistics of the Russian Federation for 2007 and 2008. A total of 371 enterprises from the nine regions were included to create a single data array to determine output growth (Table 23.1).

Using a representative sample, we calculated output growth rates and production profitability according to gross profit and efficiency coefficients of resource potential usage for 2007 and 2008. These coefficients were calculated as an enterprise's actual output divided by the calculated one according to the Cobb–Douglas regression equation based on production factors. Then, the correlation coefficients for output growth rates, profitability, subsidies and efficiency coefficients were calculated for 2007 and

Table 23.1. The number of enterprises in NFD used to calculate the effect of subsidies in 2007 and 2008. (From author's calculations.)

Region (oblast, republic)	Number of farms with data for 2007 and 2008
Arkhangelsk	37
Vologda	76
Kaliningrad	24
Leningrad	113
Murmansk	11
Novgorod	20
Pskov	34
Karelia	16
Komy	40
Total	371

2008, as well as the main production factors that characterized the amounts of resources used, such as material inputs, depreciation, labour and size of agricultural lands, in 2008 (Appendix Table A23.1a,b). Depreciation is used as a substitute for the fixed capital, for which no data were available. Descriptive statistics of the variables used in Appendix Table A23.1a and b are presented in Appendix Table A23.2.

Output growth correlates significantly at the 0.01 level or higher (correlation coefficients are within parentheses), with only enterprise profitability for the current (0.168) and previous (0.112) years, but does not correlate with either enterprise size (the maximum correlation coefficient, taking into account material inputs, was 0.023) or subsidies (–0.033) (Appendix Table A23.1a,b). To a certain extent, this justifies the approach based on the impact of profitability on growth rates. However, the correlation coefficient between output and profitability in 2008 was quite low at only 0.168. Thus, profitability, subsidy amounts and other factors listed above offer a very weak explanation of the differences in output growth rates.

Profitability for 2007 correlates significantly and crucially (0.797) with profitability for 2008, with efficiency for 2007 (0.217) and 2008 (0.376) and with enterprise sizes (Appendix Table A23.1a,b). Among the four mentioned size factors, profitability for 2008 correlates mostly with labour (correlation ratio of 0.208), but does not correlate significantly with agricultural lands (0.076). The latter is due partly to the presence of poultry farms and specialized vegetable farms in the sample that require little agricultural land. The subsidies correlate significantly with enterprise size but correlate mostly with material inputs (0.756).

In a further analysis of the effect of subsidies, it is assumed that it is appropriate to consider: (i) the impact of subsidies on agricultural output in a given year; and (ii) the impact of marginal subsidies on output increments in the same period (i.e. incremental efficiency of subsidies). While this approach does not consider directly the time lag of the impact of subsidies and other factors on production output, a model is suggested that may be reformulated with lags taken into account. Moreover, the time lag is evidently partially considered when assessing incremental efficiency. This author's experience when implementing this model has shown that without the time lag, the model possesses a rather high statistical value because it produces reliable estimations due to the complete consideration of all factors.

Basic Agricultural Output Equations in Russia

Since the 1970s, several Soviet researchers have actively used the linear production function and the Cobb–Douglas function for assessing agricultural production output. These functions were used for the analysis and the planning of the activities of agricultural enterprises to determine how efficiently the production potential of enterprises was used and to allocate the planned production tasks and the limited resources. The models followed the early work in the USA of Heady and Dillon (1965).

In Russia, Tregubov (1969) was probably the first to use the production power function for agricultural planning and analysis. Production functions have been actively used since the early 1970s (Krastin, 1971), followed by other researchers such as Rivzha and Tray (1987). A substantial contribution to the development of this approach was made by Kudinov *et al.* (1990) and Andriychuk (1983).

In the early 1980s, it was suggested using the sum of material inputs instead of the sum of current assets, as it depicted most fully the aggregate impact of different assets (Epstein, 1988), and that material inputs boost the statistical significance of the equations. This parameter is now used most often in agricultural production output equations. In the early 1990s, output regression applications were developed for solving various problems of the regulation of agricultural production for countries transitioning from a planned economy to a market economy (Epstein, 1993).

Accordingly, a basic equation is expressed as a Cobb–Douglas function of product output or gross output of an agricultural enterprise as dependent on the four basic production factors (labour, agricultural land, fixed capital and material inputs)

$$B = CM^m F^f P^p N^n, \qquad (23.1)$$

where B is the revenue from sales of produced goods, C is a constant, M is material inputs, F is fixed capital, P is agricultural land area, N is labour and m, f, p and n are the exponents of the corresponding factors. This equation allows one to compute the production function coefficients with sufficiently high accuracy, as long as reliable data are available, to evaluate statistically the impact of different factors on production output (the determination coefficient of the equation for all enterprises of an oblast is usually greater than 0.9).

Modification of the basic equation

Since a subsidy is an additional production factor, it may be included as one of the factors. If subsidies S are channelled to increase the material inputs, M_b can be singled out as that portion of material inputs that would have existed if no subsidies had been received ($M_b = M - S$). This allows us to include subsidies as an independent factor in Eqn 23.1, thus obtaining the basic Eqn 23.2, which takes into account the impact of subsidies as

$$B = CM_b^m F^f P^p N^n S. \qquad (23.2)$$

Furthermore, we can check whether or not this equation is adequate when revealing the impact of subsidies and other factors on output by using the NFD enterprise data.

Output growth rate equation

To derive an output growth rate equation from Eqn 23.2, the incremental effect of subsidies is also evaluated. By indexing with zero the values for the output and all the factors of the primary year (preceding the one being analysed), and indexing with one the values of the one being analysed, the output growth rate is B_1/B_0. This amount via the values of the main factors can be expressed, assuming that the coefficients m, f, n, p and s are invariable during the period being analysed

$$\begin{aligned} B_1/B_0 &= C_1 M_{1b}^m F_1^f P_1^p N_1^n S_1^s / C_0 M_{0b}^m F_0^f P_0^p N_0^n S_0^s \\ &= (C_1/C_0)\ (M_1/M_0)^m\ (F_1/F_0)^f\ (P_1/P_0)^p \\ &\quad (N_1/N_0)^n (S_1/S_0)^s. \qquad (23.3) \end{aligned}$$

By doing so, the formula for the output growth rate, analogous to that of the Cobb–Douglas production function, is deduced. It ties together the output growth rate with the growth rates of the factors. This equation explains why the actual output increase rate (according to the data of the NFD enterprises) correlates weakly with the size of the enterprises (production factors). They must correlate with the values of their growth but not with the values of the factors. Take the logarithm of both sides of Eqn 23.3 to deduce a linear equation that ties the logarithms of output growth rates with the logarithms of factors growth rates

$$\begin{aligned} \ln B_1/B_0 = {} & \ln(C_1/C_0) + m \ln(M_1/M_0) \\ & + f \ln(F_1/F_0) + p \ln(P_1/P_0) \\ & + n \ln(N_1/N_0) + s \ln(S_1/S_0). \qquad (23.4) \end{aligned}$$

Note that all the monetary data for year one (1) should be transformed (deflated) to the price level of year zero (0) using data on price indices for agricultural products and the means of production in the region.

The results of the basic output and output growth equations

What are the actual results of applying Eqn 23.2 and Eqn 23.4? In Table 23.2, only the coefficient of determination D and the F-criterion value are given, which are the criteria for the deduced equations and the coefficients at significant factors (all parameters are presented in Appendix Tables A23.3 and A23.4 for 2008).

In 2008, the first factor of production was material inputs (which was revealed by the beta coefficients in Table 23.2), which represented

Table 23.2. Output regression coefficients and output growth rate regression coefficients in 2008, taking into account the subsidies factor. (From author's calculations.)

	Output regression $D = 0.927$, $F = 777.8$		Output growth rate regression $D = 0.470$, $F = 128.1$	
	Unstandardized coefficients	Standardized coefficients (beta)	Unstandardized coefficients	Standardized coefficients (beta)
Labour	0.472	0.298	0.599	0.334
Depreciation	–	–	0.069	0.139
Subsidies	0.037	0.053	0.022	0.094
Material inputs minus subsidies	0.666	0.694	0.363	0.422
Agricultural land	–0.100	–0.072		

almost 70% of the output change and had the greatest impact on output and output increment. The second factor of production was labour, which represented about 25% of the output change. The third factor of production was subsidies, which represented only 5% of the output change. The coefficient related to subsidies was positive and equalled 0.037, which was a statistically significant positive effect (Table 23.2).

Based on the values of the determination coefficient D and F-criterion, the deduced equations do adequately reveal economic tendencies, and the growth rate equation shows that the output growth rate depends on the production factor growth rates, including subsidies.

The differences in coefficients related to the factors in these two equations are explained by the fact that they change annually, as distinguished from the assumed invariability of the production function coefficients in 2007 and 2008 (Appendix Table A23.1a,b). That is why the coefficients of the revenue growth rate differ from the coefficients of revenue. The coefficients of revenue are also affected by the fact that in the increase rate equation there are some missing additional significant economic variables.

Additional economic variables

Consider the following as a hypothesis. Eqn 23.1 contained only the basic constituents of the inputs as production factors. Therefore, it is tacitly accepted that output is determined by the inputs only. However, this assumption is not quite accurate. With the same inputs (accounting for differences in structure and absolute size), the more profitable enterprise will produce more output, for one or more reasons. First, its resource input per output unit is smaller. Second, the goods it produces are more readily accepted by the trade network or, alternatively, a profitable enterprise can sell its goods directly and get higher prices. Third, the more profitable enterprise has better-developed sales mechanisms. In short, higher profitability rests on some additional material or attitudinal factors that are reflected partially in the inputs only. Which reporting indicators can reflect them? The best choice would be to use gross or net profit. But for the unprofitable enterprises, the negative values would prevent using logarithms. Moreover, introducing profit into the equation would have nearly turned this equation into a mathematical identity, since many of the input components provide cost of production (revenue = cost + profit). That is why it is more appropriate to try to add to the basic equation the rate of recoupment of expenses (i.e. the ratio of revenue to input).

The management quality index is important in practice. Earlier, it was represented by the *coefficient of efficiency of production potential usage* (i.e. the relationship of real output to the one calculated in accordance with the output regression equation). Profitability reveals management quality, and to some extent depends on it. Still, it is not identical to management quality, as it depends on the enterprise's areas of specialization, on the quality of the supply and its distribution channels and on the existing level of wages and costs in different industries. For example, the correlation coefficient between profitability and coefficient of efficiency of production potential usage for the analysed NFD enterprises was

0.244 in 2007 and 0.376 in 2008 (Appendix Table A23.1,a). This confirms the hypothesis that management quality is an independent parameter in the equation, to be considered on par with profitability.

Accordingly, Eqn 23.2 with added factors yields

$$B = CM_b^m F^f P^p N^n S^s r^c k^d, \qquad (23.5)$$

where c and d are unknown parameters of the impact of the corresponding profitability r (in the form of the recoupment of expenses coefficient) and efficiency coefficient k factors.

The output growth rate equation in the logarithmic form is

$$\begin{aligned} \ln B_1/B_0 = \ln(C_1/C_0) + m\ln(M_1/M_0) \\ + f\ln(F_1/F_0) + p\ln(P_1/P_0) \\ + n\ln(N_1/N_0) + s\ln(S_1/S_0) \\ + c\ln(r_1/r_0) + d\ln(k_1/k_0). \quad (23.6) \end{aligned}$$

Results from basic equations with additional variables

The computed parameter values for Eqns 23.5 and 23.6 are listed in Table 23.3. Other parameters are presented in Appendix Tables A23.5 and A23.6.

Obtained Eqn 23.5, which takes into account the impact of additional subsidies on production, is noticeably better than Eqn 23.2, which has no added factors, where the determination coefficient is 0.927 in Appendix Table A23.3 but is 0.955 in Table 23.3. As a determinant of the management quality in 2008 (Appendix Table A23.2), the efficiency coefficient for 2007 was adopted (Table 23.3), and this factor was found to be statistically significant. That is, our assumption that this coefficient was relatively stable was justified.

As the beta-coefficients show, this factor is responsible for 16.6% of the change in output. The regression coefficient for the subsidies factor equals 0.052 and accounts for 7.2% of the output dynamics in 2008, which is higher than the impact of profitability (0.037 and 5.3%, respectively) (Table 23.2). This means that a subsidy increase of 1% leads to a revenue increase of 0.052%. In particular, a 1 rouble increase in subsidies leads to an output growth of more than 1 rouble for enterprises where the relationship of subsidies to the revenue is less than 0.052; this would apply to 152 out of 371 enterprises.

This new equation included the depreciation factor, which was not present before. Evidently, the profitability factor in this case is responsible for only 4.9% of output dispersion. In general, a more accurate and informative output equation is obtained that defines more accurately the contribution of subsidies in the output. We now turn to the output growth regression equation in 2008 as related to 2007 (the right-most half of Table 23.3). The determination coefficient for the output growth rate equation has grown from 0.470 in Table 23.2 to 0.632 in Table 23.3 after the addition of two new factors. This is a rather significant increase, as now almost two-thirds of the output growth dispersion is defined by the newly added factors. All the factors are included with positive signs. According to the output growth rate regression equation, the growth of material inputs has the greatest impact on the output growth rate. The second, in order of importance, is an increase of the efficiency coefficient; the third is the labour increase; and the fourth is the profitability increase. The impact of the increased subsidies on the output growth rate is significantly smaller in comparison to the other factors. The coefficient related to the subsidy increased rate (0.025) has decreased by a factor greater than 2, in comparison to the coefficient related to subsidies in the output equations. This may be explained by: (i) the statistical imprecision of the equation; (ii) the share of subsidies in input and output is relatively small; and (iii) the increase in subsidies over the year plays a smaller role than does the increase of the other factors.

Because practitioners in Russia are not very tolerant of non-linear models, a linear model of output growth was developed depending on the increase of the main factors, adding the profitability increase factors and the efficiency coefficients.

An equation with a high determination coefficient (0.611) is obtained, when the coefficient with subsidy equals 1.032. For each 1 rouble increase in subsidy, revenue will grow by slightly more than 1 rouble. The fundamentally important implication is that an increase in subsidies leads to a statistically significant increase in output. Therefore, the computations show that the subsidies in the NFD in 2008 were effective.

Table 23.3. Output regression coefficients and output growth rate regression coefficients with consideration of cost pay-off and management quality factors. (From author's calculations.)

	Output regression $D = 0.955$, $F = 727.539$				Output growth regression $D = 0.632$, $F = 57.5$			
Factors	Unstd. coefficient	Standardized coefficient (beta)	*t*-Test	Sig.	Unstd. coefficient	Standardized coefficient (beta)	*t*-Test	Sig.
Labour	0.509	0.326	10.848	0.000	0.482	0.305	6.698	0.000
Depreciation	0.056	0.069	2.565	0.011	0.059	0.134	3.253	0.001
Subsidies	0.052	0.072	3.662	0.000	0.025	0.112	2.715	0.007
Material inputs minus subsidies	0.530	0.557	16.080	0.000	0.346	0.412	9.157	0.000
Agricultural land	–0.069	–0.052	–3.005	0.003	0.098	0.070	1.716	0.088
Recoupment of expenses	0.277	0.049	2.415	0.017	0.476	0.223	5.174	0.000
Efficiency coefficient 2007	0.662	0.160	9.166	0.000	0.330	0.367	8.681	0.000

Conclusions

The accurate evaluations of the efficiency of subsidies – whether for the whole country or for an individual region – may be derived from regression analysis, and in particular from the basic output equation and the output growth rate equation. This approach may be used with respect to particular crops or kinds of products, provided that reliable data are available on the resource inputs for that kind of production (labour, land, material inputs, basic production assets, subsidies, etc.).

To obtain more accurate evaluations, the basic output equations must be taken into account. These output equations account for the input components, the profitability index and the management quality factor in the form of resource potential efficiency coefficient or in the form of technical efficiency.

References

Andriychuk, V.G. (1983) Effektivnost' ispol'zovaniya proizvodstvennogo potentziala, Ekonomika, Moskva, Russia. In: Andriychuk, V.G. (ed.) *The Efficiency of Production Potential Usage*. Economics, Moscow.

Bezlepkina, I. (2004) Microeconomic analysis of Russian agricultural enterprises with special reference to subsidies and debts. PhD thesis, Wageningen University, the Netherlands.

Epstein, D.B. (1988) Normative and resource methods at the state foodstuffs purchase planning. In: Epstein, D.B. (ed.) *Normative Methods of Planning and Labour Evaluation on the Basis of Final Results*. Economics, Leningrad, Russia.

Epstein, D.B. (1993) *State Regulation of Agricultural Production in Conditions of Transition to Market. Theory and Methods*. State Agrarian University, St Petersburg, Russia.

Heady, E. and Dillon, J. (1965) *Agricultural Production Functions*. Progress, Moscow.

Just, R., Hueth, D. and Schmitz, A. (2004) *Applied Welfare Economics and Public Policy*, 2nd edn. Prentice-Hall, Englewood Cliffs, New Jersey.

Krastin, O.P. (1971) *Agroekonomicheskie funktzii*. Minvuz Latv. SSR, Riga, Russia. [*Agro-economic Functions*. Ministry of High Education of Latvian SSR.]

Kudinov, V.I., Kapitonov, A.A. and Korneev, V.F. (1990) Normativno-resursnyi metod planirovaniya. Rosagropromizdat, Moskva. [*Normative and Input Methods of Planning*. Rosagropromizdat, Moscow.]

Petrick, M. and Zier, P. (2011) Regional employment impacts of Common Agricultural Policy measures in Eastern Germany: a difference-in-differences approach. *Agricultural Economics* 42, 83–193.

Pindyck, R.S. and Rubinfeld, D.L. (1992) *Microeconomics.* Economics, Moscow.

Rivzha, B.A. and Tray, B.A. (1987) *Production Potential of Agricultural Enterprises and Analysis of its Usage*. LSHA, Elgava, Russia.

Rossi, F., Schmitz, A. and Schmitz, T. (2007) US cotton subsidies: drawing a fine line on the degree of decoupling. *Journal of Agricultural and Applied Economics* 39(1), 135–149.

Schmitz, A. and Seckler, D. (1970) Mechanized agriculture and social welfare: the case of the tomato harvester. *American Journal of Agricultural Economics* 52(4), 569–577.

Shaykin, V.V., Akhmetov, R.G. and Kovalenko, N.Y. (2001) *Agricultural Markets*. Kolos, Moscow.

Tregubov, K.G. (1969) Agricultural production functions. PhD thesis, Odessa National University, Odessa, Ukraine.

Uzun, V.V. and Gataulina, E.A. (2009) *Methodological Approaches to the Estimation of Efficiency of the State Support of Agricultural Producers*. GNU VIAPI, Moscow.

Voelken, E. (1999) *Introduction to the Theory of Agrarian Market*, 3rd edn. Informagrotekh, Moscow.

Appendix

Table A23.1a. The matrix of correlation coefficients of output growth and main factors (agricultural enterprises NFD, 2008). (From author's calculations.)

Variables	Output growth 2008	Profitability 2008	Profitability 2007	Efficiency coefficient 2008	Efficiency coefficient 2007
Output growth 2008					
Profitability 2008	0.168[b]				
Profitability 2007	0.112[a]	0.797[b]			
Efficiency coefficient 2008	0.011	0.376[b]	0.300[b]		
Efficiency coefficient 2007	−0.031	0.217[b]	0.244[b]	0.831[b]	
Subsidies 2008	−0.033	−0.023	0.027	−0.132[a]	−0.045
Material inputs 2008	0.023	0.150[b]	0.246[b]	−0.080	−0.003
Depreciation 2008	−0.013	0.189[b]	0.216[b]	−0.057	−0.046
Labour 2008	−0.056	0.208[b]	0.226[b]	−0.054	0.019
Agricultural land 2008	−0.002	0.076	0.043	−0.089	−0.118[a]

Notes: [a]Correlation is significant at the 0.05 level (two-tailed); [b]correlation is significant at the 0.01 level (two-tailed).

Table A23.1b. The matrix of correlation coefficients of output growth and main factors (agricultural enterprises NFD, 2008). (From author's calculations.)

Variables	Subsidies 2008	Material inputs 2008	Depreciation 2008	Labour 2008	Agricultural land 2008
Output growth 2008					
Profitability 2008					
Profitability 2007					
Efficiency coefficient 2008					
Efficiency coefficient 2007					
Subsidies 2008					
Material inputs 2008	0.756[b]				
Depreciation 2008	0.412[b]	0.673[b]			
Labour 2008	0.540[b]	0.512[b]	0.725[b]		
Agricultural land 2008	−0.028	0.053	0.362[b]	0.350[b]	

Notes: [a]Correlation is significant at the 0.05 level (two-tailed); [b]correlation is significant at the 0.01 level (two-tailed).

Table A23.2 Descriptive statistics of the variables. (From author's calculations.)

	Mean	Standard deviation	Mean	Standard deviation
Variable	Data 2007		Data 2008	
Output (1,000 roubles)	95,675	20,6240	12,2183	2,684,912
Labour in agriculture, persons	175	143	168	139
Material inputs (1,000 roubles)	71,333	127,741	74,468	142,545
Material inputs minus subsidies (1,000 roubles)	66,943	12,6960	66,759	133,362
Depreciation (1,000 roubles)	4,786	7,655	5,976	7,319
Agricultural land (hectares)	2,920	2,749	2,975	2,655
Subsidies, (1,000 roubles)	4,391	7,112	8,735	14,727
Profitability (%)	3.0	2.1	2.1	23.3
Recoupment of expenses	1.0	0.2	1.0	0.3
Management quality (efficiency coefficient) factor	1.2	1.3	1.1	0.6

Table A23.3. Output regression coefficients in 2008 with consideration of subsidies factor. (From author's calculations.)

Adjusted R^2	F-criterion		Significance of F	Standard error of estimate	Number of farms
0.927	777.8		0.000	0.302	246
	Unstandardized coefficients		Standardized coefficients		
	B	Standard error	Beta	t-Test	Sig.
Constant	2.023	0.221		9.170	0.000
Labour	0.472	0.060	0.298	7.917	0.000
Agricultural land	–0.100	0.029	–0.072	–3.493	0.001
Subsidies	0.037	0.017	0.053	2.193	0.029
Material input minus subsidies	0.666	0.032	0.694	20.636	0.000

Table A23.4. Output growth rate regression coefficients in 2008. (From author's calculations.)

Adjusted R^2	F-criterion		Significance of F	Standard error of estimate	Number of farms
0.470	128.1		0.000	0.17	238.0
	Unstandardized coefficients		Standardized coefficients		
	B	Standard error	Beta	t-Test	Sig.
(Constant)	–0.070	0.394		–0.177	0.859
Labour growth	0.599	0.095	0.334	60.305	0.000
Depreciation growth	0.069	0.024	0.139	20.921	0.004
Subsidies growth	0.022	0.011	0.094	10.980	0.049
Material inputs minus subsidies growth	0.363	0.045	0.422	70.996	0.000

Table A23.5. Output regression coefficients in 2008 with added profitability (recoupment of expenses) and management quality (efficiency coefficient) factors. (From author's calculations.)

Adjusted R^2	*F*-criterion	Significance of *F*	Standard error of estimate	Number of farms
0.955	727.5	0.000	0.224	238

	Unstandardized coefficients		Standardized coefficients		
	B	Standard error	Beta	*t*-Test	Sig.
(Constant)	2.491	0.207		12.006	0.000
Labour	0.509	0.047	0.326	10.848	0.000
Depreciation	0.056	0.022	0.069	2.565	0.011
Subsidies	0.052	0.014	0.072	3.662	0.000
Material inputs minus subsidies	0.530	0.033	0.557	16.080	0.000
Agricultural land	–0.069	0.023	–0.052	–3.005	0.003
Recoupment of expenses	0.277	0.115	0.049	2.415	0.017
Efficiency coefficient 2007	0.662	0.072	0.160	9.166	0.00

Table A23.6 Output growth regression coefficients in 2008 with added profitability (recoupment of expenses) and management quality factors. (From author's calculations.)

Adjusted R^2	*F*-criterion	Significance of *F*	Standard error of estimate	Number of farms
0.632	57.521	0.000	0.121	230

	Unstandardized coefficients		Standardized coefficients		
	B	Standard error	Beta	*t*-Test	Sig.
(Constant)	2.491	0.207		12.006	0.000
Labour growth	0.482	0.072	0.305	6.698	0.000
Depreciation growth	0.059	0.018	0.134	3.253	0.001
Subsidies growth	0.025	0.009	0.112	2.715	0.007
Material inputs minus subsidies growth	0.346	0.038	0.412	9.157	0.000
Agricultural land growth	0.098	0.057	0.070	1.716	0.088
Recoupment of expenses growth	0.476	0.092	0.223	5.174	0.000
Efficiency coefficient growth	0.330	0.038	0.367	8.681	0.000

Index

Page numbers in **bold** type refer to figures and tables.